The Biology of Baculoviruses

Volume II
Practical Application for Insect Control

Editors

Robert R. Granados, Ph.D.
Virologist and Program Director
Biological Control Program
Boyce Thompson Institute for
Plant Research
Cornell University
Ithaca, New York

Brian A. Federici, Ph.D.
Professor of Entomology
University of California
at Riverside
Riverside, California

CRC

CRC Press, Inc.
Boca Raton, Florida

Library of Congress Cataloging-in-Publication Data

The Biology of baculoviruses.

   Includes bibliographies and index.
   Contents: v. 1. Biological properties and molecular
biology -- v. 2. Practical application for insect
control. I. Baculoviruses. I. Granados, Robert R.
II. Federici, Brian A.
QR327.B56   1986   576'.6484   86-12929
ISBN 0-8493-5987-2 (v. 1)
ISBN 0-8493-5988-0 (v. 2)

Direct all inquiries to CRC Press, Inc., 2000 Corporate Blvd., N.W., Boca Raton, Florida, 33431.

© 1986 by CRC Press, Inc.

International Standard Book Number 0-8493-5987-2 (v. 1)
International Standard Book Number 0-8493-5988-0 (v. 2)

Library of Congress Card Number 86-12929
Printed in the United States

PREFACE

The baculoviruses constitute one of the largest known groups of viruses, and because of their complex properties, widespread occurrence among economically important insect pests, and potential economic importance in insect control programs, they have been the subject of serious study by microbiologists, virologists, pathologists, and entomologists for more than a century. Prior to the 1940s, most studies focused on identification of the specific agent that caused baculovirus disease. During the 1940s, Dr. G. Bergold working in Germany, demonstrated convincingly that baculovirus diseases were caused by rod-shaped virions, and he developed techniques for the purification of these virions. Bergold's pioneering studies and the advent of centrifugation techniques and electron microscopy, stimulated many groups in Europe, North America, and Japan to study the basic biology of baculoviruses in more detail and to develop them as microbial control agents. These initial studies on the histopathology, ultrastructure, biochemical characterization, and epizootiology were reviewed in 1963 in a series of chapters in the classic two-volume series *Insect Pathology—An Advanced Treatise*, edited by the late Edward Steinhaus.

Since the publication of *Insect Pathology—An Advanced Treatise*, there has been a dramatic proliferation in both basic and applied research on baculoviruses, facilitated in part by such advances as the development of insect tissue culture and recombinant DNA technology, and the willingness of governmental agencies to permit baculoviruses to be used as insecticides. Our intention in the present two-volume series is to review the significant body of literature that now exists on baculoviruses, with emphasis on studies carried out over the past two decades, and to provide a single reference that can be used to gain access to most of the available literature on these viruses.

The division of the treatise into two volumes reflects the uniqueness of baculoviruses, first as model systems for basic research in virology and molecular biology and in biotechnology in general, and secondly, as potentially important microbial pesticides for use in insect pest management programs. Volume I is concerned with the general properties and molecular biology of baculoviruses. Volume II deals with more practical aspects of baculoviruses and their use for insect control.

The contributors to the present work are among the leaders in insect virology and each have prepared a comprehensive and well-documented review of the assigned subject matter. The editors have enlisted the talents of leading scientists from various countries giving these volumes an international scope. In addition, the authors are a blend of senior, well-established investigators as well as younger but highly qualified scientists.

This two-volume treatise will be of interest and a source of information to researchers, students, and administrators at universities, experiment stations, and industrial centers. This treatise will be of particular interest to scientists engaged in microbial control of pests in agriculture, forestry, veterinary, and medical areas, to biotechnology industries who are using baculoviruses as expression vectors of foreign gene products, and to individuals who are interested in the development of viral pesticides on an industrial scale.

We wish to express our sincere gratitude to the contributors for their cooperation and excellent contributions, to Kathleen Williams and Johanna Granados who assisted the editors in the editing of the manuscripts, and to the staff of CRC Press, Inc. for their part in the production of these volumes.

Robert R. Granados
Ithaca, N.Y.
Brian A. Federici
Riverside, CA.

THE EDITORS

Robert R. Granados, Ph.D., is Virologist and Program Director of the Biological Control of Insects Program at the Boyce Thompson Institute for Plant Research, Cornell University of Ithaca, New York.

Dr. Granados received his B.S. degree in Entomology and Parasitology from the University of California, Davis, in 1960. He obtained his M.S. and Ph.D. degrees in Entomology in 1963 and 1965, respectively, from the Department of Entomology, University of Wisconsin, Madison. Since 1965 he has served in positions of Assistant, Associate, and presently full rank Virologist at Boyce Thompson Institute for Plant Research, Cornell University. It was in 1977 that he assumed his present position.

Dr. Granados is a charter member of the Society for Invertebrate Pathology and the American Society for Virology. He is a member of the Entomological Society of America, American Association for the Advancement of Science, American Society for Microbiology and New York Academy of Sciences, and the honorary societies of Alpha Zeta and Phi Kappa Phi. He has been the recipient of a National Science Foundation Fellowship (1965), University of Wisconsin-awarded Visiting Professorship (1976), and is a member of the Scientific Advisory Board of Ecogen, Inc., Langhorne, PA. He served as a member of the editorial board of the *Journal of Invertebrate Pathology* (1978-1980) and *Phytopathology* (1974-1975).

Dr. Granados has been the recipient of research grants from the National Institutes of Health, the National Science Foundation, the Rockefeller Foundation, the U.S. Department of Agriculture, and private industry. He has published more than 65 scientific papers. His current research interests are in insect cell culture and the molecular basis of baculovirus pathogenesis.

Brian A. Federici, Ph.D., is Professor of Entomology in the Division of Biological Control, Department of Entomolgy, at the University of California, Riverside.

Dr. Federici obtained his undergraduate training at Rutgers University where he received a B.S. degree in 1966. From there he went on to the University of Florida, Gainesville, where he received the M.S. degree (1967) and Ph.D. degree (1970), both in Medical Entomology. He served as an NIH postdoctoral fellow at the Boyce Thompson Institute for Plant Research, Yonkers, N.Y., from 1972 to 1974. In 1974, he moved to the University of California, Riverside, where he became Professor of Entomology in 1983.

Dr. Federici is a member of the American Association for the Advancement of Science, American Society for Microbiology, Society for Invertebrate Pathology, Entomological Society of America, American Mosquito Control Association, and Sigma Xi. He is a member of the World Health Organization's Expert Committee on Vector Biology and Control (1981 to 1987), and the Steering Committee on the Biological Control of Vectors for WHO's Research and Training Program for Tropical Disease (1979 to 1986). He has served as a member of the editorial board of the *Journal of Invertebrate Pathology* (1982 to 1985), and a trustee of the Society for Invertebrate Pathology (1984 to 1988). He has been the recipient of several research grants from the National Institutes of Health and the World Health Organization.

Dr. Federici is the author of more than 80 papers. His current major research interest is in the basic biology of pathogens that cause diseases in insects.

This volume is dedicated to
CARLO M. IGNOFFO
whose intellect and enthusiasm
paved the way for the registration
in the U.S. of the first Baculovirus insecticide

CONTRIBUTORS

Frederick S. Betz, Ph.D.
Biologist
Hazard Evaluation Division
U.S. Environmental Protection Agency
Arlington, Virginia

G. T. Bohmfalk, Ph.D.
Project Manager — Research and
 Development
Zoecon Corporation
Douglas, Arizona

D. T. Briese, Ph.D.
Senior Research Scientist
Division of Entomology
Commonwealth Scientific and Industrial
 Research Organization
Canberra, Australia

Hugh F. Evans, D. Phil.
Principal Entomologist
Forestry Commission
Forest Research Station
Farnham, Surrey, England

James D. Harper, Ph.D.
Professor of Entomology
Department of Zoology — Entomology
Auburn University
Auburn, Alabama

Jürg Huber, Ph.D.
Insect Virologist
Biologische Bundesanstalt
Darmstadt, Federal Republic of Germany

Patrick R. Hughes, Ph.D.
Insect Pathologist
Boyce Thompson Institute for Plant
 Research
Cornell University
Ithaca, New York

Martin Shapiro, Ph.D.
Research Entomologist
Agricultural Research Service
U.S. Department of Agriculture
Otis Air Force Base, Massachusetts

James L. Vaughn, Ph.D.
Research Leader
Insect Pathology Laboratory
Plant Protection Institute
Agricultural Research Service
U.S. Department of Agriculture
Beltsville, Maryland

Stefan A. Weiss, M.S.*
Senior Research Associate
Head, Cell Culture Laboratories
Southwest Foundation for Research and
 Education
San Antonio, Texas

H. Alan Wood, Ph.D.
Virologist
Boyce Thompson Institute for Plant
 Research
Cornell University
Ithaca, New York

W. C. Yearian, Ph.D.
Professor
Department of Entomology
University of Arkansas
Fayetteville, Arkansas

Seth Y. Young, III, Ph.D.
Professor
Department of Entomology
University of Arkansas
Fayetteville, Arkansas

* Present address: CETUS Corporation, Emeryville, California.

TABLE OF CONTENTS

Volume I

TABLE OF CONTENTS

Volume II

Chapter 1

IN VIVO AND IN VITRO BIOASSAY METHODS FOR BACULOVIRUSES

P. R. Hughes and H. A. Wood

TABLE OF CONTENTS

I. INTRODUCTION

A. General Bioassay

1. What is Bioassay?

A chemical or biological product can be defined in many ways, including by its chemical or physiochemical properties or, in the case of a microorganism, by its morphological or ultrastructural characteristics. However, in many cases such definitions are either insufficient or impractical, and the more desirable alternative is to consider the agent's biological properties through bioassay. Finney[1] defines the biological assay (or bioassay) as "an experiment for estimating the nature, constitution, or potency of a material (or of a process) by means of the reaction that follows its application to living matter." Two general classes of bioassay can be identified based on the overall objective of the test. Historically, bioassay has been concerned primarily with measurement, such as the amount of a chemical or viable counts of microorganisms, with the aim of comparing the potencies of different preparations or of determining the titer of organisms. A second type of quantitative experiment that should be included in any general consideration of bioassay is one in which the object is to determine the intrinsic nature of a biological mechanism by examining the way in which it acts.[2] Within each of these general classes there may be several subclasses that differ according to their specific purposes.

2. Types of Bioassays

For reasons of logic and statistical analysis, several types of quantitative biological assays should be distinguished. First, such assays can be classed either as *direct* or *indirect*, depending on how the dose producing a specified response is determined. In direct assays the dose is measured directly, whereas in indirect assays it is mathematically estimated from the responses to a range of doses. Secondly, assays can be classed as *quantitative* or *quantal*, depending on the type of response being measured. Quantitative responses, also termed *graded responses*, are those that can assume any of a series of values, such as lesion size or time to appearance of response. Quantal responses, on the other hand, can assume only one of two alternatives, such as live vs. dead, and for this reason they are also referred to as *all-or-none* responses.

The field of bioassay has been dominated by procedures designed to determine the potency of a preparation, and large fluctuations from test to test in the estimated potency of a single sample have been observed frequently, presumably due to variability inherent in the biological system. This variability has led to the widespread use of comparative techniques and terminology particularly related to these methods. The ratio of equivalent potencies (i.e., doses producing equivalent responses) of two preparations has been found to be far more consistent than the values from which it is obtained. Therefore, the common practice is to designate a *standard preparation* or *reference standard* and to estimate the potency of a *test preparation* relative to this standard by simultaneously testing the two. Ideally, the standard is homologous with the test material, uniform and homogeneous within itself, and sufficiently stable to be usable over a long period of time. Preferably, the standard consists of a single active constituent, but if it is a mixture then the various active elements should be present in the same proportion as in the test preparation(s).[1,3-5]

Frequently, several test preparations are compared simultaneously against a single standard in a *multiple assay*. This permits more equal allotment of the test subjects among the treatments and more precise estimation of the regression coefficient than possible when each test preparation is compared singly against the standard in many *simple assays*.

By far the most common type of indirect assay is the *dilution assay*, in which the test preparation behaves as though it is simply a dilution or concentration of the standard in an inert diluent. Finney[1] distinguishes between two types of dilution assays. If none of the constituents of the test preparation except one affects the response being measured and the test preparation behaves as a dilution or concentration of the standard preparation, the assay is termed an *analytic dilution assay*. In many cases, the test preparation is known to be different from the standard material but still behaves as though qualitatively it is the same effective constituent; an assay with such a preparation is termed a *comparative dilution assay*. As discussed by Finney[1] the choice of assay technique, subject, response, and experimental conditions should have no effect on the potency estimate from an analytic dilution assay; on the other hand, the results of a comparative dilution assay may be greatly altered by any of these factors, making the estimate obtained from such an assay valid only under the specific conditions used.

The potency ratio must be constant in valid analytic or comparative dilution assays. Therefore, if linearization is achieved by the regression of response against the untransformed dose, the lines will have a common intercept but the slopes will differ. In this case, the test is called a *slope-ratio assay* because the relative potency is estimated by the ratio of the slope of the test preparation to that of the standard preparation;[1,4,6] to be fundamentally valid, the lines must intersect at zero dose. This type of relation is most commonly (but not exclusively) encountered in microbiological assays.[1]

If the best fit of the dose-response data to a straight line is obtained by regressing response against the logarithm of dose, then the intercepts will be different for the preparations but the slopes should be the same. This type of test is called a *parallel line assay,* and the relative potency is estimated as the ratio of a dose of the standard preparation to that of the test preparation producing the same response; generally the median effective dose (ED_{50}) has been used because it can be estimated more precisely than the more extreme percentage levels.[7] This relation between dosage and response is most commonly found in bioassays of chemicals and is generally attributed to a log normal distribution of the tolerances of individual test animals to the material being assayed.

B. Bioassay of Insect Viruses
1. Role in the Development of Viral Pesticides
In vivo assay is central to both basic and applied research on insect viruses. In the commercialization of microbial pesticides, it is the only means available by which the potency of a preparation can be established and monitored in order to maintain a product of known, constant quality. In vivo assay also plays essential functions in designing and evaluating control strategies using baculoviruses. To be most effective, reliable, safe, and cost efficient, such strategies must be based on an understanding of the virus-host interactions responsible for infection and pathogenicity, and this, in turn, requires quantitative information on the dose-response relationship and the replication characteristics of viruses in pest insects. Since in vivo assay is the only means by which the combined effect of all the factors determining potency can be measured, it also plays a key role in the identification of more effective isolates and formulations. In vitro titrations are required for many types of basic studies of the physical and biological properties of the viruses, particularly the nonoccluded form of baculoviruses. For example, accurate assay procedures are required in the determination of the kinetics of infection, growth curves, responses to physical or chemical agents, viral interference, and neutralization assays. In addition, in vitro assays have made possible the isolation and characterization of individual clones from wild-type and mutagenized populations.

2. Historical Perspective
The early uses of in vivo bioassay with insect viruses were primarily to determine or compare potencies of preparations in given-aged larvae or to determine the effect of larval age on the potency of a given preparation. Most of these early dose-response assays were accomplished by contaminating food with a calculated quantity of virus and allowing the

insects to ingest the virus as they fed. Usually four or more doses of a given isolate were tested and the response was scored as the number of individuals dead at a particular time after initiation of the test; the median lethal dose was calculated from these responses and used to compare potencies of different preparations. Thus, these early quantitative assays were indirect, based on a quantal response. However, the actual dose received by the insects was unknown and varied from individual to individual in an unpredictable manner that depended on feeding behavior.

The relation between dose and response in these assays was found to be linear when probit analysis, which was already widely used to analyze data from pesticide assays, was applied to the results. Biologically, this conformity to the probit model could be rationalized as resulting from the same sort of individual variation in susceptibility to the virus as was well established for the chemical pesticides. This conformity of the results to an established method, the apparent logical explanation for the conformity, and the availability and relative ease of the analytical procedure led to the adoption of probit analysis for the routine processing of data from dose-response assays of insect viruses.

The relation between time and response has also been measured in many studies.[8-11] However, little importance has been attached to the data, and generally they have not been subjected to statistical analysis.

In vitro assays of baculoviruses were first used to determine growth curves. These assays were quantal in nature and required large numbers of samples for accuracy. With the advent of quantitative procedures, the ease and accuracy of titrations improved dramatically and many new areas of investigation have opened, particularly in viral molecular biology.

II. IN VIVO ASSAY OF BACULOVIRUSES

A. Design and Planning of the Bioassay
1. General Considerations

A general treatment of the theory of experimental design can be found in any of many texts.[12-15] The application of these principles to the design of bioassay is discussed by Finney[1] and Hewitt,[5] and Meynell and Meynell[2] consider the subject specifically for microbiological assay. A detailed discussion of the subject is beyond the scope of this chapter. However, some points that are basic to most bioassays are considered below; the concepts and terminology are taken primarily from Finney,[1] and the reader is referred to this text for a more complete explanation of the points discussed.

Generally, a bioassay is comprised of three components: a *stimulus,* such as a chemical or microorganism, that is applied to a *subject,* such as an insect, to produce a characteristic, identifiable *response,* such as mortality. The magnitude of the stimulus is referred to as the *dose,* and all interest lies in determining, describing, and using the relationships between these components. The type and design of bioassay used may vary according to its immediate purpose and special considerations such as technical constraints. The design of dose-response tests may be *symmetric* or *unsymmetric.* A symmetric dose-structure, termed a k,k-point design for assays involving two preparations such as a standard and a test preparation, is one that has k doses of each preparation such that the successive doses of either preparation are a constant ratio to one another and each dose has the same number of subjects. Symmetric designs are preferred to unsymmetric designs and avoid complications in analysis of results.

In considering and analyzing dose-response relationships, a distinction must be made between the *fundamental validity* of the assay and the *statistical validity* of the analysis of the data obtained. Jerne and Wood[16] emphasized that behind the design of any assay and the conclusions drawn from it are a number of chemical, biological, and mathematical assumptions. Some of these assumptions are essential to the fundamental validity of the assay (i.e., the validity of the assay as a means of describing a relationship), whereas others relate to the statistical validity of the analysis following it (i.e., the correctness of the mathematical process applied to the data). In parallel line assays based on quantitative response, linearity is a test of statistical validity, and, when two or more preparations are compared, parallelism is a test of fundamental validity. In slope ratio assays based on quantitative response, again linearity is a test of statistical validity. However, when two or more preparations are compared, the test of fundamental validity is the intersection of the regressions at zero. In assays based on quantal response, the chi-square test for homogeneity is equivalent to the test for linearity.

One of the fundamental assumptions that is essential to recognize when comparing preparations is the *condition of similarity*. This condition, also referred to as the *hypothesis of similarity,* requires that the response evoked by a test preparation is due solely to the same effective constituent (or proportion of effective constituents) as that responsible for evoking the characteristic response from the subjects to the standard preparation. The importance of this assumption is underscored by Jerne and Wood:[16]

As Finney[17] says, if the data cannot be adequately described by the same form of (function) for both preparations, the basic assumption that only the same effective constituents were concerned in both must be false . . . The assay is therefore invalid — and it might be added that the whole idea of assaying that particular (test preparation) against that particular (standard preparation) becomes absurd.

It is now recognized that the assay is not necessarily invalid or useless when the assumption of similarity is false, but in such situations any conclusions must be drawn carefully and only in the context in which the test was performed.

2. Assays of Baculoviruses

With the exception of the time-response assay, in vivo bioassays of baculoviruses have all been based on quantal responses, and a detailed discussion of the design and planning of such assays, as opposed to those based on a quantitative response, is provided by Finney.[1] The discussion is directed particularly towards the standard assay in which a test preparation is to be compared with a standard, but the principles apply generally to all quantal response assays. The purpose of this section, which is based strongly on Finney,[1] is to summarize the most important points to be considered when planning this type of assay.

The first step in such an assay is to conduct a preliminary test, usually involving few subjects distributed equally among many doses. This test is used to investigate the form of the response equation for the regression of response on dose, to indicate what transformations are likely to be suitable, to assess if homoscedasticity can be assumed, and to determine the linear region of the response function. In addition, the information obtained will be used to estimate the doses to be used in the final test.

Whenever possible, the test itself should be of a symmetric design with successive doses a constant ratio to each other and the doses on a log scale symmetrically distributed about the median effective dose (ED_{50}). The optimal spacing of the doses to achieve the most precise estimate of the potency of the preparation is affected by many considerations, including the total number of subjects in the test (N), the sum of the weighting coefficients associated with the responses (Snw), and the sum of the squared deviations of the doses from the mean dose (S_{xx}). The precision of the estimate is increased as each Snw and S_{xx} increase, but the former is maximized by responses close to the ED_{50} while the latter is highest at extreme dose values, thus requiring a compromise between these opposing effects to attain the greatest precision. Tables 1 and 2 show how the effective variance of M, the estimated log potency, varies according to the number of subjects at each dose (n) and the selected responses for a (3,3)- or (4,4)-point assay in which the doses are symmetric about the ED_{50}. From the tables it is apparent that for an assay with three doses and from 45 to 150 subjects the best precision is obtained with doses that produce about 20, 50, and 80% response. In an assay with four doses and from 80 to 200 subjects the best precision is

Table 1
GUIDE FOR THE SELECTION OF OPTIMAL RESPONSES IN A SYMMETRIC, (k, k)-POINT ASSAY FOR THE SPECIFIED NUMBER OF SUBJECTS AT EACH DOSE USING PROBITS AND THE 95% SIGNIFICANCE LEVEL FOR THE FIDUCIAL LIMITS

Responses (%)			$b^2 V_E(M)$ at $P = 0.95$ for specified no. of subjects per dose[a]									
			10	15	20	25	30	35	40	50	60	200
4	50	96	0.5990	0.3430	0.2403	0.1849	0.1503	0.1266	0.1094	0.0859	0.0708	0.0204
7	50	93	0.4980	0.2876	0.2022	0.1559	0.1268	0.1069	0.0924	0.0727	0.0599	0.0173
12	50	88	0.4536	0.2547	0.1770	0.1357	0.1100	0.0925	0.0798	0.0626	0.0515	0.0148
18	50	82	0.5101	0.2542	0.1693	0.1269	0.1015	0.0845	0.0724	0.0563	0.0461	0.0130
27	50	73	—	0.4202	0.2188	0.1479	0.1117	0.0898	0.0750	0.0565	0.0453	0.0120
33	50	67	—	—	0.5546	0.2454	0.1575	0.1160	0.0918	0.0648	0.0500	0.0120
38	50	62	—	—	—	—	—	—	0.4012	0.1396	0.0845	0.0129
44	50	56	—	—	—	—	—	—	—	—	—	0.0326
4	27	73	96	0.3589	0.2159	0.1544	0.1202	0.0984	0.0832	0.0722	0.0570	0.0471	0.137
7	31	69	93	0.3158	0.1890	0.1348	0.1048	0.0857	0.0725	0.0628	0.0496	0.0409	0.0119
12	34	66	88	0.2964	0.1724	0.1215	0.0938	0.0764	0.0645	0.0558	0.0439	0.0362	0.0104
18	38	62	82	0.3286	0.1736	0.1179	0.0893	0.0719	0.0601	0.0517	0.0404	0.0331	0.0094
27	42	58	73	—	0.2596	0.1467	0.1023	0.0785	0.0637	0.0536	0.0406	0.0328	0.0088
33	44	56	67	—	—	0.2894	0.1535	0.1045	0.0792	0.0637	0.0459	0.0358	0.0088
38	46	54	62	—	—	—	—	—	—	0.1778	0.0842	0.0552	0.0095

[a] b is the regression coefficient of response on dose and V_E (M) is the effective variance of M, the estimate of log potency, as described by Finney.[1] P is the significance level for the fiducial limits.

obtained with doses producing about 20, 40, 60, and 80% response. With fewer subjects the doses (and therefore the responses) should be more widely spaced, and with more subjects they could be more closely spaced. These tables also provide some indication of the relative change in precision for a particular change in n or N. The equations by which the optimal spacing can be determined for any combination of doses, total number of subjects, and significance level can be found in Finney (pp. 406-407).[1]

Other formulas useful in determining the number of subjects to use in an assay based on quantal response have been described by DeArmon and Lincoln.[18] If an estimate of the slope is available, such as from the preliminary assay, an approximation of the number of subjects required to achieve a specified confidence interval (CI) of the $\log_{10} ED_{50}$ can be obtained from the equation

$$N = (10z^2)/(L^2b^2) \qquad (1)$$

where N is the total number of subjects (to be divided equally among the doses), L is the length of the CI, b is the slope of the regression line, and z is the normal deviate for the specified probability (e.g., 1.96 for the 95% CI or 2.58 for the 99% CI). The length of the CI for a given N can be estimated from the equation

$$L = (3.17z)/(N^{1/2}b) \qquad (2)$$

In choosing the number of doses to be used, Finney points out the need to distinguish between a study of the response curve, which is done best by spreading the subjects over many dose levels, and determining the most precise estimate of the ED_{50}, which is done better with fewer doses containing more subjects per dose. For estimating potency, a (2,2)-

Table 2
GUIDE FOR THE SELECTION OF OPTIMAL RESPONSE IN A SYMMETRIC, (k, k)-POINT ASSAY FOR THE SPECIFIED NUMBERS OF SUBJECTS AT EACH DOSE USING PROBITS AND THE 99% SIGNIFICANCE LEVEL FOR THE FIDUCIAL LIMITS

Responses (%)				$b^2V_E(M)$ at P = 0.99 for specified no. of subjects per dose[a]									
				10	15	20	25	30	35	40	50	60	200
4	50	96		—	0.4459	0.2867	0.2112	0.1672	0.1384	0.1180	0.0912	0.0743	0.0207
7	50	93		0.7794	0.3633	0.2369	0.1757	0.1397	0.1159	0.0990	0.0767	0.0626	0.0175
12	50	88		0.5410	0.2801	0.1889	0.1426	0.1144	0.0956	0.0821	0.0640	0.0524	0.0149
18	50	82		—	0.3838	0.2184	0.1526	0.1173	0.0952	0.0802	0.0609	0.0491	0.0132
27	50	73		—	—	0.5480	0.2491	0.1612	0.1191	0.0945	0.0668	0.0517	0.0124
33	50	67		—	—	—	—	—	0.3708	0.2102	0.1051	0.0711	0.0129
38	50	62		—	—	—	—	—	—	—	—	—	0.0151
44	50	56		—	—	—	—	—	—	—	—	—	—
2	26	74	98	0.5727	0.2927	0.1966	0.1480	0.1186	0.0990	0.0850	0.0662	0.0542	0.0153
4	27	73	96	0.4796	0.2545	0.1732	0.1312	0.1056	0.0884	0.0760	0.0594	0.0487	0.0138
7	31	69	93	0.4316	0.2251	0.1523	0.1150	0.0924	0.0772	0.0664	0.0518	0.0424	0.0120
12	34	66	88	0.4478	0.2146	0.1411	0.1051	0.0837	0.0696	0.0595	0.0462	0.0377	0.0106
18	38	62	82	0.6878	0.2397	0.1452	0.1041	0.0811	0.0665	0.0563	0.0431	0.0349	0.0095
27	42	58	73	—	—	0.2860	0.1548	0.1061	0.0804	0.0652	0.0470	0.0368	0.0091
33	44	56	67	—	—	—	—	0.3574	0.1708	0.1122	0.0665	0.0473	0.0094
38	46	54	62	—	—	—	—	—	—	—	—	0.1981	0.0108

[a] b is the regression coefficient of response on dose and $V_E(M)$ is the effective variance of M, the estimate of log potency, as described by Finney.[1] P is the significance level for the fiducial limits.

point assay (i.e., two doses of the standard material and two doses of the test material) would be optimal in terms of precision but does not allow testing of statistical validity and should be used only when this validity is unquestionable. The use of (3,3)- or (4,4)-point designs permits testing of validity and need not decrease the precision significantly if the doses are properly chosen. Rarely is a larger k required, but when little is known about a preparation, additional higher or lower doses might be included with the intention of discarding the data if they introduced marked nonlinearity.[1]

A completely randomized design is preferred, but when constraints, such as the number of tests that can be conducted in 1 day, prevent this, any of various blocking systems might be used to reduce heterogeneity of the results. The simplest is the randomized complete block design, in which every treatment appears in each block. When this is not possible, other designs can be used, and many of these alternatives are discussed by Finney.[1] Most often, assays of baculoviruses can be accomplished with the simpler designs.

In summary, one would begin the investigation of the potency of an isolate with a preliminary assay. Table 1 or 2 or the formulae of DeArmon and Lincoln[18] could then be used to select the desired response rates based on the number of doses and the number of subjects in the test, and the preliminary assay results would be used to estimate the appropriate doses to achieve these rates. Finally, the doses would be adjusted slightly so that they are equally spaced on a log scale.

If the exponential rather than the log normal model was suspected to be applicable, a preliminary assay with many doses would be conducted to examine the response function. The modified Weibull plot and statistical test described by Gart and Weiss[19] could be used to test for deviation from the model. If no significant deviation was detectable, the data could be used to obtain an initial estimate of the slope, p, and to estimate the doses to be used in the final assay.

Whereas the response curve is best determined by a number of doses, the parameter p is estimated most precisely by only a few doses. Ideally, the best estimate of p would be obtained by a single dose containing all of the subjects and producing about 20% survival. However, if little is known about p or if linearity is not unquestionable, then more doses should be used and the best estimate will be obtained by the use of a few doses grouped closely around that expected to produce 20% survival. The precision of the estimate will not be decreased appreciably if the proportion surviving is kept between 10 and 35%, but doses producing high survivor rates should be avoided.[20]

Time-mortality assays are most often conducted in the same manner as the dose-mortality assay, and frequently the data for both are collected from a single experiment simply by recording the mortality at regular time intervals throughout a dose-mortality test. A fundamental distinction should be made between tests in which the subjects are continuously exposed to the virus and those in which the subjects receive the inoculum only at the beginning of the assay. Analysis of the data is the same in both cases, but the median response time is termed the *median lethal time* (LT_{50}) in the former case and the *median survival time* (ST_{50}) in the latter case.[21]

B. Bioassay Procedures
1. Surface Contamination
Contamination of food surfaces in a quantitative manner is by far the most widely used means of administering baculoviruses to insects. The methods used can be divided into two important categories based on whether the test insects are left on the treated food for the entire period of the assay or are given only a specified time in which to acquire the virus and then transferred to untreated diet. In the former case, the generalized procedure consists of applying a measured volume of aqueous virus suspension to the surface of the food, spreading it over the surface with a glass rod or by swirling, allowing the water to evaporate, and then placing the larvae on the treated material.[9] Diet treated in the same manner with sterile water serves as the controls. The volume of food treated must be great enough to permit the larvae to feed for the duration of the test; the dosage of virus is generally expressed as the amount of virus per unit of surface area, such as the number of occlusion bodies applied per square millimeter of surface.[10,11] The number of individuals dead after a prespecified time period is recorded for each treatment and each larva is examined to verify virus as the cause of death. The use of only one larva per container has been recommended in order to eliminate the possibility of complications caused by interaction of the larvae. However, when preliminary tests show this precaution to be unnecessary, some simplification can be gained by placing several larvae per container.[22]

The surface contamination technique has been used as the basis for procedures to assess activity and standardize potency of nuclear polyhedrosis viruses of the Douglas-fir tussock moth (*Orgyia pseudotsugata*), the western spruce budworm (*Choristoneura occidentalis*), and the gypsy moth (*Lymantria dispar*).[22-25] The parameters of these procedures are summarized in Tables 3 and 4. The procedures have proven very reliable and reproducible, yielding precision values (calculated as the ratio of the upper and lower 95% confidence limits of the median effective dose) near or below 2.0.[24,27]

The advantages of such surface contamination procedures include the simplicity of the methods, their close approximation to the natural means of acquisition, and their applicability to any age of larvae. However, the disadvantages include no control of dose, the inability to quantitate the dose, and the long, uncontrolled acquisition time during which the virus is ingested. Also, acquisition is dependent on larval feeding behavior, which can be quite variable within as well as between species. The inoculum cannot be applied evenly to the surface in many cases because of surface tension, irregularities in the surface, etc., and great care must be taken to make the surfaces even and of the same size. Also, care must

Table 3
PARAMETERS OF THE BIOASSAY PROCEDURE FOR *BACULOVIRUS*
ISOLATES PATHOGENIC FOR *CHORISTONEURA OCCIDENTALIS* AND
***ORGYIA PSEUDOTSUGATA*[23-25]**

Test insect	*Choristoneura occidentalis*, strain COC-FS-01	*Orgyia pseudotsugata*, strain GL-1
Age and sex	Post-diapause larvae, fourth instars (4 to 24 hr after molt), fed, both sexes	Second instars (1 to 17 hr after molt), fed, both sexes
Weight (mean ± SD)	2.6 ± 0.2 mg	?
Confinement	1 larva per standard 2-mℓ analyzer cup with perforated cap	Same
Diet and volume	Modified McMorran diet, 1 mℓ per cup	Formulation/65-W-PEN, Forestry Service Laboratory, Corvallis, Ore., 1 mℓ per cup
Rearing temperature	30°C ± 0.5°C	Same
Light regimen	Constant darkness	Same
Holding period	14 days	12 days
Exposure to the virus	Diet-surface treatment, continuously exposed	Same
Inoculum volume	25 μℓ per cup (diet surface area: 109 mm^2)	Same
Group size	30 larvae per group, nonfeeding larvae removed at 48 hr	20 larvae per group
Replication	Two groups (60 larvae) per dose level	Two groups (40 larvae) per dose level
Dose levels	Four	Five
Serial dilution factor	1/3	1/2
Control	One group (30 larvae)	Two groups (40 larvae)
Total No.	270 per assay	240 per assay
Observations	Periodically record no. dead larvae; record no. living on day 14; verify presence of virus by microscopic examination of dead larvae	Daily record no. dead larvae; record no. living on day 12; verify presence of virus by microscopic examination of dead larvae
Evaluation	Maximum-likelihood logistic regression analysis (LC$_{50}$, LC$_{90}$, LC$_{99}$, confidence limits, precision, and slope)	Same
Units	Nanograms (dry weight) virus powder per cup; picoliters of stock virus suspension per cup; polyhedral inclusion bodies per cup	Same

be exercised to avoid (or consider in any conclusions) the possibility of interference with the assay by chemicals on the food surface. Most of these shortcomings can manifest themselves as variation in the results of the assay.

To reduce variability, a variety of means have been devised that confine the larvae to a small area of treated diet or leaf, thereby enabling a measured amount of virus to be administered and the time over which it is acquired to be defined. One of the simplest techniques is to give each insect only a small piece of food to which the virus has been applied; both artificial diet[28-32] and detached leaf discs[33-37] have been used in this way. McEwen and Hervey[38] and Drake and McEwen[39] administered a nuclear polyhedrosis virus to the cabbage looper *(Trichoplusia ni)* by inserting an 8-mm leaf disc of broccoli halfway in a slit in a cork, depositing a measured amount of virus suspension on the outer portion of the disc, and using the cork with the disc to stopper a vial containing one larva.

Chauthani[40] described a method in which diet was dispensed into 2.5-dram disposable plastic vials, a 3-mm O.D. glass tube was inserted about 5 mm into the diet, and melted paraffin was poured onto the surface of the diet to a depth of about 1 cm. The glass tube

Table 4
TEST PARAMETERS OF STANDARD BIOASSAY PROCEDURES FOR
LC_{50} AND LD_{50} DETERMINATIONS FOR THE NUCLEAR
POLYHEDROSIS VIRUS OF THE GYPSY MOTH[22]

Test insect	*Lymantria dispar*, New Jersey strain (F_{15})
Larval weight and sex	Newly molted second-instar larvae (5 to 8 mg) both sexes
Exposure to the virus:	
LC_{50} test	Virus incorporated into diet at 52°C
LD_{50} test	Virus applied by microapplicator with a magnetic stirrer operating
Inoculum volume:	
LC_{50} test	1 mℓ stock NPV to 99 mℓ diet. Two 1.25 cm³ diet cubes per sterile plastic Petri dish (100 × 15 mm)
LD_{50} test	1 μℓ to a diet plug cut by a no. 1 cork borer and trimmed to 4 mm in height
Dose unit:	
LC_{50} test	Nanograms per milliliter of diet and polyhedra per milliliter of diet
LD_{50} test	Polyhedra per diet plug
Confinement:	
LC_{50} test	Ten larvae per petri dish
LD_{50} test	One larva per 1 oz plastic creamer containing moistened cotton
Group size	50 larvae per group
Replication	One group per dose level, 5 doses
Control	One untreated group
Rearing temperature	24 ± 2°C
Light regime	16-hr photoperiod followed by 8 hr of darkness daily
Holding period	14 days (48 hr on virus diet)
Data evaluation	Berkson's logit chi-square method[26]

was removed as the paraffin hardened to leave a 7-mm² area of exposed diet to which 0.01 mℓ of distilled water or virus suspension was applied. One larva averaging 20 mg in weight was added to each vial, the vials were capped and placed in a constant-temperature room until the assay was terminated (7 days). Different sizes of glass tubing could be used to accommodate different sizes of test insect.

At least two methods have been described by which the insects are confined to small, treated areas of intact leaves. Peach et al.[41] and Allen and Ignoffo[42] applied 0.01-mℓ aliquots of virus suspensions around the margins of bean leaves, allowed the material to dry, and confined individual larvae to each of the treated spots by means of small clip-on leaf cages. The virus dilutions were prepared with 0.5% hydroxy ethyl cellulose solution to retain a homogeneous suspension and red food coloring to aid in identification of the treated areas on the leaves. A piece of acetate inserted under the bottom prong of the cage prevented the larva from escaping after the leaf area was consumed. The size of the vial used as the cage body could be changed to accommodate different sizes of larvae so that all exposed leaf tissue could be consumed within 12 hr.

In a similar approach, Evans[43] constructed feeding chambers that exposed a sufficiently small area of the leaf that larvae could consume the food and virus within a period of 24 hr or less. The chambers for use with first- and second-instar larvae were constructed by sandwiching a cabbage leaf between flat Perspex® templates (15 × 5 cm) with opposing holes approximately 3.5 mm in diameter (Figure 1). Microtiter plates (Titertek or Flow Laboratories, Ltd.) with the bases removed were used for the larger larvae. The virus suspension was applied in 3-μℓ aliquots to the center of the exposed leaf surfaces and air-dried. This was easily accomplished with the larger larvae, but the small diameter of the holes in the plastic templates required application to be achieved differently for first- and

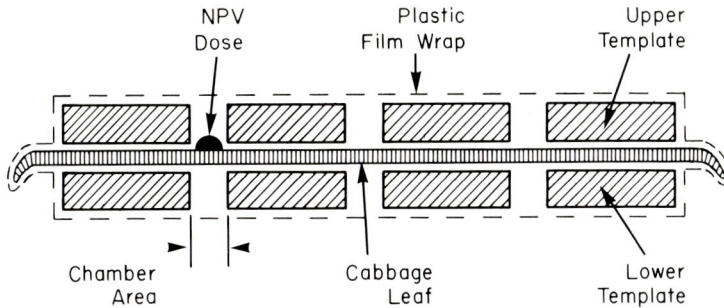

FIGURE 1.   Schematic of a leaf feeding chamber for administering measured amounts of virus to larvae over a short time period. (From Evans, H. F., *J. Invertebr. Pathol.*, 37, 101, 1981. With permission.)

second-instar larvae. For first-instar larvae, the top template was removed and a thin adhesive plastic film with 6 mm diameter holes corresponding in position to those on the plastic template was applied to the leaf strip; this film was firmly anchored to the lower template. The virus suspension was mixed with an equal part of 0.5% agar containing a minute amount of Tween® 20, and 3-$\mu\ell$ aliquots were spread evenly over each exposed leaf area. For second-instar larvae, the leaf was attached to the lower template by plastic film with 6 mm holes as above. The upper template was then replaced and the center of each exposed leaf area marked by a small pin prick. The upper template was again removed and 3-$\mu\ell$ aliquots of virus suspension were applied carefully to the leaf at each pin prick. When the suspension dried, the upper template was replaced. In all cases, larvae were placed in the feeding chambers when the virus suspension dried, and the entire assembly was covered with thin plastic film; the units were placed in the dark for 24 hr, after which those larvae that had consumed the entire area were transferred to individual containers of artificial diet and held under the same conditions for the duration of the test.

Modifications of this type that enable a measured amount of virus to be administered over a relatively short time are effective in reducing some variability, and the more stringint methods allow a better estimation of quantitative relations. However, even with these techniques the acquisition period is long relative to the residence period of material in the midgut of the insect. Thus, they provide only coarse estimates of quantitative dose relations, which are adequate for some applications but not others and are not optimal for time-response studies.

2. Diet Incorporation

Virus can also be administered by incorporating it into the diet as described by Ignoffo.[8] Artificial diet is prepared and cooled to 35 to 38°C for subsequent use. A total of 95 mℓ of the diet is added to 5 mℓ of virus suspension in a beaker and the medium thoroughly mixed and poured into a Petri dish. After the diet solidifies, cylindrical plugs are cut with a cork borer and placed in 30-mℓ plastic cups for use in the assay. This general procedure has been used as the basis for a standardized procedure (Table 5) for determining activity and relative potency of the ELCAR nuclear polyhedrosis virus registered for use against *Heliothis zea* and *H. virescens* on cotton.[25] The parameters of two other such procedures are summarized in Table 6.

The details of the diet incorporation procedure may be varied to suit individual situations, but care must be taken to cool the diet adequately before adding the virus and to mix the combination thoroughly before using. In general, this method provides more reproducible results than surface contamination techniques and it does not require the careful control of surface size and conditions. Like the surface contamination procedures, it is applicable to

Table 5
PARAMETERS OF THE BIOASSAY PROCEDURE FOR *BACULOVIRUS* OF *HELIOTHIS* SPP.[25]

Test insect	Heliothis zea, strain initiated at Brownsville ca. 1970, continuously
Larval age and sex	used since then in laboratory for estimated 250 generations
	Neonatal unfed larvae or 24-hr-old fed larvae, both sexes
Confinement	One larva per cubicle of a compartmented tray with Mylar® film cover
Diet and volume	Modified Brownsville formulation without formalin, 3 to 5 mℓ per cubicle
Rearing temperature	30°C ± 1°C
Light regimen	Variable (no special requirement)
Holding period	6 days
Exposure to the virus	Diet incorporation, larvae continuously exposed
Diet temperature when virus added	60°C
Inoculum volume	10% of diet
Group size	50 larvae per tray
Replication	Two trays (100 larvae) per dose level
Dose levels	Three
Dilution factor	1/10
Control	Two trays (100 larvae)
Total number of larvae	400 per assay
Observations	Record number of dead larvae on the 6th day
Evaluation	Probit analysis (LC$_{50}$, 95% confidence limits, and slope)
Unit	Nanograms (dry weight) of virus preparation per milliliter of diet

Table 6
PARAMETERS OF TWO BIOASSAY PROCEDURES USING THE DIET INCORPORATION METHOD

Ref.	Thomas et al.[44]	Payne[45]
Test insect(s)	*Trichoplusia ni*	*Cydia pomonella*
Age and sex	Neonate (under 12 hr), unfed, both sexes	Neonate, unfed, both sexes
Virus	Nuclear polyhedrosis	Granulosis
Confinement	1 larva per 1 oz clear plastic cup	1 larva per well, microtiter plates
Diet and volume	Modification of Ignoffo diet, quantity not specified	Modification of Hoffman et al.,[46] 200 μℓ per well
Rearing temperature	26.7 ± 0.5°C	26°C
Light regimen	Not specified	Constant darkness
Holding period	Until pupation	14 days
Exposure to the virus	Continuously exposed	Some continuously, some transferred to noncontaminated diet after 24 hr
Group size	Not specified	20 to 40 larvae per group
Temperature of agar when virus added	50°C	45°C
Data analysis	Probit analysis[47]	Maximum likelihood procedure (MLP) program

larvae of any age. If only a single species of insect is being used, the results are not as affected by differences in larval feeding behavior as they are with the surface techniques. However, behavioral differences can strongly affect the results if different species are being used in a single assay, such as when the potency of a particular isolate is compared in two or more species. The method also suffers from several of the other shortcomings described for the surface contamination techniques. The results are still relatively variable, control of the dose received has been poor under the best of conditions, and it is applicable only to insects for which an artificial diet has been developed.

3. Direct Feeding

Several methods have been developed by which the virus suspension can be administered directly to the insects. These methods share the great advantage of permitting a measured, known dose of virus to be administered to each test insect.

a. Per Os Injection

Martignoni[48] used microneedles and a microinjector to administer virus preparations directly into the mouths of larvae. More recently, Paschke et al.[26] described in detail a similar p.o. method used to administer nuclear polyhedrosis and granulosis viruses to fourth-instar larvae of *T. ni*. Microinjection needles were made from glass capillary tubes and coupled to a calibrated 0.25-mℓ glass syringe.[49] Using these syringes and an Isco model M microapplicator (Instrument Specialties Co., Lincoln, Neb.), each test larva was force-fed 2.0 $\mu\ell$ of either a virus suspension or sterile water (controls). Injections were performed while observing through a stereo dissecting microscope to verify that the entire dose was received and not regurgitated.

This method is more accurate and reproducible than the diet contamination methods and allows quantitation of the dose. The major disadvantages are that it is more tedious and is applicable only to late (large) instar larvae. Because of the length of time required to treat a single larva, the number of larvae that can be treated for a single test is limited and the acquisition of virus is not synchronous among the individuals. Also, some care must be taken to ensure that the virus remains suspended and thoroughly mixed in the syringe.

b. Measured Drop Feeding Method for Late Instar Larvae

The larvae of many species of insects will actively seek and drink free water. This behavior has been utilized to administer virus in aqueous suspensions. Bird and Whalen[50] determined the median lethal dose of a nuclear polyhedrosis virus in the European pine sawfly *(Neodiprion sertifer)* by starving the larvae for 24 hr and then allowing them to imbibe 0.5 $\mu\ell$ droplets of virus suspension containing an estimated 5, 50, 500, or 5000 purified polyhedra. The age of the larvae at the time they were treated was not specified but must have been between 24- and 96-hr.

This method was described in greater detail by Klein[51] with the Egyptian cotton worm, *Spodoptera littoralis*. Using Parafilm membrane as the substrate, fifth-instar larvae that had been starved for the previous 24 to 48 hr were each given 5.0 $\mu\ell$ of either a virus suspension or distilled water (controls). Only those larvae that ingested the entire volume were used in the assay, and these larvae were returned to their diet 24 hr after the inoculation. The number of larvae treated per dose group was not specified, nor was the end point of the assay. Preliminary tests showed that very few of the polyhedra remained on the Parafilm after the drops were consumed but large numbers remained when plastic was used as the substrate.

The measured-drop technique permits careful quantitation of the dose administered; although insufficient information is available to fully evaluate the precision and accuracy of this procedure, they are most certainly greater than those of the contaminated-food methods. The major disadvantages are that the insects must be starved for a relatively long period of time prior to treatment, the treatment itself allows only three or four larvae to be treated per minute, and it is not practical for small larvae. Care must be taken to eliminate subjects that do not ingest all of the inoculum or that walk through it and to avoid stimulating regurgitation when the treated larvae are transferred. Also, the procedure is not well suited for investigations that require synchronized treatment of large numbers of insects, such as some time-response studies.

c. Droplet Feeding Method for Neonate Larvae

Recently, a droplet feeding method for neonate larvae was developed as a partial solution to many of the difficulties concerning quantitation of dose received by individual larvae, synchrony of acquisition of the virus, and reproducibility and precision of the assay.[52] As applied to *T. ni*, eggs laid on paper toweling are sterilized by sandwiching the towels between two 28 × 28-cm pieces of 1.6-cm hardware cloth and submerging them in 3% formalin for 30 min. They are then washed for 30 min in running water, air-dried, and placed in polyethylene bags in an incubator. Rather than refrigerate the eggs at 4°C for 18 hr followed by incubation at 28°C as originally described, egg hatch is now timed to occur on the third morning after collection by incubating constantly at 23°C. Larvae that are 3 to 5-hr-old are preselected for vigor by allowing them to climb a ramp made of blotter paper. The selected larvae are gently knocked onto discs 15 cm in diameter cut from plastic-coated picnic plates, a disposable substitute for the 15 × 15-cm Teflon® squares originally used. The discs are previously ringed with a 2:1 mixture of heavy mineral oil and vaseline dispensed from a 50-cc syringe to prevent the larvae from escaping. The virus suspension is dispensed by means of 15-cm disposable glass Pasteur pipettes, the tips of which have been drawn into a small diameter. The suspension, containing 0.5 mg of FD&C Blue No. 1 per milliliter of water used to make the dilutions, is drawn into the pipette, the bulb removed, and the virus dispensed near each larva as a small droplet by touching the tip of the pipette to the plate. When the larvae have finished ingesting the suspension, as evidenced by their moving away from the droplets, they are transferred individually by means of a vacuum pencil to 30-mℓ plastic cups containing approximately 2 mℓ of diet.[53] Only larvae that contain the blue coloring and therefore have ingested the virus suspension are used. The cups are placed in an incubator at 28°C in constant darkness. The larvae are checked for individuals that were killed or injured by handling 24 hr after initiation of the test, and these are replaced by extras. If not used, the extras are discarded. Mortality is recorded on the 10th day in dose-mortality studies, by which time all unaffected larvae have pupated, and at 6-hr intervals over the appropriate time period in time-mortality studies. Originally, the virus suspension contained 5 or 20% (w/v) sucrose. However, this has been found to be unnecessary for *T. ni* as well as several other species. Also, the dispersing agent Darvan No. 2® (R. T. Vanderbilt Co., Los Angeles, Calif.) was used at the rate of 10 mg/mℓ of stock solution to minimize clumping of the occlusion bodies,[54] but this practice has been replaced by the use of bath sonication.

Modifications of the mechanics of the assay can be made to accommodate the habits and peculiarities of different species. For example, the hairy larvae of *Estigmene acrea* tend to break the surface tension and become entrapped in the droplets. This difficulty was overcome by pouring the virus suspension onto the center of the plate, covering it with a piece of Parafilm that had numerous perforations made with dissecting needles, and allowing the larvae to drink through these perforations. This species also requires about 20 min to complete feeding, so the plates were covered with plexiglas boxes containing blotter paper moistened with saturated K_2SO_4 to prevent excessive evaporation from the virus suspension. The critical consideration when redesigning the assay is to present the virus suspension in a manner that permits the insects to approach, ingest, and leave the preparation unhindered.

Dose is measured as the mean number of occlusion bodies per larva, which is determined as the product of the mean number of occlusion bodies per microliter of suspension and the mean number of microliters of suspension ingested. If an estimate of the mean number of virus particles per occlusion body is available, then the dose can be expressed in terms of the mean number of virus particles per larva.

The volumes of suspension ingested by neonate larvae of several species have been shown to be quite constant by the use of solutions of ^{32}P, suspensions of ^{32}P-labeled occlusion bodies, and/or the difference in weight between treated (fed) and untreated (unfed) larvae. *T. ni* larvae ingested an average of 0.006 (± 0.001) $\mu\ell$ of 20% (w/v) sucrose[52] and 0.013 (± 0.003) $\mu\ell$ of water preparations without sucrose.[55,56] Mean volumes of water suspensions ingested by other species, based on gravimetric determinations from various assays, were 0.014 (± 0.004) $\mu\ell$ *(H. zea)*, 0.049 (± 0.006) $\mu\ell$ *(E. acrea)*, and 0.006 (± 0.002) $\mu\ell$ *(Spodoptera frugiperda)*.[54,56]

This droplet feeding method offers many advantages over the previously described methods. It permits the synchronous treatment of large numbers of larvae with a uniform, definable dose of virus. The method eliminates variability due to differences in feeding behavior between individual test subjects, and the use of neonate larvae ensures test subjects that are very homogeneous physiologically, uniform in size and development, sensitive to the virus, and readily and cheaply available. Also, the technique is simple and its reproducibility and precision make possible the use of very few insects in many applications. The major disadvantages are that it is not very practical for other instars and the synchrony in response obtained by this method requires that mortality checks in time-response studies be made at relatively short intervals for fast-acting viruses.

d. Egg-Dip Method

Neonate larvae of many species of insects ingest the shells of their eggs during or immediately after eclosion. Taking advantage of this behavior, Ignoffo et al.[57] used an egg-dip procedure to infect neonate *H. zea* larvae with a cytoplasmic polyhedrosis virus. Eggs were placed in a 50 × 12-mm Petri dish and immersed in either a suspension of virus in sterile deionized water or the water alone. After 15 min the excess fluid was removed by means of a Pasteur pipette and the eggs air-dried. The eggs were then incubated at 30 ± 1°C until hatch occurred. The hatching larvae were transferred to cells containing artificial diet, placing one larva per cell, and reared to the fourth instar at 30 ± 1°C. They were then examined for virus infection.

A modification of this technique was described by Williamson[58] to infect *Phthorimaea operculella* (potato moth) with a granulosis virus. Sectors of filter paper containing eggs close to eclosion were randomly allocated to sterile plastic Petri dishes for treatment; sectors containing densely aggregated egg masses were excluded. Each virus dilution was poured over the appropriate group of eggs, completely immersing them. The control solution consisted of BSA (0.2% w/v) and NH_4CO_3 (0.001% w/w). After 15 min, the papers were removed, drained, and dried in open Petri dishes under a laminar flow hood. The dried sectors were placed in separate sterile containers and stored at 28°C until egg hatch. The neonate larvae were transferred to prepared slices of potatoes within 4 hr after hatching,[59] and held at 28°C. After 17 days, response was scored as failure to pupate and viral infection was confirmed by microscopy.

e. Topical Treatment of Pharate Larvae

Many lepidopterous larvae also ingest their own exuvium immediately after molting. We have used this behavior successfully to treat third- through fifth-instar larvae of *T. ni* with nuclear polyhedrosis virus (NPV). The readily identifiable pharate larvae are placed individually in plastic 30-mℓ cups containing a single disc of filter paper on which the larva can attach. A measured amount of virus suspension containing FD&C Blue No. 1 is applied to the dorsum immediately behind the head by means of a microsyringe, and the larvae are placed in an incubator at 30°C. After 4 hr, those larvae that have molted and ingested the virus, as evidenced by blue coloration within the larva or a blue fecal pellet, are transferred to diet for use in the assay.

4. Hemocoelic Injection

Liberated or nonoccluded virus can be administered to larvae by direct injection into the hemocoel.[60-62] Bergold[61] injected early fourth-instar larvae (approximately 3 to 5 mm in length) of the spruce budworm *(Choristoneura fumiferana)* with 1 μℓ quantities of dilutions of virus particles liberated from a nuclear polyhedrosis virus. The microneedle was inserted into the tip of a proleg so that the muscle would close the wound and prevent bleeding after injection. Three concentrations of virus were used in each test with 20 to 30 insects per dose.

The applications of this method are very limited. It is effective only for liberated or nonoccluded virus, it is tedious, and it is abusive to the larvae. Care must be taken to prevent secondary infections and to keep suspensions well mixed. Perhaps its most useful applications are in studies of host range and barriers to infection (e.g., Granados et al.[63]).

C. Analysis of Results

A detailed description of the mathematical bases and computations of the various statistical procedures used to analyze bioassay data is beyond the scope of this chapter. Their omission is further justified, in part, by the belief that most researchers will not be interested in the details of the procedures beyond what is needed to interpret the results and understand their limitations. Accordingly, the discussion will concentrate on the uses of particular methods, some of the assumptions upon which they are based, their strengths or shortcomings, and the major differences and/or similarities between the methods. Where appropriate, the reader will be directed to specific references describing the mathematical calculations.

1. Dose-Mortality Assays

When considering the analysis of data from dose-mortality assays, a distinction must be made between assays conducted to compare the potencies of two or more preparations and those in which the purpose is to determine the relationship between a particular virus and a given host. Assays conducted to determine the relative potencies of two or more preparations can be analytic dilution assays (e.g., tests of different batches of a given virus), comparative dilution assays (e.g., tests of different isolates in a given host), or simply comparative biological assays (e.g., tests of a given isolate in different hosts). The statistical procedures most commonly used apply in a strict sense only to analytic dilution assays, but they can be used for comparative dilution assays providing the limitations regarding conclusions drawn from the results are clearly understood. If the agent or combination of agents responsible for producing the characteristic response is not the same in each preparation but only appears to be acting in the same manner (i.e., it is a comparative dilution assay), then the results might change if different hosts, methods, response, or experimental conditions were used; in an analytic dilution assay, the potency estimation (but not necessarily the precision of this estimation) should be unaffected by such differences.

Frequently in baculovirus research, isolates that are known to be quite different are compared within a given host in comparative dilution assays. Data from such assays can be analyzed by the standard methods and statistical validity tested as linearity. However, the condition of similarity cannot be assumed and any deviation from parallelism signals the need for additional caution in drawing conclusions from the results.

a. Log-Normal Model

The most widely used method for analyzing the results of dose-mortality assays of insect viruses is probit analysis. The procedure is described in great detail by Finney,[1] and numerous

Table 7
COMPUTER PROGRAM SCORES BY SELECTED CRITERIA[64]

		Program				
Criterion[a]	BMD 03S	Probit analysis	Potency probit analysis	LOCSAN	POLO	SAS probit
I	2	4	4	3	1	2
II	3	2	2	2	4	1
III	2	1	1	1	4	3
IV	3	2	2	2	4	3
V	3	2	2	1	2	3
VI	1	3	3	4	3	2
VII	1	1	4	3	5	2

[a]   I = Portability (1-4); II = handling of natural mortality (1-4); III = ease of data entry (1-5); IV = quality of printing (1-5); V = dose transformations available (1-3); VI = effective doses and fiducial intervals (1-4); VII = statistical tests for validity of model and hypotheses (1-5). Higher numbers denote greater versatility or completeness of capability.

computerized programs have been developed to accomplish the mechanics of the analysis. The widespread availability of these programs makes probit analysis the method of choice under several circumstances. If the only interest is an estimate of the median effective dose or if inferences with respect to absolute numbers of virus particles cannot be made, then the ease and availability of the procedure make it the likely selection. If the larvae are known to vary considerably in susceptibility to the virus, then the probit model likely will be the most appropriate in terms of fundamental as well as statistical validity. If the dose administered to each larva is not carefully controlled, the results are likely to fit the probit model because the variation in doses received by individuals is statistically indistinguishable from individual variation in susceptibility. Major assumptions of this model are (1) for each host there exists an individual effective dose (IED) and receipt of that many or more particles will invariably cause a response whereas any fewer will not, and (2) the IEDs within a population are log-normally distributed. The infective particles are assumed to act cooperatively to produce the response. With microorganisms this can be interpreted in a broader sense of acting cooperatively to overcome host defenses rather than to initiate infection, or in the case of organisms such as baculoviruses that must be ingested and pass barriers in the alimentary canal to get to suitable sites, the individual variability in susceptibility could reside in these "preinfection" barriers rather than in a requirement for cooperative action.

Several of the computer programs for probit analysis have been compared by Mr. G. S. Walton of the U.S. Forest Service, Hamden, Conn. (Table 7).[64] The programs were scored on the basis of seven criteria, i.e., portability, handling of natural mortality, ease of data entry, quality of printing, dose transformations available, effective doses and fiducial intervals, and statistical tests for validity of model and hypotheses. The programs evaluated were BMD 03S, Probit Analysis,[47,65] Potency Probit Analysis,[47,65] POLO,[66] and SAS Probit.[67] A versatile program not included in the above evaluation is PARLIN.[68]

For a relatively fast solution to the dose-response curve that does not require a computer or extensive computations, Litchfield and Wilcoxon[69] provide a graphic method. By the use of nomographs, simple solutions can be obtained that are numerically equal to probit analysis but do not require transformation of the data. Unlike the various approximate methods, this procedure permits tests for parallelism and heterogeneity and calculation of confidence limits, slope of the regression line, relative potency, etc. The ready availability of computers and calculators makes this type of solution far less useful now than when it was first proposed, but it still finds valuable application under some conditions.

Heterogeneity is tested in a quantal response assay of a single preparation by chi-square given as

$$\chi^2 = S \ [(r - nP)^2]/[nP(1 - P)] \tag{3}$$

where n is the number of subjects used at a dose, r is the number responding to the given dose, and P is the expected probability of occurrence of response taken from the response curve at the particular dose. The number of degrees of freedom is two less than the number of doses (i.e., $k - 2$). The test presupposes that none of the expected frequencies is very small; if nP or $n(1 - P)$ is less than about five for any dose, the chi-square should be calculated after combining the observed and expected frequencies for that dose with the next dose, and the degrees of freedom should be reduced by one. This chi-square also could be calculated as

$$\chi^2_{k-2} = S_{yy} - [(S_{xy})^2/S_{xx}] \tag{4}$$

where $S_{xy} = Snwxy - [(Snwx)(Snwy)/Snw]$, $S_{yy} = Snwy^2 - (Snwy)^2/Snw$, n is the number of subjects per dose, w is the weighting coefficient, and y is the working probit.[1]

When two or more preparations are compared by a quantal response assay, chi-square for heterogeneity can be calculated as

$$\chi^2 = \Sigma S_{yy} - \Sigma[(S_{xy})^2/S_{xx}] \tag{5}$$

where Σ denotes summation over data for the different preparations and S denotes summation over doses within a preparation.[1] If the number of preparations is denoted by a, the degrees of freedom can be calculated as $a(k - 2)$.

If a large chi-square for heterogeneity is observed, the data should be examined such as by a plot to determine if the significance is produced by random or by systematic deviation from the expected. This constitutes a check of statistical validity equivalent to that for deviation from linearity used in assays based on quantitative responses, and systematic variation would indicate that the model is invalid. If the deviation is random, the *heterogeneity factor* can be calculated as

$$s^2 = \chi^2/f \tag{6}$$

where f, the number of degrees of freedom, is two less than the number of doses for an assay of a single preparation or four less than the total number of doses tested in an assay of two preparations.

If no heterogeneity is detected, parallelism (and thus fundamental validity) can be tested when two preparations are compared in an assay based on quantal response by chi-square given as

$$\chi^2 = [\Sigma((S_{xy})^2/S_{xx})] - [(\Sigma S_{xy})^2/\Sigma S_{xx}] \tag{7}$$

with one degree of freedom. If heterogeneity is found and the heterogeneity factor is used, a variance ratio test should be used to test parallelism instead of the chi-square. F would be taken as the ratio of the above chi-square value to s^2 with one and f degrees of freedom.

b. Logistic Model (Minimum Chi-Square)

Berkson[70-73] described and applied the use of the logit transformation and minimum chi-square solution to bioassay results. The method was considered to be an easier solution to accomplish, which is no longer a consideration with the availability of computers. Finney[1] concluded that maximum likelihood is the preferable principle of estimation, since minimum chi-square has no theoretical advantages, can be troubled if class frequencies are small, and is more restricted in its applications. Nonetheless, if the potential difficulties are recognized, the minimum logit chi-square method can be applied successfully to the bioassay of baculoviruses, and a computer program to do so, called LOCSAN, has been described and used by Paschke et al.[26]

c. Exponential Model

Unlike chemicals, self-replicating agents such as bacteria and viruses are capable of acting independently of each other to produce a response in a subject. When this occurs and if the hosts are reasonably uniform in susceptibility to the pathogen, the dose-response relation can be more properly described and interpreted through an exponential model.[74-77] This model, also referred to as the independent action model, describes a stochastic process (as opposed to the deterministic process described by the probit model) in which the lethal dose is the same for all individuals and the response rates reflect the chance of receiving this dose.

The model has received considerable attention in work on pathogens of vertebrates[20,78-80] and plants.[81] Only recently has it been advanced seriously to describe the relations between some baculoviruses and insects.[52,55,82] The main assumptions of the model are that the virus particles act independently and the hosts are uniformly susceptible to the virus. However, in practice the model is not too perturbed by even moderate variability in host susceptibility.[74] Quantitative features of an infection predicted by this model include (1) an exponential relation between dose and the proportion of hosts responding, with the slope of the line estimating the proportion of the virus effective in producing an infection, (2) a critical concentration of organisms at which response occurs, (3) the same proportion of hosts responding even when the total dose is split into several smaller doses, (4) the probability that a host given a particular dose will respond due to the multiplication of one, two, or any number of inoculated organisms, and (5) a mean response time that tends to become constant at doses less than the ED_{50} and is inversely related to log of dose at higher concentrations. Using the notation of Peto,[20] the model is in the form

$$\ln S = -pn \tag{8}$$

where lnS is the natural log of the proportion surviving, n is the number of organisms administered to each subject (i.e., the dose), and p is the probability of any one organism killing a subject (also referred to variously as the proportion of the inoculum that is effective in initiating an infection or the susceptibility of the host). A maximum likelihood solution is provided by Peto[20] and can be readily computerized; two such programs, one in APL and one in Basic language, are available from the authors upon request. Linearity can be tested with chi-square as in probit analysis according to the equation

$$\chi^2 = \text{Sum } [(r - mS)^2/mS(1 - S)] \tag{9}$$

where r is the number of subjects surviving, m is the number of subjects tested, and S is the expected proportion surviving. The number of degrees of freedom is one less than the number of doses used, since only one parameter (i.e., p) is being estimated.

d. Other Methods for Estimating the ED_{50}

Many approximate methods have been proposed to permit estimation of the ED_{50}, including many graphic and nonparametric procedures that are listed by Finney.[1] Dougherty[83] describes in detail with examples the application of some of these methods to virus research. Two that have been applied to studies of baculoviruses are the Spearman-Karber method and the Reed-Muench method. The primary attribute of the Spearman-Karber method is that for an appropriately designed assay it can give an estimation equivalent to that obtained by maximum likelihood but without requiring an assumption of the form of the tolerance distribution. The primary disadvantages include the lack of provision for validity tests, the need for a specialized experimental design, and the requirement for a greater number of doses than previously described methods.[1] There are strong objections to the Reed-Muench method, including the known frequent invalidity of the basic assumption that all subjects reacting to a given dose would have reacted to a higher dose and the inability to calculate estimates of error.[83] Regarding this method, Finney[1] states that "except as statistical history, (it) should be forgotten."

2. Time-Mortality Assays

Time-response can be studied (1) to estimate the relationship between time and the proportion of subjects responding, (2) to determine virus titer, or (3) to examine the nature of the virus-host interaction.

Sampford[84-86] provides a comprehensive treatment of the estimation of time-response distributions for a variety of situations. The manner in which the data are grouped, the completeness of the sample, and, if the sample is not complete, the reason for the truncation are all important in determining the method by which the data from a time-response assay are to be analyzed. Using the terminology of Sampford, the data are *ungrouped* if observations are made continuously throughout the experiment. If the observations are made periodically, then the groupings can be *regular* (i.e., the observations made at equal intervals on the scale of the time metameter), *transformed regular* (i.e., taken at regular intervals in the original time scale and transformed to another scale for analysis), or *irregular* (i.e., any grouping other than described above). The sample can be *complete* (all individuals respond) or *incomplete*. Incomplete data can result from *survival*, in which some individuals fail to respond because of immunity, recovery, or failure to receive the inoculum, or *truncation*, in which the responses of some individuals are intentionally or accidentally not recorded. Space limitations preclude a detailed consideration in this chapter of the various situations, but a reference guide to many situations encountered and the methods for analyzing the data from each is given by Sampford.[86]

Litchfield[87] described a graphic solution for time-response curves that has been used frequently in baculovirus research. The method permits calculation of response times, standard errors, confidence limits, reaction time ratios, and a test for parallelism. In addition, it accommodates both complete and truncated data. Bliss[88] described a mathematical solution, and an interactive computer program that provides maximum likelihood ST_{50} values based on a logit version of Bliss's probit model is available[54] (Figure 2). The program is started by entering the name "logit", after which the experimenter is asked to make entries of the first time check at which mortality was observed, the number of dead subjects found in each time interval (not cumulative), and the length of the time interval at which checks were made; as written, the program assumes that response is not artificially truncated and requires that all time intervals be the same duration. The program then calculates the slope and median response time and their respective standard errors. An example is given in Figure 2 of the analysis of data from an assay of a nuclear polyhedrosis virus in neonate *Heliothis zea*. The first mortality was observed at 66 hr, the number of larvae dead at the 66-, 72-, 78-, 84-, and 90-hour checks were 2, 4, 25, 8, and 1, respectively, or a total of 31 larvae responding. Checks were made at 6-hr intervals, and the ST_{50} was 75.4 hr with a standard error of 0.7 hr.

Although time-mortality data cannot be analyzed validly by the same methods as used for dose-mortality assays, from the standpoint of practicality a method such as probit analysis

```
     ∇LOGIT[□]∇
      ∇ LOGIT
[1]   'ENTER BIOASSAY IDENTIFIER'
[2]   R←□
[3]   'ENTER THE FIRST CHECK TIME (IN HOURS) WHEN MORTALITY OCCURRED'
[4]   X1←●T←□
[5]   'ENTER (THE ARRAY OF) NUMBER OF DEATHS IN THIS AND IN EACH'
[6]   'SUCCEEDING INTERVAL UP TO AND INCLUDING THE LAST DEATH(S)'
[7]   ⍴THIS ARRAY IS THEN BRACKETED BY A ZERO COUNT AT EACH END AND
[8]   ⍴SURVIVAL TIME IS ASSUMED TO BE CENSORED OUTSIDE THIS TIME SPAN
[9]   L0←●((X÷(N-X←S[1])),X÷((N+÷/S)-X←(+\S)[(K+⍴S←□)-1]))
[10]  'ENTER THE INTERVAL LENGTH (IN HOURS) BETWEEN MORTALITY CHECKS'
[11]  PA←φ((●((X2,-X1)+.×L0÷(D×P1))),P1←(-/L0)÷D×X1-X2←●(T+(K-2)×L←□))
[12]  LT←●(((T-L×(T>L))-L)+L×⍳K+⍴S←(T=L)+(0,S,0))
[13]  WT:W←(PF←(F,1)-(0,F÷1÷1+×Z1←-Z←(P1+PA[1])×(LT-●PA[2])))*⁻0.5
[14]  P←((W×(((Z,0)×A)-(0,Z)×B)÷P1),W×((B+(0,G))-A←((G+F×(1-F)),0))×(÷/PA))
[15]  PA←PA+DEL←(V←⊞(N×PX+.×⍉PX))+.×((PX←(2,K+1)⍴P)+.×(S,0)×W))
[16]  →(~I,I←((+/|DEL)<1E⁻10))/WT,ANS
[17]  ANS:'THE SLOPE AND ST50 ARE:  ',⍕PA
[18]  'WITH STANDARD ERRORS:     ',⍕(V[1;1],V[2;2])*0.5
      ∇
```

Example:

```
      LOGIT
ENTER BIOASSAY IDENTIFIER
BIOASSAY 120
ENTER THE FIRST CHECK TIME (IN HOURS) WHEN MORTALITY OCCURRED
□:
      66
ENTER (THE ARRAY OF) NUMBER OF DEATHS IN THIS AND IN EACH
SUCCEEDING INTERVAL UP TO AND INCLUDING THE LAST DEATH(S)
□:
      2 4 25 8 1
ENTER THE INTERVAL LENGTH (IN HOURS) BETWEEN MORTALITY CHECKS
□:
      6
THE SLOPE AND ST50 ARE:  32.27523342 75.37229625
WITH STANDARD ERRORS:    4.762284049 0.7136832902
```

FIGURE 2.   An interactive computer program in APL language for the analysis of time-mortality data and an example of its use. (Program provided courtesy of Dr. D. S. Robson,[89] Department of Plant Breeding and Biometry, Cornell University, Ithaca, N.Y.)

will generally give good approximations of the ST_{50} or LT_{50} if such an estimate is all that is desired.

III. IN VITRO ASSAY OF BACULOVIRUSES

A. In Vitro Infection

Titration assays of viruses in vitro are in many ways simplified as compared to whole-animal assays. The experimenter has more control over the conditions of the assay, many of which significantly affect titration results. Of prime importance are the state of the cell, composition of the diluent, density of cells, and cell culture medium. These factors can affect the efficiency of plating, virus replication, and/or development of infection foci.

Since viruses are not like bacteria where most cells can replicate, the number of virus particles per sample cannot be determined directly through titration procedures. This raises the question of whether the initiation of infection occurs through cooperation of two or more virus particles or if a single particle is capable of initiating an infection. The answer to this question can be obtained through plaque assay data from a dilution series. The number of plaques obtained at each lower dilution will increase by the dilution factor to the power of the number of virus particles required to initiate an infection. For instance, a twofold increase in inoculum concentration will result in a twofold (2^1) increase in the number of plaques if a single particle can initiate an infection. However, if two particles are required, the number of plaques will increase fourfold (i.e., 2^2).

This relationship is expected only if the virions or particles are relatively rare in occurrence and randomly distributed within the inoculum, and, therefore, their occurrence in a sample should follow the Poisson distribution described by the equation

$$p_r = (s^r e^{-s})/r! \tag{10}$$

where s is the average number of virus particles per sample, r is the actual number of virus particles per sample, p_r is the probability of having exactly r virus particles in a given sample, and e is the base of the natural logarithm system. By this relationship, the probability of a sample containing no particles (i.e., r = 0) is given by

$$p_0 = e^{-s} \tag{11}$$

The probability of the sample containing exactly one particle is

$$p_1 = se^{-s} \tag{12}$$

and that of the sample containing exactly two particles is

$$p_2 = (s^2 e^{-s})/2 \tag{13}$$

Agreement of the frequency distribution of plaques with this distribution indicates that the virions are randomly distributed within the original and diluted samples and that one virion is capable of initiating an infection. A plot of data conforming to these conditions presents a linear relationship between the numbers of plaques and the concentrations of virus inoculum. The standard deviation of mean plaque number at a particular dilution can be estimated from the square root of the variance (the sum of squares of the difference between each replicate value and mean divided by one less than the total number of replicates). Since the variance of the Poisson distribution is equal to the mean, the standard deviation calculated in this manner should not differ greatly from the square root of the mean.

Although all plaque assays of insect viruses analyzed to date indicate single-hit Poisson kinetics, this does not mean that every virus particle succeeds in initiating an infection. Some particles may be inactivated, attach to dead cells, or simply never come in contact with an attachment site on a susceptible cell. Viral titers are measures only of the average number of successful virions under a particular set of conditions. For example, virus titrations and particle counting by electron microscopy have provided estimates that there are 266 ± 177 particles per plaque forming unit (PFU) for the *Spodoptera frugiperda* nuclear polyhedrosis virus.[90] In another study, Volkman et al.[91] estimated that there were 128 particles per PFU for the *Autographa californica* nuclear polyhedrosis virus (AcNPV). Since the number of PFUs does not decrease significantly when inoculum is transferred from one set of cells onto another,[92] the high ratios of physical particles to plaque forming units appear to arise through lack of contact of the virus with attachment sites on competent cells.

B. Qualitative Assay

Plaque assay procedures have been developed for several nuclear polyhedrosis viruses replicating in *T. ni* (TN-368),[93-95] *S. frugiperda* (IPLB-SF 21),[90,96,97] and *H. zea* (IPLB-HZ 1075)[98] cell lines. The most common procedures rely on the presence of virus occlusion bodies within infected cells to detect plaques. However, alterations in cell morphology and inhibition of cellular division following infection can also be used to easily discern plaque development. For instance, virus mutants lacking the ability to produce occlusion bodies cause cell lysis within 1 to 2 days.[99] Cell lysis also occurs following productive infections with the nonoccluded Hz-1 virus in all lepidopterous cell lines tested.[100]

The first step for assaying any insect virus in vitro is to obtain indicator cells in the logarithmic phase of growth. Lepidopterous cells are most susceptible during this phase.[94] The cells can be seeded into multiwell plates, Petri plates, or on microscope slides (for immunological procedures). The appropriate cell densities are determined by the doubling time of the cells and the replication rate of the virus under the conditions used. For example, the Hz-1 virus replicates faster than the AcNPV. Therefore, the initial cell density used for best results in Hz-1 virus plaque assays is higher than that used in assays of the AcNPV. Although the desired physical contact between the virus and cells can be increased by increasing cell density, excessively high cell densities may interfere with virus replication. A cell contact inhibition phenomenon that inhibits baculovirus replication has been demonstrated with all lepidopterous cell lines tested.[101,102] Therefore, several cell densities should be tested to obtain optimum plaque numbers and size.

The cells are allowed to attach to the vessel surface by centrifugation or by allowing them to incubate undisturbed for 2 to 24 hr. After attachment of the cells to the culture vessel, the tissue culture medium is removed and virus inoculum is added in a sufficient volume of medium to prevent drying. Inoculation of the cells is usually allowed to proceed for 1 to 2 hr. This is accomplished with a rocker platform or by tilting the plates every 15 min. When feasible (e.g., with cell lines such as TN-368 and IPLB-SF 21) the plates can be centrifuged for 1 hr at $1000 \times g$. Centrifugation increases the sensitivity of the assay approximately tenfold[94] by concentrating the virus and cells in the bottom of the plate and thereby optimizing virus-cell contact.[92] Dougherty et al.[103] indicated that adsorption of the AcNPV to TN-368 cells was optimized at pH 7 and by pretreatment of the cells with poly-*l*-lysine or heparin.

Removal of the virus inoculum from the cells is sometimes a critical step. If an agarose overlay is to be used, excessive amounts of residual medium can result in a thin monolayer in which cells and/or progeny virions can move under the overlay, resulting in undefined plaque areas. Efficient removal of the inoculum medium without loss of an excessive number of cells can be accomplished by placing an aspiration device with a small orifice at the edge of the plate and removing the medium as capillary action moves it up the side of the vessel. If a large number of samples is being handled, this must be done quickly in order to avoid desiccation of the cells before the overlay is applied. In multiwell plates handled in a horizontal laminar flow hood, desiccation can be identified by a crescent-shaped area of dead cells in the region farthest from the sterile air source during handling.

The purpose of using an overlay in the plaque assay is twofold. First, the amplification of the infection foci allows for readily visible infection units that are easily enumerated when the appropriate virus dilutions are used. Secondly, the delineation of the infection site allows for plaque purification of virus isolates. If the second criterion is not of interest, minimal or no overlays may be needed with NPV infections, and the primary infections can still be identified in many cases by the presence of occlusion bodies in a single cell prior to their production in secondarily infected cells.

The first plaque assay overlay procedure for use with NPVs in cell culture was published by Hink and Vail in 1973[93] and was later modified by Hink and Strauss.[95] The procedure calls for a 0.6% (w/v) methyl cellulose (4000 CP) overlay in tissue culture medium containing $0.1 M$ 2-(N-morpholino) ethanesulfonic acid (MES) adjusted to pH 6.15 with KOH. The addition of MES reduces the formation of salt precipitates, allowing for longer incubation periods and easier plaque identification. Volkman and Goldsmith[104] reported a similar assay in which cells were grown on slides, inoculated under methyl cellulose, and then stained with an immunoperoxidase procedure. Although these procedures are excellent for the enumeration of infectious units, they are not suitable for plaque purification because progeny virions are not sufficiently localized in the viscous overlay or they are inactivated.

The agarose overlay procedure is generally the superior method for obtaining a measure of infectious units and cloning virus isolates. Several types of agarose have been used successfully. However, Seaplaque™ agarose (Marine Colloids, Rockland, Maine) has been used most extensively because it has a gelling temperature below 30°C, which eliminates exposure of the cells to inactivating temperatures. As with the methyl cellulose overlay, salt precipitates can interfere with visualization of the plaques, but this can be eliminated by autoclaving the agarose in the basal medium (i.e., Grace's medium). We use the basic salt solution of the Grace's medium for dissolution, swirl the tubes after autoclaving to suspend any precipitates, and then cool to room temperature. Prior to use, the $2\times$ agarose solution is melted in a boiling water bath, mixed with complete medium, and allowed to cool in a 37°C water bath. Although the overlay contains half-concentration medium, cell division and virus development still proceed normally.

The final concentration of the agarose which can be used is dependent on the gel strength and can vary from 0.3 to 2% (w/v). For instance, Seaplaque™ agarose has a low gel strength and is usually used at 1.5 to 2%, whereas Ultrapure™ agarose (Bio-rad Labs, N.Y.) has a high gel strength and can be used at 0.3%. The optimum gel strength may vary with different viruses. In our assays of the AcNPV and Hz-1 virus in TN-368 cells, optimal plaque development is obtained with 1.5 and 1% Seaplaque™ agarose, respectively. The reason for this difference may be that the Hz-1 virus is large and therefore diffuses more slowly.

After applying the overlay, the assay plates are sealed to prevent desiccation and incubated undisturbed for 5 to 8 days. Plaques containing cells with virus occlusion bodies usually can be viewed with the unaided eye and counted with the aid of a binocular microscope or Bellco plaque viewer (Vineland, N.J.). Stains have been employed to enhance the visualization of plaque areas. Visualization of lytic plaques has been enhanced with 0.5% (w/v) trypan blue in PBS.[100] Solutions of 2% (w/v) 2-(p-iodophenyl)-3-(p-nitrophenyl)-5-phenyltetrazolium or 0.1% (w/v) neutral red in PBS have been used to enhance the visibility of plaques that have cells containing NPV occlusion bodies.[96]

C. Quantal Assay

Typically, quantitation of insect viruses in vitro is accomplished by the inoculation of cells dispensed in multiwell plates with a series of virus dilutions. After 5 days, the plates are scored daily for positive responses until no further responses are noted. Typical responses used include the presence of viral inclusion bodies, swelling of cell nuclei, and/or cell lysis. The data are then used to calculate the median tissue culture infective dose ($TCID_{50}$), which is the dose that produces positive responses in 50% of the inoculated wells. Two methods have been used most frequently for the analysis of data from in vitro assays — the Reed-Muench method and the Spearman-Karber method.

1. Analysis by the Reed-Muench Method

The procedure for estimation of the median lethal dose (LD_{50}) devised by Reed and Muench[105] is used quite widely in insect virology. It is based on the assumption that cells becoming infected at a given virus dilution would also be infected at a lower dilution and, conversely, that cells failing to be infected at a given dilution would also not be infected at a higher dilution. Accordingly, the estimate of the LD_{50} is based on the cumulative frequencies of positive responses and extrapolation to the dilution that would produce 50% response. An example of the application of this method is given in Table 8. The $TCID_{50}$ can be calculated by the formula

$$\text{Log } TCID_{50} = \text{Log of the dilution} + [\text{the proportionate distance of}$$
$$50\% \text{ from the observed response greater than or less}$$
$$\text{than } 50\% \text{ times the log of the dilution factor}] \qquad (14)$$

For the example given in Table 8, using the observed response greater than 50%, this is

Table 8
EXAMPLE DATA FOR DETERMINATION OF THE TCID$_{50}$ BY
THE REED-MUENCH METHOD[a]

Log virus dilution	Infections per no. inoc.	Cumulative infections	Cumulative healthy	Infection ratio	Percent infection
−4	24/24	51	0	51/51	100
−5	22/24	27	2	27/29	93
−6	5/24	5	21	5/26	19
−7	0/24	0	45	0/45	0

[a]   TCID$_{50}$ is the mean tissue culture infective dose.

$$\text{Log TCID}_{50} = -5.0 - [((93 - 50)/(93 - 19))(1)]$$

$$= -5.58$$

Therefore, the TCID$_{50}$ in this assay is the antilog of -5.58 or $1/(3.8 \times 10^5)$. Since this represents the dilution factor, its reciprocal divided by the volume of inoculum used will give the number of TCID$_{50}$ units per milliliter of sample. In this example, if each well was inoculated with 0.1 mℓ of sample, there would be $(3.8 \times 10^5)/0.1$ or 3.8×10^6 TCID$_{50}$ units per milliliter of sample.

This method has been very popular due to the simplicity of the assay, the ease of analysis, and because, for most purposes, it gives usable results. However, one should be aware that there is no method of deriving an estimate of error from the observed data and the assumption that virus infections at a particular dose would have occurred at higher doses is not always correct.[106] Also, the method assumes that the responses are distributed symmetrically about the ED$_{50}$, which is not true for the exponential relationship described by the first term of the Poisson distribution.[1]

2. Analysis by the Spearman-Karber Method

The Spearman-Karber method is an alternative to the Reed-Muench method and is recommended by Finney[1] over other approximate methods because it performs well, does not require assumptions about the form of the distribution, and permits estimates of error to be made. Properly applied, the method requires equal numbers of subjects at each dose, equal and narrow spacing of doses on a logarithmic scale, and a long series of doses. This method is illustrated in Table 9 using the same example as used for the Reed-Muench method. The TCID$_{50}$ can be calculated from the formula

$$\text{Log TCID}_{50} = x_{p-1} + (1/2 \text{ d}) - d\Sigma p \qquad (15)$$

where x_{p-1} is the highest log dilution giving all positive responses, d is the log dilution factor, p is the proportion positive at a given dose, and Σp is the sum of values of p for x_{p-1} and all higher dilutions.[83] For the example in Table 9

$$\text{Log TCID}_{50} = -4 + 1/2(1) - (1)(2.13)$$

$$= -5.63.$$

Therefore, the TCID$_{50} = 1/(4.3 \times 10^5)$ and the sample titer is 4.3×10^6 TCID$_{50}$ units per milliliter. The standard error (SE) of the Log TCID$_{50}$ can be calculated from

Table 9
EXAMPLE DATA FOR DETERMINATION OF THE TCID$_{50}$ BY THE SPEARMAN-KARBER METHOD[a]

Log virus dilution (x)	No. inoc. (n)	No. positive	Proportion positive (p)	(1-p)
−4	24	24	1.00	0
−5	24	22	0.92	0.08
−6	24	5	0.21	0.79
−7	24	0	0	1.00

[a] TCID$_{50}$ is the mean tissue culture infective dose.

$$SE = [d^2\Sigma(p(1 - p)/(n - 1))]^{1/2} \qquad (16)$$

where n is the number of inoculated samples. Thus, for this example

$$SE = [1^2((0.92)(0.08)/(24 - 1)) + ((0.21)(0.79)/(24 - 1))]^{1/2}$$

$$= \pm 0.10.$$

It should be noted that, just as with plaque assay results, TCID$_{50}$ titrations of insect viruses will vary depending upon the density of cells seeded per well and the incubation period of the virus required to identify infection.[94]

IV. CONCLUDING REMARKS

The importance of bioassay in determining the potency of preparations and monitoring the quality of commercially produced microbial pesticides is well established. Bioassay is also well entrenched as a valuable tool in research. However, even though many virus titrations appear simple and straightforward, often the reliability of the results obtained could be increased if greater attention was given to the experimental design and conditions of the assay. Also, additional information frequently could be gleaned from the results of such assays if the basis, assumptions, and applications of the biological and statistical methods used were more completely understood by researchers. Much of the potential of bioassay in elucidating the pathogen-host relationship and infection processes has been unrealized in the past, but, as the methods and level of understanding improve, this potential is certain to be more completely filled both in basic and applied research.

To help promote this understanding, we have attempted as part of this chapter to outline many of the major concepts that form the biological and statistical foundations of bioassay; additional consideration of this subject as it relates to the assay of insect pathogens can be found in the recent summary by Reichelderfer.[107]

REFERENCES

1. **Finney, D. J.,** *Statistical Method in Biological Assay,* 3rd ed., Charles Griffin & Co., London, 1978.
2. **Meynell, G. G. and Meynell, E.,** *Theory and Practice in Experimental Bacteriology,* 2nd ed., Cambridge University Press, London, 1970, chap. 6.

3. **Miles, A. A.,** The biological unit of activity, *Bull. W. H. O.,* 2, 205, 1949.

4. **Burn, J. H., Finney, D. J., and Goodwin, L. G.,** *Biological Standardization,* 2nd ed., Oxford University Press, London, 1950.

5. **Hewitt, W.,** *Microbiological Assay: An Introduction to Quantitative Principles and Evaluation,* Academic Press, New York, 1977, chap. 1.

6. **Wood, E. C.,** Calculation of the results of microbiological assays, *Nature (London),* 155, 632, 1945.

7. **Finney, D. J.,** *Probit Analysis,* 3rd ed., Cambridge University Press, London, 1971.

8. **Ignoffo, C. M.,** Bioassay technique and pathogenicity of a nuclear-polyhedrosis virus of the cabbage looper, *Trichopulsia ni* (Huebner), *J. Insect Pathol.,* 6, 237, 1964.

9. **Ignoffo, C. M.,** The nuclear-polyhedrosis virus of *Heliothis zea* (Boddie) and *Heliothis virescens* (Fabricius) IV. Bioassay of virus activity, *J. Invertebr. Pathol.,* 7, 315, 1965.

10. **Ignoffo, C. M.,** Effects of age on mortality of *Heliothis zea* and *Heliothis virescens* larvae exposed to a nuclear-polyhedrosis virus, *J. Invertebr. Pathol.,* 8, 279, 1966.

11. **Ignoffo, C. M.,** Susceptibility of the first-instar of the bollworm, *Heliothis zea,* and the tobacco budworm, *Heliothis virescens,* to *Heliothis* nuclear-polyhedrosis virus, *J. Invertebr. Pathol.,* 8, 531, 1966.

12. **Cochran, W. G. and Cox, G. M.,** *Experimental Designs,* 2nd ed., John Wiley & Sons, New York, 1957.

13. **Cox, D. R.,** *Planning of Experiments,* John Wiley & Sons, New York, 1958.

14. **Finney, D. J.,** *An Introduction to the Theory of Experimental Design,* The University of Chicago Press, Chicago, 1959.

15. **John, J. A. and Quenouille, M. H.,** *Experiments: Design and Analysis,* 2nd ed., Macmillan, New York, 1977.

16. **Jerne, N. K. and Wood, E. C.,** The validity and meaning of the results of biological assays, *Biometrics,* 5, 273, 1949.

17. **Finney, D. J.,** *Probit Analysis: A Statistical Treatment of the Sigmoid Response Curve,* Macmillan, Cambridge, 1947.

18. **DeArmon, I. A., Jr. and Lincoln, R. E.,** Number of animals required in the bio-assay of pathogens, *J. Bacteriol.,* 78, 651, 1959.

19. **Gart, J. J. and Weiss, G. H.,** Graphically oriented tests for host variability in dilution experiments, *Biometrics,* 23, 269, 1967.

20. **Peto, S.,** A dose-response equation for the invasion of micro-organisms, *Biometrics,* 9, 320, 1953.

21. **Steinhaus, E. A. and Martignoni, M. E.,** *An Abridged Glossary of Terms Used in Invertebrate Pathology,* 2nd ed., PNW Forest and Range Exp., Sta., U.S. Department of Agriculture Forest Service, Corvallis, Ore., 1970.

22. **Lewis, F. B., Rollinson, W. D., and Yendol, W. G.,** Gypsy moth nucleopolyhedrosis virus: laboratory evaluation, *USDA For. Serv. Tech. Bull.,* No. 1584, 455, 1981.

23. **Martignoni, M. E. and Iwai, P. J.,** Peroral bioassay of technical-grade preparations of the Douglas-fir tussock moth nucleopolyhedrosis virus *(Baculovirus), USDA For. Serv. Res. Pap.,* PNW-222, 1977.

24. **Martignoni, M. E. and Iwai, P. J.,** Peroral bioassay of nucleopolyhedrosis viruses in larvae of the western spruce budworm. *USDA For. Serv. Res. Pap.,* PNW-285, 1981.

25. **Martignoni, M. E. and Ignoffo, C. M.,** Biological activity of *Baculovirus* preparations: *in vivo* assay, in *Characterization, Production and Utilization of Entomopathogenic Viruses,* Ignoffo, C. M., Martignoni, M. E., and Vaughn, J. L., Eds., American Society for Microbiology and National Science Foundation, Washington, D.C., 1980, 138.

26. **Paschke, J. D., Lowe, R. E., and Giese, R. L.,** Bioassay of the nucleopolyhedrosis and granulosis viruses of *Trichoplusia ni, J. Invertebr. Pathol.,* 10, 327, 1968.

27. **Martignoni, M. E. and Iwai, P. J.,** Activity standardization of technical preparations of Douglas-fir tussock moth *Baculovirus, J. Econ. Entomol.,* 71, 473, 1978.

28. **Stairs, G. R.,** Dosage-mortality response of *Galleria mellonella* (Linnaeus) to a nuclear-polyhedrosis virus, *J. Invertebr. Pathol.,* 7, 5, 1965.

29. **Magnoler, A.,** Bioassay of a nucleopolyhedrosis virus of the gypsy moth, *Porthetria dispar, J. Invertebr. Pathol.,* 23, 190, 1974.

30. **Magnoler, A.,** Bioassay of nucleopolyhedrosis virus against larval instars of *Malacosoma neustria, J. Invertebr. Pathol.,* 25, 343, 1975.

31. **Boucias, D. G. and Nordin, G. L.,** Interinstar susceptibility of the fall webworm, *Hyphantria cunea,* to its nucleopolyhedrosis and granulosis viruses, *J. Invertebr. Pathol.,* 30, 68, 1977.

32. **Boucias, D. G., Johnson, D. W., and Allen, G. E.,** Effects of host age, virus dosage, and temperature on the infectivity of a nucleopolyhedrosis virus against velvetbean caterpillar, *Anticarsia gemmatalis,* larvae, *Environ. Entomol.,* 9, 59, 1980.

33. **Jaques, R. P.,** The Ecology of Polyhedrosis of the Cabbage Looper, *Trichoplusia ni* Huebner, Ph.D. thesis, Cornell University, Ithaca, N.Y., 1960.

34. **Jaques, R. P.,** The influence of physical stress on growth and nuclear polyhedrosis of *Trichoplusia ni* (Huebner), *J. Insect Pathol.,* 3, 47, 1961.

35. **Jaques, R. P.,** Tests on protectants for foliar deposits of a polyhedrosis virus, *J. Invertebr. Pathol.,* 17, 9, 1971.
36. **Ignoffo, C. M., Hostetter, D. L., and Smith, D. B.,** Gustatory stimulant, sunlight protectant, evaporation retardant: three characteristics of a microbial insecticidal adjuvant, *J. Econ. Entomol.,* 69, 207, 1976.
37. **Potter, K. N., Jaques, R. P., and Faulkner, P.,** Modification of *Trichoplusia ni* nuclear polyhedrosis virus passaged *in vivo, Intervirology,* 9, 76, 1978.
38. **McEwen, F. L. and Hervey, G. E. R.,** Control of the cabbage looper with a virus disease, *J. Econ. Entomol.,* 51, 626, 1958.
39. **Drake, E. L. and McEwen, F. L.,** Pathology of a nuclear polyhedrosis of the cabbage looper, *Trichoplusia ni* (Huebner), *J. Insect Pathol.,* 1, 281, 1959.
40. **Chauthani, A. R.,** Bioassay technique for insect viruses, *J. Invertebr. Pathol.,* 11, 242, 1968.
41. **Peach, M. J., Allen, G. E., and Brazzel, J. R.,** A technique for confining insects to leaf surfaces and its importance in the bioassay of pathogens, *J. Econ. Entomol.,* 62, 1227, 1969.
42. **Allen, G. E. and Ignoffo, C. M.,** The nucleopolyhedrosis virus of *Heliothis:* quantitative in vivo estimates of virulence, *J. Invertebr. Pathol.,* 13, 378, 1969.
43. **Evans, H. F.,** Quantitative assessment of the relationships between dosage and response of the nuclear polyhedrosis virus of *Mamestra brassicae, J. Invertebr. Pathol.,* 37, 101, 1981.
44. **Thomas, E. D., Reichelderfer, C. F., and Heimpel, A. M.,** Accumulation and persistence of a nuclear polyhedrosis virus of the cabbage looper in the field, *J. Invertebr. Pathol.,* 20, 157, 1972.
45. **Payne, C. C.,** The susceptibility of the pea moth, *Cydia nigricana,* to infection by the granulosis virus of the codling moth, *Cydia pomonella, J. Invertebr. Pathol.,* 38, 71, 1981.
46. **Hoffman, J. D., Lawson, F. R., and Yamamoto, R.,** Tobacco hornworms, in *Insect Colonization and Mass Production,* Smith, C. N., Ed., Academic Press, New York, 1966, chap. 34.
47. **Daum, R. J.,** Revision of two computer programs for probit analysis, *Bull. Entomol. Soc. Am.,* 16, 10, 1970.
48. **Martignoni, M. E.,** Uber zwei Viruskrankheiten von Forstinsekten im Engadin, *Mitt. Schweiz. Entomol. Ges.,* 27, 147, 1954.
49. **Martignoni, M. E.,** Preparation of glass needles for microinjection, *J. Insect. Pathol.,* 1, 294, 1959.
50. **Bird, F. T. and Whalen, M. M.,** A virus disease of the European pine sawfly, *Neodiprion sertifer* (Geoffr.), *Can. Entomol.,* 85, 433, 1953.
51. **Klein, M.,** An improved peroral administration technique for bioassay of nucleopolyhedrosis viruses against Egyptian cotton worm, *Spodoptera littoralis, J. Invertebr. Pathol.,* 31, 134, 1978.
52. **Hughes, P. R. and Wood, H. A.,** A synchronous peroral technique for the bioassay of insect viruses, *J. Invertebr. Pathol.,* 37, 154, 1981.
53. **Bell, R. A., Owens, C. D., Shapiro, M., and Tardif, J. R.,** Development of mass rearing technology, *USDA For. Serv. Tech. Bull.,* No. 1584, 599, 1981.
54. **Hughes, P. R., Gettig, R. R., and McCarthy, W. J.,** Comparison of the time-mortality response of *Heliothis zea* to 14 isolates of *Heliothis* nuclear polyhedrosis virus, *J. Invertebr. Pathol.,* 41, 256, 1983.
55. **Hughes, P. R., Wood, H. A., Burand, J. P., and Granados, R. R.,** Quantification of the dose-mortality response of *Trichoplusia ni, Heliothis zea,* and *Spodoptera frugiperda* to nuclear polyhedrosis viruses: applicability of an exponential model, *J. Invertebr. Pathol.,* 43, 343, 1984.
56. **Hughes, P. R. and Wood, H. A.,** unpublished data, 1981.
57. **Ignoffo, C. M., Hostetter, D. L., Sikorowski, P. P., Sutter, G., and Brooks, W. M.,** Inactivation of representative species of entomopathogenic viruses, a bacterium, fungus, and protozoan by an ultraviolet light source, *Environ. Entomol.,* 6, 411, 1977.
58. **Williamson, W. E. P.,** Aspects of the Standardization and Stability of the Potato Moth Granulosis Virus, Master of Pharmacy thesis, Victorian College of Pharmacy, Ltd., Parkville, Victoria, Australia, 1981.
59. **Griffith, I. P., Smith, A. M., and Williamson, W. E. P.,** Raising potato moth larvae (*Phthorimaea operculella* [Zeller]: Lepidoptera: Gelechiidae) in isolation, *J. Aust. Entomol. Soc.,* 18, 348, 1979.
60. **Bergold, G.,** Eine Mikroinjektionsspritze und Mikroburette bis zu 0.1 cmm, *Biol. Zentralbl.,* 61, 158, 1941.
61. **Bergold, G. H.,** The polyhedral disease of the spruce budworm, *Choristoneura fumiferana* (Clem.) (Lepidoptera: Tortricidae), *Can. J. Zool.,* 29, 17, 1951.
62. **Plus, N.,** Étude de la multiplication du virus de la sensibilité au gaz carbonique chez la drosophile, *Bull. Biol. Fr. Belg.,* 88, 248, 1954.
63. **Granados, R. R., Nguyen, T., and Cato, B.,** An insect cell line persistently infected with a baculovirus-like particle, *Intervirology,* 10, 309, 1978.
64. **Walton, G. S.,** unpublished data, 1982.
65. **Daum, R. J. and Killcreas, W.,** Two computer programs for probit analysis, *Bull. Entomol. Soc. Am.,* 12, 365, 1966.
66. **Russell, R. M., Robertson, J. L., and Savin, N. E.,** POLO: a new computer program for probit analysis, *Bull. Entomol. Soc. Am.,* 23, 209, 1977.

67. **Helwig, J. and Council, K. A., Eds.,** *SAS User's Guide, 1979 Edition,* SAS Institute, Cary, N.C., 1979, 357.

68. **Finney, D. J.,** A computer program for parallel line bioassays, *J. Pharmacol. Exp. Ther.,* 198, 497, 1976.

69. **Litchfield, J. T., Jr. and Wilcoxon, F.,** A simplified method of evaluating dose-effect experiments, *J. Pharmacol. Exp. Ther.,* 96, 99, 1949.

70. **Berkson, J.,** Application of the logistic function to bio-assay, *J. Am. Stat. Assoc.,* 39, 357, 1944.

71. **Berkson, J.,** Approximation of chi-square by "probits" and by "logits", *J. Am. Stat. Assoc.,* 41, 70, 1946.

72. **Berkson, J.,** Minimum χ^2 and maximum likelihood solution in terms of a linear transform, with particular reference to bio-assay, *J. Am. Stat. Assoc.,* 44, 273, 1949.

73. **Berkson, J.,** A statistically precise and relatively simple method of estimating the bioassay with quantal response, based on the logistic function, *J. Am. Stat. Assoc.,* 48, 565, 1953.

74. **Armitage, P. and Spicer, C. C.,** The detection of variation in host susceptibility in dilution counting experiments, *J. Hyg.,* 54, 401, 1956.

75. **Meynell, G. G.,** Inherently low precision of infectivity titrations using a quantal response, *Biometrics,* 13, 149, 1957.

76. **Meynell, G. G.,** The applicability of the hypothesis of independent action to fatal infections in mice given *Salmonella typhimurium* by mouth, *J. Gen. Microbiol.,* 16, 396, 1957.

77. **Meynell, G. G. and Stocker, B. A. D.,** Some hypotheses on the aetiology of fatal infections in partially resistant hosts and their application to mice challenged with *Salmonella paratyphi-B* or *Salmonella typhimurium* by intraperitoneal injection, *J. Gen. Microbiol.,* 16, 38, 1957.

78. **Druett, H. A.,** Bacterial invasion, *Nature (London),* 170, 288, 1952.

79. **Moran, P. A. P.,** The dilution assay of viruses, *J. Hyg.,* 52, 189, 1954.

80. **Moran, P. A. P.,** The dilution assay of viruses. II, *J. Hyg.,* 52, 444, 1954.

81. **Lauffer, M. A. and Price, W. C.,** Infection by viruses, *Arch. Biochem.,* 8, 449, 1945.

82. **Huber, J. and Hughes, P. R.,** The quantitative bioassay in insect pathology, *Bull. Entomol. Soc. Am.,* 30(3), 31, 1984.

83. **Dougherty, R. M.,** Animal virus titration techniques, in *Techniques in Experimental Virology,* Harris, R. J. C., Ed., Academic Press, New York, 1964, chap. 6.

84. **Sampford, M. R.,** The estimation of response-time distributions. I. Fundamental concepts and general methods, *Biometrics,* 8, 13, 1952.

85. **Sampford, M. R.,** The estimation of response-time distributions. II. Multi-stimulus distributions, *Biometrics,* 8, 307, 1952.

86. **Sampford, M. R.,** The estimation of response-time distribution. III. Truncation and survival, *Biometrics,* 10, 531, 1954.

87. **Litchfield, J. T., Jr.,** A method for rapid graphic solution of time-per cent effect curves, *J. Pharmacol. Exp. Ther.,* 97, 399, 1949.

88. **Bliss, C. I.,** The calculation of the time-mortality curve, *Ann. Appl. Biol.,* 24, 815, 1937.

89. **Robson, D. S.,** personal communication, 1982.

90. **Knudson, D. L. and Tinsley, T. W.,** Replication of a nuclear polyhedrosis virus in a continuous cell culture of *Spodoptera frugiperda:* purification, assay of infectivity, and growth characteristics of the virus, *J. Virol.,* 14, 934, 1974.

91. **Volkman, L. E., Summers, M. D., and Hsieh, C.-H.,** Occluded and nonoccluded nuclear polyhedrosis virus grown in *Trichoplusia ni:* comparative neutralization, comparative infectivity, and *in vitro* growth studies, *J. Virol.,* 19, 820, 1976.

92. **Wood, H. A.,** unpublished data, 1977.

93. **Hink, W. F. and Vail, P. V.,** A plaque assay for titration of alfalfa looper nuclear polyhedrosis virus in cabbage looper (TN-368) cell line, *J. Invertebr. Pathol.,* 22, 168, 1973.

94. **Wood, H. A.,** An agar overlay plaque assay method for *Autographa californica* nuclear-polyhedrosis virus, *J. Invertebr. Pathol.,* 29, 304, 1977.

95. **Hink, W. F. and Strauss, E. M.,** An improved technique for plaque assay of *Autographa californica* nuclear polyhedrosis virus on TN-368 cells, *J. Invertebr. Pathol.,* 29, 390, 1977.

96. **Brown, M. and Faulkner, P.,** A plaque assay for nuclear polyhedrosis viruses using a solid overlay, *J. Gen. Virol.,* 36, 361, 1977.

97. **Knudson, D. L.,** Plaque assay of baculoviruses employing an agarose-nutrient overlay, *Intervirology,* 11, 40, 1979.

98. **Gettig, R. R.,** *Heliothis* spp. Baculoviruses: Genetic Characterization of Geographical Isolates, *In Vitro* Replication and Molecular Cloning, Ph.D. thesis, The Pennsylvania State University, University Park, 1983.

99. **Smith, G. E., Fraser, M. J., and Summers, M. D.,** Molecular engineering of the *Autographa californica* nuclear polyhedrosis virus genome: deletion mutations within the polyhedrin gene, *J. Virol.,* 46, 584, 1983.

100. **Burand, J. P., Wood, H. A., and Summers, M. D.,** Defective particles from a persistent baculovirus infection in *Trichoplusia ni* tissue culture cells, *J. Gen. Virol.,* 64, 391, 1983.

101. **Wood, H. A., Johnston, L. B., and Burand, J. P.,** Inhibition of *Autographa californica* nuclear polyhedrosis virus replication in high-density *Trichoplusia ni* cell cultures, *Virology,* 119, 245, 1982.

102. **Wood, H. A. and Johnston, L. B.,** unpublished data, 1981.

103. **Dougherty, E. M., Weiner, R. M., Vaughn, J. L., and Reichelderfer, C. F.,** Physical factors that affect *in vitro Autographa californica* nuclear polyhedrosis virus infection, *Appl. Environ. Microbiol.,* 41, 1166, 1981.

104. **Volkman, L. E. and Goldsmith, P. A.,** Baculovirus bioassay not dependent upon polyhedra production, *J. Gen. Virol.,* 56, 203, 1981.

105. **Reed, L. J. and Muench, H.,** A simple method of estimating fifty percent endpoints, *Am. J. Hyg.,* 27, 493, 1938.

106. **Luria, S. E. and Darnell, J. E., Jr.,** *General Virology,* 2nd ed., John Wiley & Sons, New York, 1967.

107. **Reichelderfer, C.,** Assays with insect pathogens, in *Handbook of Natural Pesticides: Methods,* Mandava, N. B., Ed., CRC Press, Boca Raton, Fla., 1985.

Chapter 2

IN VIVO PRODUCTION OF BACULOVIRUSES

Martin Shapiro

TABLE OF CONTENTS

I. INTRODUCTION

Within the past 30 years much effort has been devoted to the study of insect pathogenic viruses, especially the baculoviruses, in hopes of utilizing these microbes as safe and effective insecticides.[1] Because of these efforts, four baculoviruses have been registered as microbial control agents by the U.S. Environmental Protection Agency: the nucleopolyhedrosis viruses (NPVs) of the cotton bollworm *(Heliothis zea)*, the Douglas-fir tussock moth *(Orgyia pseudotsugata)*, the gypsy moth *(Lymantria dispar)*, and most recently, the European pine sawfly *(Neodiprion sertifer)*. Moreover, it is anticipated that more baculoviruses, both nucleopolyhedrosis viruses (NPV) and granulosis viruses (GV) will be registered as microbial insecticides within the next several years.

Despite great advances in the field of insect tissue culture during the past 20 years, it is still anticipated that mass production of baculoviruses will be dependent upon the living, susceptible host insect for the next several years.[2-5] During the past 20 years, the development of mass rearing technology, as pioneered by Vanderzant et al.,[6,7] led the way to the development and subsequent commercialization of in host baculovirus production.[8-11]

In this chapter, it is the author's hope to discuss in vivo production of baculoviruses not as an art form but as a legitimate branch of both insect pathology and insect ecology, and to demonstrate the impact of production technology not only on the cost of the microbial but also on its effectiveness.

II. OPTIMIZATION OF VIRUS PRODUCTION

The goal of virus production is to obtain the greatest amount of biologically active virus at the lowest cost, while conforming to quality control standards.[12,13] In other words, we want to utilize host tissue as efficiently as possible for maximal virus replication[11] and maximal biological activity. In order to achieve this goal, it is necessary to determine and then optimize those factors in the virus-host-environment interrelationship that influences both the quantity *and* quality of the virus being produced.

III. FACTORS INFLUENCING PRODUCTION

A. The Host
1. Wild vs. Colonized Insects

Field-collected insects are used commonly as starting material for colony establishment, as well as for mass production of insect viruses. Feral insects are used because of their genetic variability, relative availability, relatively low cost of collection, and because sufficient numbers of colonized insects are not available. A serious drawback, however, may be the presence of parasites, pathogens, or contaminants in or on these insects.

Virus diseases may occur in field populations and cause natural epizootics, e.g., nucleopolyhedroses of the alfalfa butterfly *(Colias eurytheme)*,[14] the Douglas-fir tussock moth *(Orgyia pseudotsugata)*,[15] the gypsy moth *(Lymantria dispar)*.[16] Douglas-fir tussock moth larvae, from field-collected egg masses, had about 40% incidence of nucleopolyhedrosis.[15] Moreover, egg hatch was very low and viruses (nucleo- and cytoplasmic polyhedrosis) occurred among the larvae.[17] Virus mortalities from field-collected gypsy moths varied from year to year, i.e., 35% from the 1969—1970 sample and 95% from the 1970—1971 sample.[16] A high incidence of NPV among larvae from dense populations was noted, whereas virus-induced mortality was low in sparse populations.[18,19] To minimize these problems, gypsy moth egg masses were obtained from lightly infested areas.[20,21] Although spruce budworm

(Choristoneura fumiferana) egg masses were collected from sparse populations in the fall of the year, virosis often occurred.[22]

Because of disease problems, it was concluded that laboratory-reared, colonized insects must be used for virus production.[11,17] These insects are available on a year-round basis, growth and development are both predictable and vigorous, response to virus is predictable, and these insects are more healthy than field-collected insects.[23]

Regardless of the source of insects used, strict sanitation procedures (including insect disinfection) must be utilized to reduce or eliminate microbial contaminants and/or pathogens.

2. Sanitation and Decontamination

The importance of strict sanitation measures to minimize or eliminate pathogenic or non-pathogenic microorganisms cannot be overemphasized and is a keystone to rearing.[11,24-26] Sanitation includes not only room sanitation but equipment sanitation. All equipment that can be autoclaved, should be. Items that cannot be autoclaved should be soaked in such disinfectants as formalin[24,27] or sodium hypochlorite.[25,28] Recently, the feasibility of ethylene oxide to decontaminate facilities and equipment has been demonstrated.[29,30] Refer to the excellent reviews[24,31] for more detailed information on this subject.

a. Disinfection of the Insect

Two chemicals have been used most often for disinfection, i.e., sodium hypochlorite and formalin. Sodium hypochlorite was considered an ideal, general disinfectant because of its wide microbicidal spectrum of activity, good solubility in water, stability in aqueous solutions, nontoxicity to humans and insects, availability, and low price.[32] In addition, since treatment of eggs with hypochlorite results in partial dechorionation, eggs are susceptible to dessication and mechanical injury. In general, two treatments with variations, have been utilized for disinfection: (1) eggs are soaked in hypochlorite and rinsed in water[33] or (2) eggs are soaked in hypochlorite, treated in sodium thiosulfate to neutralize the chlorine, and rinsed in water.[32]

Formaldehyde solution (= formalin) has also been employed as an egg disinfectant.[15,34,35] Temperature is very important for the action of formaldehyde, as this disinfectant is more effective at warm temperatures.[24] Other chemicals that have been utilized as disinfectants include mercuric chloride,[36,37] quaternary ammonium salts,[38] cupric sulfate,[39] *para*-hydroxymethyl-benzoate,[40] and sorbic acid.[41]

Rearing of the cabbage looper was severely hampered by viral disease (= nucleopolyhedrosis), until the development of egg sterilization techniques.[32,33] When eggs of the forest tent caterpillar *(Malacosoma disstria)* were treated with 0.5% sodium hypochlorite, no significant virus infection was observed.[22] In the case of the gypsy moth, field-collected egg masses were disinfected with sodium hypochlorite to minimize or eliminate NPV.[16,36-38]

Formaldehyde (10%) was recommended as an egg disinfectant, because of its viricidal activity.[14] Many chemicals were tested and formalin was found most effective in reducing the incidence of nucleopolyhedrosis in the silkworm *(Bombyx mori)*.[34] Although natural virus incidence from field-collected Douglas-fir tussock moths was high, i.e., 30 to 100%, formalin treatment was very effective in eliminating the virus from the treated eggs.[15,39] At our laboratory, sodium hypochlorite was used for disinfection of gypsy moth egg masses, but NPV was still present in the rearing stock. Formalin and sodium hypochlorite were evaluated for disinfection of field-collected and laboratory colonized egg masses. While both materials were effective in suppressing the NPV, formalin had greater activity. Formalin treatment (2.5%/5 min) was used successfully to eliminate GV from the Indian meal moth *(Plodia interpunctella)*.[40] Formalin treatment (0.04%), however, did not eliminate the occurrence of granulosis among *Pieris brassicae* larvae after egg treatment. When the formalin concentration was increased to 0.05%, all virus was inactivated.[41]

3. Host Biology

The biology or behavior of the insect must be taken into account, especially when determining optimal host density. Because of their aggressive, cannibalistic nature, the cotton bollworm *(Heliothis zea)* and the tobacco budworm *(H. virescens)* were reared in individual containers.[9,10,42] Since codling moth larvae *(Cydia pomonella)* are solitary by nature,[43] they were reared individually for production of GV.[44]

Some cannibalism occurred among variegated cutworm larvae *(Peridroma saucia)*, when ten larvae were reared per 240-mℓ container for NPV production.[45] Large Douglas-fir tussock moth larvae were cannibalistic when more than six were placed in a 120-mℓ container. For large-scale NPV production, four larvae were reared per container.[15] Since cabbage looper larvae are gregarious by nature, these insects may be reared in aggregate for virus production. NPV yield was optimal at a host density of 300 larvae per container.[46] Although the total yield increased as larval density increased from 100 to 400 per container, the yield of virus per larva decreased from 9.2×10^8 polyhedral inclusion bodies (PIB) (at 100) to 3.3×10^8 PIB (at 400 per container).

For virus production, the gypsy moth has been reared gregariously, i.e., in groups of 10 in 100×15-mm petri dishes[48] or 12 in 480-mℓ ice cream containers or waxed cups.[38,49] During our 1977 pilot-scale production of *L. dispar* NPV, fourth-instar larvae were used at a density of ten per 180-mℓ container. This density was chosen because larval growth and development were normal and crowding was minimal. As larval density increased, the amount of NPV produced per larva decreased, but the total virus yield per container increased. Apparently space or food was not a limiting factor, even at the highest host density used. For the 1979 large-scale production, fifth-instar larvae were used at a density of ten per 180-mℓ container. As larval density increased (to 14), virus production per insect decreased. Space became a limiting factor for these insects and the total yield per container remained constant at densities of 5 (only females), 10 (males and females), and 14 (males and females). For virus production, ten larvae were reared per 180-mℓ container, as total NPV yields were excellent, and labor costs were minimal (in comparison to the other two density systems).

4. Age, Stage, and Sex

The objective of virus production is to utilize host tissue as efficiently as possible and to obtain the maximal yield of biologically active virus per insect or per container.[11,13,49] Cabbage looper larvae 6- to 7-days-old were used; by this time 70% of their larval development had been completed. The use of later instar larvae accomplished several objectives. The ingestion of a lethal dose occurred within a short period, maximum utilization of larval tissues was obtained, and various phases of virus production were easily systematized.[11] In later mass production of *Baculovirus heliothis*, 5-day-old larvae were infected for maximal NPV production, i.e., an increase of 10,000-fold over the inoculum dose.[50]

Since the greatest virus yield occurred among fifth-instars, bird-cherry moth *(Hyponomeuta evonymellus)* larvae were infected during second instar, while the spindle-tree moth *(H. cognatellus)* and the apple moth *(H. malinellus)* were infected as third-instar larvae.[49] In order to ensure that maximal size and subsequent virus yield were obtained, variegated cutworms were infected as mid fifth-stage larvae,[45] while late fifth and early sixth-instar spruce budworm larvae *(Choristoneura fumiferana)* were utilized for NPV production.[51] While third-instar *Pieris rapae* larvae were used to produce GV,[52] late fourth or early fifth-stage European pine shoot moth larvae *(Rhyaciona buoliana)* were used to produce its GV.[53]

Production of the gypsy moth NPV is an interesting case in point, as some thorough research has been done relating virus yield and larval stage. When late third-instar larvae were used, an average yield of approximately 9×10^8 PIB per larvae was obtained.[38] The greatest virus yield per insect was obtained when larvae weighing between 400 to 499 mg were fed on a diet containing 1.5×10^5 PIB per milliliter of diet. The average yield per insect was 1.64×10^9 PIB, with a maximum of 2.23×10^9.

Table 1
VIRUS PRODUCTION FROM MALE AND
FEMALE GYPSY MOTH LARVAE[a]

Larval stage when inoculated	Mean PIBs per larva ($\times 10^9$)		PIBs per milligram ($\times 10^6$)	
	Male	Female	Male	Female
Fourth	1.57	3.99	1.85	1.78
Fifth	1.40	4.96	1.73	2.10

[a] Day 10 harvest, 29°C. From Shapiro, M., Bell, R. A., and Owens, C. D., *USDA For. Serv. Tech. Bull.*, No. 1584, 633, 1981. With permission.

These authors[47] concluded that maximal production was 1.56×10^6 PIB per milligram of final larval weight. It was later found that virus yield was not as dependent upon the initial weight of the host insects as on the rate of weight gain during virosis.[54] For example, virus yield was greater among larvae having an initial weight of 100 to 199 mg than among larvae weighing 50 to 99 mg and 300 to 499 mg. Data indicated that larvae infected in the later instars produced more virus than those infected in earlier instars.[13] Since the goal of the 1977 pilot-scale production was to produce at least 1×10^9 PIB per insect, it was necessary to infect larvae in the fourth or later instars. A total of 500,000 larvae were utilized (fourth-instar) and a total of 1×10^{15} PIB were obtained ($= 2.04 \times 10^9$ PIB per larva). For every milligram of larval tissue, 2.26×10^6 PIB were produced, which was much higher than the 1.56 million PIB per milligram produced from the Pennsylvania feral insects.[47] In all likelihood, differences in larval diet as well as source of insects played critical roles in determining virus yield. In the 1979 large-scale pilot production involving 1.5 million larvae, fifth-instar larvae were used, and an average yield of 2.8×10^9 PIB per larva was obtained.[23]

In our production efforts, as well as those of others, no distinction was made between male and female larvae because physical separation would be labor-intensive and/or little difference in size (weight) existed between the sexes. Since the female gypsy moth larva has one more instar than the male and weighs considerably more (as a last-stage larva), studies were initiated to compare virus yields from males and females. During the 1977 production, ten living infected males and ten living infected females were removed daily for 38 days (Table 1). More virus was produced in females, but the female did not produce any more NPV per milligram of body weight than the male, i.e., since the female attained a greater biomass than the male, more virus was produced. These data indicated the potential advantages of female larvae for NPV production.

It was shown previously that females infected as fourth-stage larvae produced more NPV than the males. The same trends occurred in larvae infected as fifth-instar (Table 1). Total virus per insect was always higher in females, and it was not uncommon to obtain five 8×10^9 PIB per female larva. Moreover, the larger the female, the greater was the yield of virus produced per insect and 2.0 to 2.4 million PIB were produced per milligram of body weight. Although female larvae were used in these studies, it was recognized that much labor would be involved in separating them from males for virus production. Therefore, a mixed population of fifth-stage male and female larvae were used for the 1979 pilot plant production. Whether five fifth-stage female larvae (optimal density), or ten fifth-stage males and females were used, similar virus yields were produced, i.e., the final total biomass was similar in both cases. An ideal system would utilize female larvae for virus production and male larvae for sterile male production.

5. Alternate Hosts

In general, entomopathogenic viruses have been produced in their respective homologous hosts, presumably because of their host specificity. Many of these insects, however, may have undesirable qualitites for rearing and virus production, i.e., obligatory diapause, prolonged developmental time, presence of allergenic and urticarious setae, small size, cannibalistic behavior, etc. so that more "desirable" hosts might be sought.[2,11,55]

Virus produced in the alternate or factitious host may be utilized as a microbial control agent either against the original host or the factitious host, but there may be great difficulties in either approach. Most of the work on virus specificity or host range concerned the insecticidal activity of the virus. Several noctuid species were exposed to a multicapsid NPV of *Heliothis armigera*. While neonate larvae of the bollworm were most susceptible (LC_{50} = 10^3 PIB per milliliter of diet), larvae of the tobacco budworm and the beet armyworm *(Spodoptera exigua)* were less susceptible (LC_{50} = about 10^4 PIB per milliliter), larvae of the fall armyworm *(S. frugiperda)* were least susceptible (LC_{50} = about $10^5 - 10^6$ PIB per milliliter.[56] Since the bollworm NP (= *Baculovirus heliothis*) and an NPV from *H. punctigera* were infective for neonate larvae of *H. armigera* and *H. punctigera*, it was felt that *H. zea* NPV could be considered for control of the other *Heliothis* species in Australia, because of its demonstrated safety and production technology.[57]

Although the granulosis virus from the codling moth was not infective for either the bollworm of the navel orangeworm *(Paramyelois transitella)*, it was infective for the Oriental fruit moth *(Grapholitha molesta)*,[58] the European pine shoot moth,[54] as well as the pea moth *(Cydia nigricana)*.[59] In the latter case, however, the homologous host (= the codling moth) was more susceptible to the GV(LC_{50} = 1.54×10^4 capsules per milliliter of diet) than the factitious host (LC_{50} = 1.90×10^5). On the other hand, the GV from *Pieris brassice* was more active against *P. rapae* neonate larvae (LC_{50} = 5 capsules per larva) than *P. brassicae* neonates (LC_{50} = 66 capsules per larva).[60]

A GV obtained from the arctiid, *Diacrisia virginica*, was infective for fall webworm larvae *(Hyphantria cunea)*. After the virus was serially passed twice through the factitious host *(H. cunea)*, it still retained infectivity for the homologous host. Interestingly, a GV from *H. cunea* was not infective for larvae of *D. virginica*.[62]

For more than a decade, much effort has been centered on the NPV from the alfalfa looper *(Autographa californica)* because of its wide host spectrum.[46,62-70] All species initially tested, i.e., the cabbage looper *(Trichoplusia ni)*, diamondback moth *(Plutella xylostella)*, beet armyworm, bollworm, cotton leaf perforator *(Buccalatrix thurberiella)*, and the saltmarsh caterpillar *(Estigmene acrea)*, were susceptible and virus remained infective for the indicator host *(T. ni)* after passage through the alternate hosts. Since the original host *(A. californica)* was difficult to rear, it was felt that the beet armyworm and saltmarsh caterpillar might be used for in vivo production.[62] At this time it was noted that the bollworm was not easily infected. Moreover, the lethal incubation period was much longer than in the other species.[62] Later, it was demonstrated that the disease symptoms in the bollworm were atypical, and that this insect would be a poor host for *A. californica* NPV (= AcMNPV).[65] Furthermore, *Heliothis zea* larvae were at least tenfold less susceptible to the NPV than larvae of the closely related species *H. virescens*.[71]

When AcMNPV was serially passed in beet armyworm larvae, the size of the PIB and numbers of nucleocapsids per virion increased as well as infectivity for cabbage looper larvae. Similarly, *T. ni* NPV passed through beet armyworm larvae contained more nucleocapsids per virion and were more infective for cabbage looper larvae.[72] From these studies, the authors concluded that the most desirable host for virus production would be the one producing the virus with greatest biological activity. Although AcMNPV replicated in cell lines from the saltmarsh caterpillar, the gypsy moth, the cabbage moth *(Mamestra brassicae)*, the fall armyworm, and the cabbage looper, the greatest virus yield was produced from *M. brassicae* and *T. ni* cells. The cabbage looper cell line also produced the most pathogenic virus and would be a prime candidate for large-scale production.[73]

The use of hosts which are closely related to the target pest should be considered for virus production, if it is advantageous to do so. Four ermine moths, i.e., the spindle-tree moth, the bird-cherry moth, the apple moth, and the fruit moth, were exposed to an NPV from the apple moth. Although all were susceptible to the virus, the greatest amount of virus was produced in the largest insects, i.e., the spindle-tree moth and the bird-cherry moth. Moreover, other insect species were susceptible to the NPV, i.e., *Arctia caja, Argyresthia conjugella, Vanessa urticae, V. io, V. prorosa,* and are being considered for use as factitious hosts.[74]

The Eastern spruce budworm is a major forest pest in Canada and much effort has been made to utilize NPV to control it.[75] Since the host insect presents problems in large-scale rearing for NPV production, e.g., obligatory diapause, small size and subsequent low virus yields, production of silk webbing, and frequent infection with a microsporidian parasite, an alternate host was sought.[55] Initial virus-caused mortality was low when 4-day-old saltmarsh caterpillar larvae were exposed to spruce budworm NPV (CfMNPV) (0.4%). Upon subsequent passage, mortality due to virus infection increased to 48%. After one additional passage, virtually all larvae died from nucleopolyhedrosis infection. When this virus was passed through 8-day old larvae, mortality was very high (98%). Subsequent passages in 10- to 14-day old *E. acrea* larvae resulted in high mortality (97%) and high virus yield (2 $\times$ 10^9 PIB per larva). These yields were at least four times greater than those produced in the homologous host (= spruce budworm).

Although virus could be produced in the factitious host (= saltmarsh caterpillar), the NPV had to be passed back into the original host at a very high concentration (10^7 PIB per milliliter) to obtain a fatal infection in the spruce budworm. After one passage in the spruce budworm, the virus was quite active upon subsequent passage. NPV that had been passed into the spruce budworm had very little activity for saltmarsh caterpillar larvae upon subsequent challenge. Thus, complications arose that made the use of a factitious host inappropriate. The spruce budworm NPV was also infective to neonate larvae of the cabbage looper and the greater wax moth *(Galleria mellonella),* and increased its pathogenicity for these insects upon serial passage. Although viruses from these factitious hosts were still infective for spruce budworm larvae, the larvae did not "wilt" and only few PIBs were found in infected budworms. Although the "altered" virus was not considered suitable for control of the original host (spruce budworm), it was felt that the virus might be useful against the factitious hosts (wax moth, cabbage looper).[76]

The virulence of the Douglas-fir tussock moth NPV (OpMNPV) for larvae of the saltmarsh caterpillar was enhanced after successive passages in the alternate host. After six passages through *E. acrea,* virus yields of 2 $\times$ 10^9 PIB per larva were obtained, which represented at least a fourfold increase compared with yields from the homologous host. After passage in *E. acrea* larvae, the NPV was as virulent for the original host as the original virus inoculum.[55] In this case, the use of an alternate host may be successful for production purposes, since both high virus yield and high virus activity are achieved.

In the case of a virus having a wide host spectrum, i.e., AcMNPV, the virus may be produced in one host and utilized against another susceptible host. A mutant of AcMNPV was obtained,[77] which was quite active against the Douglas-fir tussock moth. At equal levels of activity (AcMNPV had one third the activity of the homologous OpMNPV), AcMNPV was as efficacious as OpMNPV. Since AcMNPV could be produced in a more easily reared insect (cabbage looper) and at least cost, it was suggested that AcMNPV could be utilized as a microbial control agent of the Douglas-fir tussock moth.[78]

In the U.S.S.R., the gypsy moth and cabbage moth larvae are used as production hosts for VIRIN-ENsh and VIRIN-EKS, respectively. The original viruses were isolated from other insects and were adapted to *L. dispar* and *M. brassicae,* respectively.[54] In this case, the alternate hosts (for production) were also the species that were controlled by the viruses.

In the case of viruses from closely related insects, it may be possible to produce one or more viruses in the target insect. NPVs from *Malacosoma americanum, M. pluviale,* and *M. disstria* were equally infective for *M. disstria* larvae, while an NPV from *M. alpicola* acted more slowly.[79] NPVs from the Eastern spruce budworm and the Western spruce budworm *(Choristoneura occidentalis)* were equally infective for larvae of the Eastern spruce budworm. In the winter of 1976—1977, 957,000 *C. fumiferana* larvae were infected with CfMNPV, while 610,000 larvae were infected with CoMNPV. Since host rearing, however, was labor intensive, the cost of production was quite high, i.e., $125./h.[75] Baculoviruses from the Douglas-fir tussock moth *(Orgyia pseudotsugata)* and the white marked tussock moth *(O. leucostigma)* were produced in larvae of the Douglas-fir tussock moth. Both viruses, when field-tested, provided excellent foliage protection against the Douglas-fir tussock moth.[80]

In the case of viruses from closely related insects, where cross transmission has been achieved, it may be possible to utilize either of the insects as a production host and/or a target insect, using either virus as inoculum. NPVs from the gypsy moth and the nun moth *(Lymantria monacha)* are related to each other serologically and are cross infective. In each instance, greater mortality is obtained with the homologous NPV (and homologous host).[81]

A certain "danger" exists in this approach, which must be borne in mind. It is imperative to determine whether the virus isolated from the susceptible alternate host is "identical" to the original virus used for inoculum. An NPV and CPV from the apple leafroller *(Adoxophyes orana)* was produced in the cabbage moth because *Mamestra* was larger, easier to rear, and could be reared on an artificial diet.[82] Later, it was found that both hosts were infected with their own virus, and the "foreign" (exogenous) virus triggered the "native" (endogenous) virus.[83]

B. The Virus Inoculum

Although virus inoculum has most often been obtained from infected and/or virus-killed insects,[8,17] primary contaminant-free inoculum may be obtained from infected tissue or cell cultures.[50,77] While it may be desirable to utilize inoculum from tissue culture, this approach may not be feasible in all cases, due to the lack of an available in vitro system. Primary inoculum of the Douglas-fir tussock moth NPV (OpMNPV), after purification and testing, was provided to the commercial producer by the U.S. Forest Service Laboratory, Corvallis, Ore.

The producer increased the amount of inoculum by serially passing the virus through mature larvae once (= secondary inoculum) or twice (= tertiary inoculum), i.e., only secondary or tertiary inoculum could be used for large-scale production.[17]

Primary inoculum of the gypsy moth NPV (LdMNPV) was provided by the U.S. Forest Service Laboratory, Hamden, Conn., for both commercial production and large-scale pilot plant production by the USDA.[84] The inoculum had passed all quality control tests and the amount was sufficient to complete a large-scale production. A large quantity of the selected *Baculovirus heliothis* (NPV) isolate was produced as inoculum, and was sufficient for several years production. Subsamples from the first large-scale passage were maintained as reference standards under optimal storage conditions.[1]

Insect tissue culture has been utilized successfully not only to provide "clean" inoculum,[50] but also to provide clones with greater insecticidal activity.[77] Hopefully, the use of in vitro systems will be utilized more often to provide quality primary inoculum for virus production.

1. Purity

Virus inoculum may be obtained from either in vitro tissue culture or in vivo host production. Tissue culture-derived virus, either nonoccluded free virions and/or PIB, represents bacteria-free, noncontaminated inoculum. In some cases, hemolymph from virus-infected insects may also be used as "clean" inoculum.

The contaminant level in host-produced virus inoculum should be minimized, since the bacterial increase during infection is partially dependent upon the original bacterial "load".[13] While it has been suggested that virus inoculum should contain less than 10^4 aerobes per gram,[50] this level may not be attainable without further processing. Roccal solution (at 250 ppm) in the inoculum suspension reduced the bacterial count to less than 10 spore formers per milliliter, without affecting virus yield.[85] Contaminant levels may be reduced greatly by centrifugation in a sucrose gradient,[45] using a zonal rotor.[86,87] The ratio of bacteria to PIB changed from 1:2.5 in a crude preparation of the European pine sawfly NPV to 1:400,000 in the purified fractions. Further reduction in the contaminant level could be realized after a second centrifugation.[88]

2. Quality Control

Certain requirements must be met before a viral preparation may be used as inoculum for in vivo production: (1) *quantitative* — contaminant levels should not exceed a certain level and (2) *qualitative* — the inoculum should be free of certain contaminants, e.g., other viruses, *Salmonella, Shigella,* etc. The battery of tests required are[17] (1) dark-field microscopy and electron microscopy; (2) virus concentration; (3) detection of extraneous insect viruses and microorganisms; (4) detection and enumeration of microbial contaminants, which includes coliforms, fecal coliforms, and pathogenic Enterobacteriaceae; and (5) safety to mice by intraperitoneal injection and per oral administration. In addition, serological[85] and biochemical identification[50] should be made, to guarantee the integrity of the viral product.

3. Activity

While it may seem apparent and axiomatic, the virus produced in living hosts (or tissue cultures) will be the *one* used as a microbial control agent against a target pest (or pests). Therefore, the inoculum used for virus production should be selected carefully to maximize both viral activity (= virulence) and viral yield. The occurrence of geographical variability among the Baculoviruses,[89] with differences in biological activity,[90-94] may be the means of obtaining optimal inocula for virus production.[12]

Five NPV isolates (LdMNPV) from France, Yugoslavia, Italy, the U.S., and Japan were compared, and greater than 1000-fold differences in activity were found.[21] The most active isolate originated from the U.S. (Connecticut), while the least active isolates were from France and Japan. Of 19 geographical isolates of LdMNPV tested, the North American isolates were generally the most active, while the Japanese isolate was least active. Five isolates were more active than the Connecticut standard,[94] and one is being tested further as a possible candidate for in vivo production.

Activities of three *Baculovirus heliothis* isolates from the U.S. (Texas) and Africa (Sudan, Ivory Coast) were compared, and less than twofold differences were found.[97] On the other hand, 56-fold differences were found among 34 *B. heliothis* isolates. These isolates could be assigned to at least two and possibly three to eight activity classes.[91] More recently, 12 isolates could be assigned to six activity groups, based upon differences in 50% survival time.[93]

Although genotypic variability exists among AcMNPV natural isolates, no phenotypic variations (including host-range alterations) have been found.[89] It may be possible, however, to enhance activity through chemical or biological selection. Following replication of a random virus clone in the presence of 2-aminopurine, a mutant (= HOB) was obtained with increased insecticidal activity. Moreover, cabbage looper infected with the HOB mutant produced more PIB than larvae infected with the wild isolate or parental clone.[77] When AcMNPV was serially passed several times in the cabbage looper, following passage in the beet armyworm, the size of the PIB increased, contained more nucleocapsids per virion, and was more pathogenic for the cabbage looper.[72]

4. Dosage

For production, a uniform response to the virus is required, i.e., 100% infection or possibly 100% mortality. The relationship between inoculum dose and virus yield must be determined for each production system; the goal is to utilize host tissue(s) as efficiently as possible and to obtain the maximal yield of biologically active virus per insect or per container.[13] If the dose is too high, larval growth and development may be retarded, resulting in less-than-optimal yields. If the concentration is too low, infection will not be maximal, many insects may pupate, and yield will also be suboptimal.

In general, inoculum concentrations have varied from 1×10^5 to 5×10^7 PIB per milliliter in all described systems, i.e., bollworm: 1.8×10^7;[11] cabbage looper: 1×10^6;[11] Douglas-fir tussock moth: 1×10^6.[15] For gypsy moth NPV productions, concentrations of 1×10,[5,47,48] 3×10^6,[38] and 5×10^6 PIB per milliliter[85] were used. The greatest production of LdMNPV occurred at the lowest dose (1.5×10^5), while high doses (1.5×10^6, 1.5×10^7) appeared to reduce virus yield.[47] In our production, an inoculum dose of 5×10^6 PIB per milliliter container was utilized. This dosage was sufficient to inoculate ten or more larvae (within a container) and still permit the desired growth and subsequent virus yield. A greater initial virus concentration did not necessarily result in greater virus yields, and doses less than 5×10^6 PIB per milliliter resulted in suboptimal yields.[13,96]

C. The Environment

Since the rearing environment may affect both the quantity and quality of virus produced, environmental parameters must be investigated and optimized for each production system.

1. Temperature

In general, virus production is carried out at temperatures between 20 and 26°C.[2,8,17,48] The relationship of temperature and humidity upon the course of virosis among gypsy moth larvae was investigated.[97] When the temperature was maintained at 26.5 to 27.5°C, the lethal infection was reduced. As the temperature was increased to 29 to 31°C, virus development was not enhanced. For routine rearing in our laboratory, *L. dispar* are maintained at 25 to 26°C. To determine the optimal temperature for NPV production, larvae were held at 23, 26, 29, or 32°C after exposure to virus. The effect of temperature was evaluated with respect to host mortality, virus yield, lethal infection time, and virulence of the virus extracted from the test insects. NPV yields and viral activity were similar from insects reared at 23, 26, or 29°C. Yields and activity of NPV produced at 32°C, however, were much lower than at the other three temperatures. As the temperature increased, the LT_{50}, i.e., the time required to achieve 50% mortality of the test population, decreased. From a production standpoint, virus-induced mortality (the sole criterion used by Yadava) was less meaningful than either virus yield or virus activity. Although virus-induced mortality was more rapid and greater at 32°C than at 29°C, yield *and* activity were much lower at 32°C. The higher temperature apparently had a detrimental effect upon the virus.[13]

NPVs of the cabbage looper and bollworm were incapable of infecting their respective hosts at 39°C or above.[98,99] At these elevated temperatures, however, the host insects were also adversely affected. When tobacco budworm larvae were reared at 36°C, the resultant adults suffered slight teratological effects. At 40°C sterility occurred, and teratological effects increased.[100] Although viral replication was slower at 15.6 and 21.1°C than at 26.7°C, total virus-induced mortalities were similar. At a higher temperature (32.2°C), however, 30 to 50% less virus-caused host mortality occurred.[101]

When larvae of the Indian-meal moth *(Plodia interpunctella)* were exposed to a GV at

22, 27, 32, and 37°C, the highest host mortality occurred at 32°C. Larvae reared at 37°C did not exhibit granulosis symptoms, but apparently contained low levels of virus.[102] Granulosis-caused mortality among codling moth larvae increased at 20 to 24°C, decreased slightly at 28°C, and decreased significantly at higher temperatures (32 to 36°C).[103] When silkworms *(Bombyx mori)* were exposed to a cytoplasmic polyhedrosis virus (CPV) at 35°C, the number of PIB was very low and PIB formation was delayed, in comparison with larvae reared at 25°C. At the elevated temperature, the accumulation of viral and polyhedral RNA was greatly inhibited.[104]

2. Humidity

An increase in humidity led to an increase in virus incidence among gypsy moth larvae.[105] Moreover, high temperature, excessive humidity, starvation, or feeding with unusual food activated occult virus (LdMNPV).[106,107] On the other hand, it was found that relative humidity had little effect upon virus replication, while temperature played a critical role.[97] In the author's laboratory, little difference occurred in either the rate of virus-induced mortality or total mortality at low (32 to 36%) or high (60 to 62%) relative humidity, at 29°C. For virus production, the humidity level in the incubation chamber was set for 40 to 45%, but reached a level of 65% within the rearing containers.[13] At these levels, the incidence of fungal contamination increased.

3. Air Flow

In general, air movement within a rearing chamber is seldom considered or controlled but may play an important role in microbial contamination and/or disease incidence. Virosis among gypsy moth larvae was enhanced when the relative humidity was increased without adequate air movement.[108] Under these conditions, excess water ingested in the food could not be eliminated and feeding stopped, causing a secondary nutritional stress. When air circulation was increased, the evaporation gradient between the larvae and the environment increased, and the incidence of polyhedrosis decreased.

In the author's virus production facility, the rearing and work area are under negative pressure to reduce the possibility of virus-laden air entering other parts of the mass rearing facility. A small fan draws air out of the rooms through a 99.95% HEPA filter at a rate of 4.2 m³/min, before blowing it outdoors. The rearing room has its own temperature and humidity control equipment, in addition to a separate air-circulating system equipped with a 99.97% HEPA filter to clean the air in the chamber. The work area and office have an air conditioning and heating system that recirculates the air every 3 min through a 99% HEPA filter.[13] For an excellent review on this subject, please refer to References 109 and 110.

4. Photoperiod

In general, the same light:dark cycle that is used for rearing is also used for virus production. The gypsy moth is commonly reared in a 16-hr light: 8-hr dark cycle.[20,21,35-37,111] The effects of photoperiod were compared by rearing insects in light: dark regimens of 16:8, 12:12, and 8:16. Virus-caused mortalities and yields were evaluated for larvae infected as neonates and reared under different light regimens, and insects reared under standard conditions (16:8) for 14 days and then maintained under different photoperiods. Little differences in larval growth, virus-induced mortality, or virus yield were observed.[13]

In a subsequent study, larvae were reared for 18 days under standard conditions (26°C, 16L:8D, 50% relative humidity), and then maintained under different photoperiods, e.g., 20:4, 12:12, 4:20. Again, few differences in mortality, larval weights, or virus yields were observed.[23] For virus production, insects were maintained at 29°C, 50 to 55% relative humidity, and a 12L:12D photoperiod.

5. Diet

In the past, foliage has been utilized to rear insects as well as to produce insect viruses.[50,112,113] In general, foliage is utilized when a semisynthetic artificial diet is not available for the insect to be reared,[62,114] and is still used for production of sawfly NPVs.[48,115-117] Of the four insect pathogenic viruses registered in the U.S. (by the U.S. Environmental Protection Agency), only the European pine sawfly NPV (= Neochek-S®) is produced from foliage-fed insects. About 20,000 acres worth of Neochek-S® were produced by the U.S. Forest Service and the virus was distributed to Christmas tree growers. At this time, the Forest Service has no plans to continue production.[118]

The use of natural foliage, i.e., pine foliage for production of sawfly NPVs, placed dependence upon the seasonal availability of food, and decreased the opportunity for a controlled, continuous production.[119] Since it was difficult to obtain virus-free foliage, it was difficult to conduct many studies of host-pathogen interactions.[120] Moreover, the use of unsuitable foliage, i.e., age or species, sometimes led to an increase in disease incidence.[120,122]

The development of artificial diets, pioneered by Vanderzant and co-workers[6,7] facilitated *continuous* mass-production of hosts and large-scale virus production.[5,13,119,123,124] Shortly thereafter, the practicality of producing an NPV from the cabbage looper was demonstrated under controlled laboratory conditions using a semisynthetic diet.[8] Since then, semisynthetic diets have been utilized for small- and large-scale virus productions,[10,11,13,15,17,38,58,75] involving such insects as the bollworm, Douglas-fir tussock moth, gypsy moth, spruce budworm, and codling moth.

Several wheat germ-casein based diets have been developed for gypsy moth rearing[20,36,37,111,125] with varying success. Some of these diets have also been used for virus studies and production, i.e., Leonard-Doane diet,[120] Odell-Rollinson diet,[38,47,48] and Magnoler diet.[21,92] These diets, as well as a lima bean diet,[126] were compared with a modified hornworm diet[127] and a simplified wheat germ-casein diet developed for the gypsy moth.[35] Virus yields were highest from larvae reared on the modified hornworm diet, while the simplified high wheat germ diet was superior to the others. Biological activity of virus obtained from larvae reared on the modified hornworm diet, high wheat germ diet, and ODell-Rollinson diet was comparable. The high wheat germ diet, however, would be most suitable, from a cost-effectiveness standpoint, for large-scale in vivo production of LdMNPV.[128]

Reduction in dietary costs would lead to a more efficient, cost-effective production and could be achieved in several ways, i.e., use of less costly dietary ingredients, reduction or elimination of nonessential ingredients, and agar substitutes. Rearing insects for virus production may require a less complex diet than for other purposes, since the objective is good larval growth and not pupal, adult, or egg production.[13]

Although the modified hornworm diet was excellent both for colony perpetuation and NPV production, certain ingredients could be deleted, e.g., cholesterol and linseed oil (source of lineolenic acid), raw sugars (carbohydrate source), and torula yeast. Sufficient sterols and fatty acids, as well as carbohydrates, were provided by increasing the wheat germ concentration from 8 to 12%. Sufficient vitamins were provided by the vitamin mix and sufficient protein(s) was provided by the wheat germ and casein.[35] These changes resulted in a simplified, less expensive diet, which could be used for both NPV production and colony perpetuation.

Blended Food Product, Child Food Supplement, Formula No. 2, commonly called CSM (corn, soy flour, and milk solids) was developed as an inexpensive, nutritious dietary supplement for protein-deficient countries. The diet was adapted for bollworm rearing,[129] and later modified further with the addition of torula yeast and ascorbic acid.[130] Subsequently, the diet was supplemented with both casein and yeast and successfully reared the tobacco hornworm *(Manduca sexta)*.[28] A modification of the CSM diet[129] and a pinto bean diet were as effective as the standard wheat germ-casein diet for NPV production, although only the standard diet was used for NPV production studies.[46]

When calcium, magnesium, potassium, mixed minor elements, choline chloride, ascorbic acid, mixed vitamins, Aureomycin®, or methyl *p*-hydroxybenzoate levels were reduced in a semisynthetic diet, the incidence of granulosis was not changed. When either the sucrose or casein content was reduced, however, a greater number of *Pieris brassicae* larvae died from granulosis.[41] The incidence of nucleopolyhedrosis was greater when sucrose was deleted from the modified hornworm diet.[131] In the author's studies, larval biomass and virus yield increased with dietary protein levels. Larvae feeding on the modified hornworm diet (6.0% protein) or on the high wheat germ diet (5.1%) produced more virus than those feeding on the ODell-Rollinson diet (3.4% protein). When NPV was extracted from these larvae and tested for virulence, little differences in activity were found. Thus, differences in viral quality were minimal, although differences in viral quantity were great.[128] The incidence of granulosis was higher when *P. rapae* larvae were fed a partially dehydrated diet (50% moisture) than when fed fresh diet (80% moisture).[132] The incidence of nucleopolyhedrosis, however, was greater when gypsy moth larvae were transferred to fresh virus-contaminated diet, as compared to 14-day-old diet. The fresh diet not only contained a different moisture level, but a different nutrient level as well.

When cabbage looper larvae were exposed to AcMNPV on diet plugs or on cabbage disks, virus was more "potent" when larvae were fed diet plugs, irrespective of virus source.[133] In this regard, we have consistently observed that less NPV is required to kill gypsy moth larvae on a semisynthetic diet ($LC_{50} = 6.4 \times 10^3$ PIB per milliliter) than on red oak seedlings ($LC_{50} = 575 \times 10^3$ PIB per milliliter).[134] Whether these observations represent inherent differences in viral activity or differences due to viral inactivation are not presently known, but these differences in activity have practical implications.

6. Dietary Ingredients

Wheat germ is a major ingredient and a source of proteins, minerals, lipids, etc. Wheat bran and wheat meal were inferior to wheat germ. Kretschmer® (Toasted) wheat germ was comparable to the raw wheat germ used at our facility, and has some advantages over raw wheat germ, e.g., the toasted and vacuum-packed wheat germ has lower contaminant levels than the raw wheat germ and greater storage stability, and can be stored at room temperature (unopened) with no loss in biological activity.

Casein has been used as a primary protein source in many insect diets,[135] and in all diets for the gypsy moth. In a series of tests, different proteins and their hydrolysates were compared to casein at equivalent protein concentrations (in the high wheat germ diet). Several proteins, namely trypticase (a casein hydrolysate), soy, phytone (a soy hydrolysate), and lactalbumin were comparable to casein. Torula yeast, a byproduct of the paper pulp industry, was almost as effective as casein, and is an inexpensive source of protein.[13] In subsequent tests, egg albumin was found to be comparable to casein. Cost and availability are important factors in the selection of protein sources for both rearing and virus production and should be borne in mind.

In general, the vitamins required for insect growth and development are the B-vitamins and ascorbic acid. These materials can be added to the diet as individual vitamins or as a prepackaged mix. A small increase in vitamin (mix) concentration resulted in little increase in virus yield, but a tenfold increase resulted in an increase from 4.4×10^9 to 5.6×10^9 PIB per female larva. Whether the increase in virus yield was due solely to the vitamins or to other materials present in the mix was not known.[13] The additional cost, however, would not seem to justify the additional amount of vitamin mix.

The pH of semisynthetic diets for lepidopterous insects is acidic, i.e., between pHs 5.3 to 5.8 for the modified hornworm and high wheat germ diets without any further adjustment.

When the pH of the hornworm diet was increased by 4 *M* KOH or decreased by 1 *M* HCl, little difference in virus yield occurred at pHs 4 to 7. As the pH increased above 6, the level of microbial contaminants increased.[13] Moreover, antimicrobial agents were ineffective in eliminating contamination at a dietary pH above 6.5.[35]

In general, agar has been used as the primary gelling agent in insect diets. Agar has been the most expensive ingredient in insect diets, accounting for up to 50% of the total cost. As the price of agar increased, greater attention was given to alternate gelling agents. Calcium alginate,[136-138] sodium alginate,[139] and carrageenan[28,46,140] were utilized successfully for the rearing of several lepidopterous insects. Moreover, it was suggested that carrageenan could be utilized as the gelling agent for production of the cabbage looper NPV.[46]

As part of our research efforts to optimize in vivo NPV production, the use of carrageenans as gelling agents was investigated, using virus yield, virus activity, and cost savings as critical parameters. From the standpoint of virus yield alone, Carrageenan® GH and Seagel® GH were superior to the other gelling agents, i.e., agar, Seakem® 202, Gelcarin® HWG, and Seagel® Pet. With respect to viral activity, Gelcarin® HWG was the best gelling agent. When both yield and activity were considered, Gelcarin® HWG was best, followed by Seagel® GH and Carrageenan® GH. Moreover, since carrageenans cost at least 40% less than agar, a significant cost saving would be realized.[141]

7. The Container

The choice of a suitable rearing container may be influenced by such factors as host biology (gregarious vs. solitary), host size, host density, economics, etc. A favorable microclimate must be obtained within the container[35] in order to obtain maximal host biomass and virus yield. Because of their aggressive behavior, bollworms and tobacco budworms were reared individually in 60 mℓ transparent containers.[9,10] Subsequently, plastic trays containing individual cells were utilized for large-scale production.[42,50] Many different containers (jars, paper cartons, waxed cartons, and plastic-coated cartons) were tested for production of the cabbage looper NPV.[8] Although paper cartons were comparable to jars, microbial contamination was greater in the paper cartons. Paraffin-coated paper bags were subsequently tested for production of the cabbage looper and alfalfa looper NPVs.[46] Container costs as well as the number of containers needed were reduced, but additional space was required in comparison with 240 mℓ cups (the standard).

Various containers (180 mℓ Dixie® containers, 30 mℓ clear plastic cups, plastic petri dishes, and cardboard trays) were evaluated for production of the Douglas-fir tussock moth NPV. The 30 mℓ cups (four larvae per cup) were superior and were used for large-scale production.[15] European pine sawfly larvae (5,000 to 10,000 third stage larvae) were placed in clean, disposable cardboard boxes, which were then subdivided to contain 2,000 to 3,000 larvae.[48,116] The gypsy moth can be reared individually, but has been maintained in groups for virus production, i.e., 10 larvae per 100 × 15-mm petri dish,[47] or 12 larvae per 480-mℓ waxed cup.[38] At the author's laboratory, the best success was achieved with plastic food cups (180 mℓ) with paper lids. The lids permit adequate gaseous exchange to occur and provide a suitable surface for attachment of larvae during molting and pupation.[35]

D. Production Methods

In host virus production is dependent upon efficient rearing of the host insect.[5,11,35,142] The development of semisynthetic diets, reliable sanitation (including disinfection) techniques, and improvements in rearing technology have made the *continuous* production of insects available for virus production *feasible* and *practicable*.

1. Inoculation of Susceptible Insects

A suspension of virus can either be incorporated into the semisynthetic diet or simply

Table 2
MORTALITY AND NPV YIELD WHEN
INOCULUM WAS APPLIED TO THE
SURFACE VS. INCORPORATED IN THE DIET

Treatment	Percent mortality[a]	PIBs per larva
Surface	28.0	1.49×10^8
Incorporation	2.8	4.86×10^8

[a] Virus-induced mortality at 10 days postinoculation; 29°C fourth-
     instar larvae. From Shapiro, M., Bell, R. A., and Owens, C.
     D., *USDA For. Serv. Tech. Bull.*, No. 1584, 633, 1981. With
     permission.

placed on the diet surface (artificial or natural), and both methods of inoculation have been used. LdMNPV was incorporated into the cooling diet at 47°C[47] or 55°C.[48] The highest yields of virus were obtained when cabbage looper NPV was incorporated into the diet.[143]

In general, most studies concerning virus production, as well as the actual production, utilize surface treatment. The cabbage looper and bollworm NPVs were sprayed on the diet surface with a hand atomizer and electric hand sprayer, respectively.[8,9] In all cases, a standard volume of virus was sprayed on the diet.[17,46,53,133] In the case of the gypsy moth, surface treatment has some advantages over diet incorporation, in that less inoculum is required (50 times less) to attain a lethal infection. Diet can be prepared in the regular diet processing area and transferred to virus production for inoculation, thus avoiding duplication of diet preparation equipment. Also, by introducing virus on the surface of the diet after cooling, as opposed to incorporation into warm diet, there is much less chance of inactivation by heat. Diet temperature can play a critical role in virus inactivation, as inactivation of LdMNPV occurred at diet temperatures greater than 58°C.[144] The effectiveness of the two methods is compared in Table 2. The mortality obtained at 10 days postinoculation and the virus yield per larva were much greater when the inoculum was applied to the diet surface.[13]

If insects are reared on foliage, virus is sprayed on the foliage (pine sawfly NPV;[115] European pine sawfly NPV;[48,116] or the foliage is dipped in a virus suspension (alfalfa looper NPV,[45] apple moth NPV,[49] and Eastern hemlock looper NPV.[114]

2. Virus Incubation

From an economic standpoint, infected insects should remain on virus-contaminated diet throughout the production period. If larvae are transferred to fresh, uncontaminated diet, additional costs are incurred because of additional diet, containers, and labor. If additional diet is presented to the infected larvae, additional costs of diet and labor are incurred. During production of bollworm and cabbage looper NPVs, larvae remained on virus-treated diets.[8,11] For the gypsy moth, old diet was removed and fresh, uncontaminated diet was added as needed.[48] Larvae were fed virus-treated diet for 48 hr at which time fresh, virus-free diet was offered.[38] In the author's 1977 pilot plant production, 14-day-old larvae were transferred to fresh virus-treated diet for the production period.[84] Because of improvements in diet preparation,[35] it became feasible to maintain larvae in the same containers (and diet) throughout the incubation period, with no reduction in virus yield.

A significant relationship was observed between the time of larval feeding on virus-treated diet and the virus concentration in the diet.[47] Virus yield was similar when larvae were exposed to virus-treated diet for 48 hr or continuously, but yields were greater when larvae fed on virus-treated diets for 96 hr. In the author's studies, no differences were found in either virus kill or virus yield when larvae fed on virus-treated diets for more than 24 hr (day 10 harvest). When the incubation period was increased to 15 days, no differences in

46 *The Biology of Baculoviruses*

virus yields or virus-caused mortalities were observed.[13] From these data, it would appear that enough virus was consumed within the initial 24 to 48 hr to ensure 100% infection.

Ideally, the time of harvest (=total virus incubation period) should be that time when virus yield and biological activity of the virus are maximal. Insects may be collected as living, infected larvae,[2,63,145] when disease symptoms are maximal,[42,64] or just prior to death.[124] Virus-infected gypsy moth larvae were collected about 24 hr prior to death, placed in clean plastic containers, and refrigerated[48] to minimize wilting and loss of recoverable virus. More commonly, larvae are usually collected after death,[38,52,114,116,117] presumably because virus yield and activity are optimal. In our 1977 production, larvae were harvested at day 10, i.e., the time of approximately 30% mortality, in order to minimize wilting of virus-killed larve and loss of recoverable virus and to minimize the "bacterial load" per larva. At this time, virus yield was comparable to that from larvae collected on subsequent days.[13]

Although preparations from virus-killed bollworm larvae contained greater numbers of bacterial contaminants than those from living, infected larvae, the activity of virus from virus-killed larvae was seven- to nine-fold greater.[146] Similarly, NPV from virus-killed gypsy moth larvae (day 15, 100% mortality) was about seven times more active than virus from living-infected larvae (day 7, 0% mortality) and approximately three times more active than virus from a mixed population of living, infected, and virus-killed larvae (day 11, 50% mortality).[147] Thus, when virus is collected at the proper time after inoculation, both virus yield *and* activity may be maximized.

3. Harvest

In vivo virus production is simple and straightforward up to the time of harvest. Insects are exposed to virus and become infected. Eventually these insects will die if left undisturbed. The problem arises at this point as how to maximize virus harvest by minimizing tissue histolysis and subsequent loss of virus by leakage into the diet (or on the plant surface). Larvae and diet could be harvested or the loss of virus into the diet could be prevented. The former method may seem desirable, but prior virus productions have centered around larval collection.[13]

Diseased, living intact cabbage looper[8] and bollworm larvae[9] were collected, transferred to liquid-tight cartons, and frozen (−20°C) until processed. Virus-infected (GV) codling moth larvae, maintained on serological, depression slides, were frozen at −75°C.[44] Gypsy moth larvae (within containers) were placed in a large freezer (−20°C) and left overnight. The following day, the frozen larvae (infected and virus-killed) were easily removed from the containers. Following harvest, the larvae were placed in plastic bags, returned to the freezer, and held for subsequent processing.[84] Virus-killed European pine sawfly larvae were handpicked from the foliage and the rearing containers.[116]

Harvesting was the most time consuming and laborious operation in virus production. Since 1700 cups (ten larvae per cup) were inoculated per day, larvae from each of the cups had to be harvested 14 days later. The manual operations of removing lids from each cup and removing larvae took more time than any other single operation. If fewer containers could be used, each with greater numbers of larvae, this step could become more efficient. Harvest could also become more efficient if the diet remaining in the container (= unconsumed diet) were either minimal or totally consumed. Moreover, if the minimal amount of unconsumed diet could also be harvested (in addition to the larvae), time would be reduced greatly.[23] Despite these shortcomings, four workers could harvest 10,000 larvae within 60 min. Although the biological activities of virus preparations from bollworm larvae plus unconsumed diet were similar to those preparations without diet, the final virus yield (PIB per gram of dried preparation) was significantly lower in the larvae plus diet product.[146] It would appear, in this case, that the virus was diluted significantly by the unconsumed diet.

Handpicking European pine sawfly larvae was the most time-consuming operation in NPV production, and one person could harvest around 8 oz of cadavers in about 8 hr. Six full-time workers, however, could not harvest all virus-killed larvae during peak production so the remaining cadavers were stored at 4°C until subsequent harvest.[116] The efficiency of manual harvest was enhanced greatly by collection of virus-killed larvae by vacuum.[1,38,133]

Harvest should occur at a time (postinoculation) when both virus yield and biological activity are maximal *and* contaminant levels conform to quality control standards. From the author's 1977 pilot-scale production, day 10 (30% kill) was selected as the optimal time based upon considerations of virus yield and bacterial contaminant levels. Since it was later demonstrated that biological activity increased significantly during the virus incubation period,[147] day 14 (75% kill) was selected as the time of harvest. Moreover, improvements in processing, e.g., freeze-drying of whole larvae, reduced the total bacterial population in the final virus powder to an acceptable level (see Section III.D.6).

In several instances, multiple harvest of infected and/or virus-killed insects took place to ensure maximal numbers of insects. Thus, NPV-infected variegated cutworm larvae *(Peridroma saucia)* were collected 6 to 9 days after inoculation.[148] Cabbage looper larvae were harvested 3 to 7 days after inoculation,[46] while Douglas-fir tussock moth larvae were harvested daily 8 to 14 days after inoculation.[15,17] Diseased spruce budworm larvae were collected after 8 days, and the remaining insects were harvested 48 hr later.[51] From an economic standpoint, however, infected and/or virus-killed insects should be harvested only once, which must be determined for each virus production system.

4. Processing
Standardized methods should be developed with these goals in mind: recovery of virus should be maximized; activity of virus should be preserved; and contaminant levels should be minimized.[13] Either only living, infected, or recently killed larvae should be used for processing.[11] Infected gypsy moth larvae were refrigerated and allowed to die,[48] or cadavers were processed after harvest.[38] In the author's production scheme, larvae were placed in plastic bags or in 1-gal ice cream containers until processed. Methods of recovery are quite diverse, but in general, involve extraction of virus from insect tissues, filtration, and concentration. After concentration, virus may be stored as suspensions or powders under refrigeration, freezing, or ambient conditions.[11]

a. Extraction of Virus
This is of utmost importance since the goal is to recover as much virus as possible from the tissues of the production insects. The classic approach has been to blend (or homogenize) the larvae in diluent (water or buffer) and then filter the material through cheesecloth and/or nylon[4,11,53,143,149] or through a series of screens.[17] Surprisingly, little attention has been paid to the blending process. Bollworm larvae were blended (one part insect to two to three parts water) for 15 to 30 sec and then filtered through cheesecloth and/or nylon.[11] For all our processing, the standard procedure is to blend 1 g of larval weight (= wet weight of whole insects) in 10 mℓ of distilled water for 60 sec (60 g would be blended in 600 mℓ H_2O, etc.). Following filtration through coarse cheesecloth, about 95% of the PIB can be recovered from infected larvae. Initial tests compared the relationship between dilution (grams larvae per milliliter H_2O) and recovery. At dilutions of 1:1 to 1:5, less than 40% of all PIB were recovered after a single blend. At a dilution of 1:10, however, recovery was increased to 94%. At greater dilutions recovery was increased only slightly. Blending time can be very important, also, as inadequate blending will not liberate enough PIB or capsules from the host tissues. At a larval weight per H_2O ratio of 1:10, a 5-sec blend resulted in 75% PIB recovery. When the blend time was increased to 15 sec, recovery was increased to 95%. Additional blending time resulted in little additional recovery.[13]

b. Concentration of Virus

This is obtained primarily by centrifugation of "crude" viral filtrate in diluent[11,38,48,54] or in sucrose gradients.[4,86,88,89] Bollworm NPV was centrifuged at 5,000 to 15,000 rpm for 15 min, while gypsy moth PIB were allowed to settle for 1 week under refrigeration (4°C) prior to continuous flow centrifugation (9000 rpm).[38] In another case, gypsy moth NPV was concentrated by differential centrifugation (750, 2000, and 5000 rpm).[48]

Following centrifugation, virus was sucrose gradient purified to remove most bacterial contaminants.[88,148] When diseased or virus-killed bollworm larvae were allowed to ferment in nutrient broth, the density gradient profiles lacked one or more spore-bacteria zones and the PIB yield was increased.[87] Activities of AcMNPV were compared after centrifugation through 40% sucrose and filtration with urea, sodium dodecyl sulfate, and sodium chloride and final centrifugation through a sucrose density gradient. The virus collected from the urea, SDS, NaCl, sucrose gradient was more virulent for cabbage looper larvae than virus through a 40% sucrose gradient.[133] In this case, the method of concentration and diluent(s) used had an effect on viral activity.

The question arises whether the centrifugation step is necessary to obtain the desired virus product. PIBs were precipitated from concentrated suspensions in lactose (4 to 6%) by adding acetone. The lactose precipitated with the polyhedra, and a dry powder was obtained. Moreover, PIBs could also be precipitated in larval homogenates. Yields of over 80% were obtained and the activity of the lactose-acetone method was comparable to the standard centrifugation-lyophilization method.[143] The ease and simplicity of this method have led to its usage, especially in the U.S.S.R.[49,54] Although there were no initial differences in activity of an aqueous preparation of virus, lyophilized virus, or acetone-extracted virus (from the bollworm), preliminary studies indicated that acetone-extracted preparations might lose activity after long-term storage.[146]

Initial tests involving elimination of the centrifugation step indicated that a suitable virus product could be obtained by direct air-drying of the larval homogenate. One problem that arose in the standard production scheme of blending, centrifugation, and air-drying was the removal of large volumes of water. Moreover, air-drying did not inhibit bacterial multiplication. Whether the suspensions were centrifuged or not, bacterial levels were similar in virus pellets and filtrates. Moreover, if initial bacterial levels were high from virus-killed insects, the final bacterial levels could exceed the acceptable limit of 10^9 aerobes per gram of final dry product. Because of these concerns, tests were initiated to compare air-drying with freeze-drying. In these tests, as in previous tests, frozen infected gypsy moth larvae (day 10 postinoculation) were blended in sterile water. Bacterial counts were taken of the filtrates, before and after drying, on trypticase soy agar plates. Freeze-drying resulted in a decrease in total aerobes, i.e., vegetative cells, from 4.6×10^7 to 3.2×10^5 per gram. Air-drying, on the other hand, led to a 16-fold increase from 4.9×10^7 to 77×10^7 per gram.[13] Because of the need to remove large volumes of water after centrifugation, much attention was focused on an alternative (simplified) method of processing that eliminated the "buildup" of bacteria during drying, and removed allergenic and urticarious setae.[13] Frozen larvae were placed in trays and were lyophilized for 22 to 24 hr. The freeze-dried intact larvae were placed in a plastic screen (6-mesh) cylinder and shaken for 5 min at 120 c/min. The setae were dislodged easily, passed through the screening, and were collected by vacuum. The dehaired larvae were then ground through a mill (75 to 100 mesh) and technical virus powder was collected in 1/2-gal ice cream containers. The resultant powder contained about 5% NPV, 95% host materials, and 1×10^{10} PIB per gram.[149]

NPV-infected cabbage looper larvae were frozen, homogenized, and virus was obtained by differential centrifugation. Maltose was added to the virus suspension until a slurry was formed; the slurry was lyophilized and ground to a powder.[64] After NPV-infected Douglas-fir tussock moth larvae were homogenized, the virus suspensions were lyophilized, weighed, and packaged under vacuum.[17] On the other hand, intact virus-infected Eastern hemlock looper larvae[114] and Eastern spruce budworm larvae[51] were lyophilized and milled, without prior blending or centrifugation. In the latter production, 1.646 million budworm larvae produced 15.7 kg of technical virus containing 1.2×10^{10} PIB per gram.[51] European pine sawfly NPV was prepared either as a liquid suspension or as a dry powder. As a liquid suspension, the virus was suspended in glycerin (50%) or mineral oil and sterile distilled water. As a powder, NPV was mixed with kaolin, bentonite, lactose, skim milk powder, methylcellulose, or polyglucide and dried in an incubator (26 to 28°C) or freeze-dried. The activity of the freeze-dried powder was about 22% greater than the air-dried material due to excessive clumping of the latter. While methylcellulose, skim milk powder, and lactose had good qualities, e.g., good adhesion and good dispersion, kaolin and bentonite were too viscous.[117]

5. Storage

In general, few guidelines exist on optimal storage conditions for insect viruses, and conditions vary from worker to worker. Virus preparations (in 50% glycerin) were stored at 2 to 4°C or were dried in a dessicator at 30 to 33°C.[49] After 6 years storage at 4°C, aqueous suspensions of NPV had lost most of their virions and pathogenicity. Polyhedra contained many empty areas where virions had been. Freeze-drying was then adopted as a method of processing, and material was stored at 4°C with little loss in activity.[114] Virus suspensions were lyophilized, milled to a fine powder, and packaged under vacuum.[15,17] The resultant powder was stored in a cool, dry place with no loss in activity (for at least 5 years).[17] Whatever the storage conditions employed, the biological activity of the virus preparation should be maintained and contaminant levels should not increase. Moreover, the virus should be packaged in airtight (nonporous) containers to prevent the accumulation of moisture which might lead to contaminant buildup and possible viral degradation or inactivation. From the author's experience, it would be ideal to package the dried virus under vacuum in airtight containers and store the material at -20°C until use.

If virus is produced on an annual basis as part of an ongoing operational program, short-term, storage, i.e., 1 year or less, would be sufficient. If production occurs on an irregular, sporadic basis where large quantities may be stockpiled for future use, long-term storage conditions should be maintained. In either case, both the biological activity and the contaminant population (both quantitative and qualitative) must be assessed at regular intervals to assure an acceptable viral product.

6. Quality Control

The virus product must be characterized on the basis of biological activity, microbial contaminants (quantitative, qualitative), and safety[1,17] and must conform to specifications. Safety of baculoviruses is a vast and important field[3,17,85,150,151] and will be discussed in great detail in Volume I, Chapter 9.

a. Activity

Biological activity of the virus product is the criterion by which the product should be measured. In order to ensure that activity of a viral product is uniform from one production lot to another (quality control), biological assays are conducted using larvae of a susceptible host.[12] The bioassay measures the response of the insect to a given pathogen under prescribed conditions. In general, the response is all-or-none (=dead or alive) or quantal, and the results are expressed as percent mortality. The end point is measured at the 50% level, or that dilution of virus which kills one half of the test animals (=LD_{50})[152] and/or the time required to kill 50% of the test population (=LT_{50}).[153]

Various methods are available for evaluation of dosage-mortality response, ranging from a graphical method(s) without transformation of the data, with transformation of the doses, with transformation of doses and percentage kill,[152] to the more sophisticated maximum likelihood estimate[154] or maximum normit X^2 estimate.[155] In general, those methods which give the LD_{50} or LT_{50}, the slope of the line, and the confidence limits are widely used. (There are several complete and detailed descriptions of bioassay.)[156-162]

Regardless of the virus being used, the following guidelines have been recommended for adoption:[162]

1.   The bioassay procedure as well as activity standardization procedure should be described in meticulous detail.
2.   An activity unit should be defined for each virus, and the activity or potency of each virus (or each production batch) should be reported as activity units per gram or milliliter.
3.   A designated insect strain (laboratory, colonized) should be employed as the standard insect strain, and should be maintained at one or more laboratories.
4.   One preparation of virus should be designated as the standard virus preparation, which will be used for determination of relative activity of production batches.
5.   Field application doses of each virus should be specified as activity units per acre or hectare.[162]

Moreover, the bollworm NPV,[162] the Douglas-fir tussock moth NPV,[17] and the gypsy moth NPV,[163] which were registered by the U.S. Environmental Protection Agency for use as microbial insecticides, specify the number of activity units (= insecticidal units; = potency units) per gram of virus preparation.

b. Identification

To ensure the integrity of the virus preparation, the inoculum as well as the technical virus must be identified by the most sensitive methods available, e.g., biochemical identification of viral nucleic acid,[50,89,164] and serological identification by enzyme-linked immunosorbent assay (ELISA).[165,166] Restriction endonuclease analysis (REN) of viral nucleic acid has been used to characterize and identify different baculoviruses as well as their genotypic variants and should be used as a standard tool for quality control of production lots. The use of immunological techniques to identify and characterize baculoviruses[167,168] has advanced our basic knowledge concerning these viruses. More recently, the development of the ELISA has proven to be an invaluable tool in the characterization and identification of insect viruses,[165] as well as in the identification of viral contaminants.[50,169]

c. Microbial Contaminants

Each production lot should be monitored to guarantee that the types and numbers of microbial contaminants do not represent a safety or health hazard. Total contaminant levels of (aerobic) bacteria represent both preharvest and postharvest populations.[13] Several workers showed that contaminants, in numbers and types, were greater in virus-killed larvae than in living-infected larvae.[87,146,170]

Maximum levels of aerobic bacteria have been established for each registered baculovirus and must not be exceeded, i.e., 10^7 colonies per gram for the bollworm NPV,[85] 10^9 colonies per gram for the Douglas-fir tussock moth NPV[17] and gypsy moth NPV.[171] The virus preparation should be free of coliforms and primary pathogenic bacteria (such as *Salmonella*, *Shigella*, *Vibrio*, etc.). Since the bacteria present in the virus preparations primarily represent bacteria from infected and/or virus-killed larvae, i.e., endogenous bacteria, it is felt that the " . . . qualitative assessment of bacterial contaminants is more important than the quantitative one,"[85] as long as no "forbidden" bacteria are present.

The virus preparation should be nontoxic to mice, as determined by intraperitoneal injection[4,85,172] of "one projected acre dose".[17] Additional tests have been performed on a variety of nontarget tissues and organisms (in vivo and in vitro), to assure nontoxicity and nonpathogenicity of the virus preparations.[17,173] No other virus shall be in the preparation, irrespective of its biological effect. For example, a cytoplasmic polyhedrosis virus (CPV) was found in cabbage looper larvae infected with an NPV.[63] The CPV was infectious to cabbage looper larvae but had low virulence. NPV preparations from field-collected larvae were shown to be contaminated with a small isometric ribonucleic acid (RNA) virus. The RNA virus was not eliminated from an NPV preparation after two cycles of differential centrifugation, but could be eliminated after one cycle of centrifugation through sucrose.[174] A small isometric RNA virus contaminant was present in CPV preparations from *Chrysodeixis eriosoma* and was infective for larvae of *C. eriosoma, Spodoptera litura*, and *Heliothis armigera*.[175] A 35-nm RNA virus from the cabbage looper was found in preparations of AcMNPV,[176] which greatly reduced larval weight, but had little effect upon the biological activities of AcMNPV in the cabbage looper and tobacco budworm.[169,177] Since the virus was virulent only at high doses, its presence could easily have been overlooked. Quality control procedures were proposed to detect the presence of these contaminants:[176] (1) extracts of NPV-infected insects should be centrifuged and examined by electron microscopy, (2) the viruses should be isolated and characterized, and antisera should be produced, and (3) a sensitive serological test, such as ELISA, should be used to confirm the presence of the contaminant virus. In reality, serological identification of the RNA contaminant virus was a standard procedure for both inoculum and commercially produced *Baculovirus heliothis*.[50] When sophisticated techniques such as monoclonal antibodies become available for different baculoviruses, more sensitive and specific quality control tools can be developed.

In summary, each production lot of technical virus must meet certain requirements before it can be released for usage. These requirements include (1) identification, (2) activity, (3) types and numbers of microbial contaminants, and (4) safety.[4,17,50,85,172]

E. The Physical Facility

The virus production facility (or area) should be separated and isolated from the rearing and diet preparation facilities (or areas).[1,17,48,85,178] In other words, the virus production area should be designated as a *quarantine* area to prevent the spread of microorganisms from a contaminated area (virus production) to a "clean" area (rearing, diet preparation, diet storage, etc.). Personnel employed in virus production should have no access to the rearing areas and vice versa. Ideally, each facility should be in a separate building,[178] but in many cases this cannot be accomplished due to space limitations. In either case, the virus production area must be considered as a source of microbial contamination and strict sanitation measures must be employed to prevent contamination of the rearing facility.[17]

1. Design

For the author's large-scale production, the prototype facility was designed as a quarantine facility, having the capacity to rear and process 17,000 gypsy moth larvae per day. The facility consisted of a rearing room (3.6 × 5.5 m) which was a modular environmental chamber. The main part of the prototype (46 m^2) is a work area where virus inoculation, harvest, and processing can be done. After harvest, the infected larvae are held in a freezer (2.5 × 3.6 m) that has the capacity of storing several days harvest, as well as the processed virus. The entrance to the facility is via a clean-air shower which workers must pass through. A 2.5 × 3-m room is provided which may be utilized as an office or laboratory. Adjacent to the office is a room equipped with a steam cleaner where all carts and trays leaving the area will be cleaned. Restrooms for men and women were included for worker cleanup before leaving. Except for the restrooms, the entire facility was comprised of modular panels from the former mass-rearing facility.[13]

The rearing room has its own temperature and humidity control equipment, in addition to a separate air-circulating system equipped with a 99.97% HEPA filter to clean the air in the chamber. The work area and office have an air-conditioning and heating system that recirculates the air every 3 min through a 99% HEPA filter.

To reduce the possibility of virus-laden air entering other parts of the mass-rearing facility, the rearing and work areas are under negative pressure. A small fan draws air out of the rooms through a 99.95% HEPA filter at a rate of 4.2 m³/min, before blowing it outdoors. The air shower should reduce the amount of contaminants carried on personnel leaving or entering the facility. High velocity, sterile, low pressure air (368 m/min) collects contaminants from the workers. The airborne particles are exhausted through floor level ducts and returned to the blower section for recycling through 99.97% HEPA filters.

The area for infecting larvae and processing virus has a hood system over the entire work counter. The air drawn into the hood is passed through HEPA filters and returned to the room, thus minimizing contamination from spraying of virus or virus processing.[13]

2. Equipment and Operation

Diet making, filling of cups, and insect rearing (up to the fifth instar) are performed in the main (separate) rearing facility. Therefore, each day a cart (0.68 × 1.2 m) holding up to 16,800 larvae will enter the virus facility through the air shower. Also, one cart will leave through the steam room coming from virus production. All other operations of virus production can be carried out in this facility, from inoculation to virus processing to storage.[13]

A special work table, a clean air hood mounted on wheels, was constructed for the protection of the workers during harvest. The tabletop is a 64-cm wide counter with long slots cut into the splash board. Plexiglass sloping down to the splash board is used as a hood. Lights are mounted over the plexiglass to improve visibility. The slots are ducted to a fan below the counter. The air is pulled into the slots at about 150 m/min to draw setae and scales away from the workers.[13] The workers wear protective clothing, disposable polyethylene gloves, and filter masks to minimize problems in handling the larvae, i.e., urticarious and allergenic setae.[17]

At 14 days postinoculation, the virus-infected and/or virus-killed larvae (about 70 to 75% host mortality) are frozen within their containers in a walk-in deep freezer (2.5 × 3.6 m). The freezer was installed so that carts coming from virus rearing could be placed in the chamber (−20°C). Frozen larvae, including wilted ones, are easily harvested by workers. During the pilot production effort, about 10,000 larvae were harvested in 1 hr by two workers. The freezer is used both for storage of harvested larvae and processed virus.

Lyophilization of intact larvae was carried out in four freeze-dryers, each having the capacity of removing 6 ℓ of water. Frozen insects were placed in trays (three trays per machine) and were dried for 22 to 24 hr. Heat was continually added to the drying chamber to facilitate drying within 24 hr. During the drying period, the temperature within the larvae did not exceed 30°C and activity was not adversely affected. Freeze-dried larvae were placed in a plastic screen (six mesh) cylinder and shaken for 5 min, at 120 c/min. The setae were dislodged easily, passed through the screening, and they were collected by vacuum. The dehaired larvae were then ground through a mill (75 to 100 mesh) and technical virus powder was collected in 1/2-gal ice cream containers and stored at −20°C.[13]

3. Scheme of Operations

The virus production system involves several operations: virus inoculation, rearing of infected larvae (incubation), harvest of virus-infected and/or virus-killed larvae, and classically, extraction, concentration, and storage of virus. Variations can occur and certain steps, e.g., extraction and concentration, may be eliminated by direct lyophilization of intact larvae. The operational procedures in production of the Douglas-fir tussock moth NPV[17]

Table 3
COMPARISON OF OPERATIONS INVOLVED IN PRODUCTION OF THE DOUGLAS-FIR TUSSOCK MOTH AND GYPSY MOTH NPVs

	Douglas-fir tussock moth	Gypsy moth
Inoculum	Secondary or tertiary	Secondary
Virus dose	10^6 PIB per milliliter	5×10^6 PIB per milliliter
Treatment	Surface of diet	Surface of diet
Insect	Fifth-stage	Fifth-stage
Temperature	25°C	29°C
Incubation	8 to 14 days	14 days
Harvest	Daily from days 8 to 14	Once at day 14
Process	Refrigerate larvae	Freeze larvae
	Blend	Lyophilize larvae
	Screen	Dehair larvae
	Lyophilize	Mill
	Package under vacuum	Package in ice cream containers
	Store in a cool, dry place	Store at −20°C

and the gypsy moth NPV[13] will be summarized to highlight both the similarities and the differences (Table 3). In both cases, the investigators had concluded that the procedures were quite sufficient and resulted in efficient systems.

F. Production Efficiency

The economics of in vivo baculovirus production are a combination of costs of production of larvae prior to virus inoculation, costs involved from inoculation to harvest, and operations from processing to storage. In addition, quality control and safety testing must also be considered as necessary and vital adjuncts to virus production. The estimated cost of bollworm NPV production, excluding overhead, was calculated at 7 cents per larva or 1 cent per billion PIBs. Labor costs were higher than both rearing containers and diet combined.[11] It was felt that costs could be reduced through more efficient rearing techniques (automation of egg collection and larval infest) and virus production techniques (automated virus inoculation). Further cost reductions could be realized by optimization of containers and diet. Initially, insects were reared in plastic containers (60 mℓ), which involved much time and labor. Later, clear polystyrene trays, containing 50 cubicles (15 mℓ per cubicle) were utilized.[42] Media simplification reduced costs by about tenfold, and the use of an alternate, noncannibalistic host would further reduce costs to half or one tenth of the present amount.[2]

Gypsy moth NPV, utilizing third-instar larvae, was produced at the cost of 8 cents per larva or 10 cents per 10^9 PIBs.[38] During our 1977 production, NPV was produced in fourth-instar larvae (2 cents per larva) and processed (about 1.5 cents per larva) for about 3.5 cents per 10^9 PIBs because of improvements in the insect stock and in the diet. Subsequent improvements in diet preparation[35] as well as a simplified diet[35] and virus production methodology (insect stage, incubation time, processing) further reduced costs to about 1 to 1.5 cents per 10^9 PIBs.[149] Further reductions in dietary costs were realized by the use of carrageenans as gelling agents.[141] Production costs would be reduced further by improvements in rearing, e.g., egg infestation, mechanization of virus inoculation, larger rearing containers, and mechanization of harvest.

Although the Douglas-fir tussock moth NPV was produced for several years,[15,17] more work still needs to be done to rear the insect economically.[15] Since the cost of producing the virus was quite high, i.e., $40/ha and since no commercial interest was expressed in producing the virus, research was initiated to explore the possibility of using a "foreign" virus to control the Douglas-fir tussock moth.[78] The HOB isolate of AcMNPV[77] was found to be virulent for *Orgyia* larvae. Moreover, since AcMNPV could be produced more efficiently (in cabbage looper larvae) than OpMNPV, it was felt that a cost-savings could be realized if AcMNPV were utilized against *Orgyia*. Although the spruce budworm NPV (= CfMNPV) had been produced,[51,75] the cost was quite high, i.e., $150 to 250/ha. Not only was production labor-intensive, but the virus yield per larva (about 1×10^8 PIB) was quite low. Because of these concerns,[51] an effort was made to find alternate hosts for NPV production. Although the cabbage looper[76] and saltmarsh caterpillar[55] could be infected with CfMNPV, the viruses obtained did not possess high activities for the spruce budworm. Since CfMNPV passed through the cabbage looper was very active for the heterologous host (= the cabbage looper), the authors felt that the virus might be usable against the cabbage looper.[76]

The use of heterologous hosts for virus production offers an exciting approach for production, as well as basic information on host specificity (or host range). Though it may be possible to utilize a heterologous host and obtain virus highly virulent for the original host,[55,63] virus obtained from the new host may have been altered in such a way as to render it less virulent for the original host.[55,76] A more direct and feasible approach may be to utilize the ''unnatural'', susceptible host as the target insect. In this regard, *L. dispar* and *M. brassicae* were infected with NPVs from different host species, and these viruses were produced for control of *L. dispar* and *M. brassicae*, as VIRIN-ENsh and VIRIN-EKS, respectively.[54] This approach would appear to be worthy of further investigation and could result in more potent viruses for control of insect pests.

It would appear that the greatest single expenditure in virus production is the construction of a production facility. In order to minimize this expense, the facility could be made of inexpensive materials,[178] or the virus production facility could have the capability of producing more than one insect virus.[2] Or, perhaps regional production facilities could be constructed to serve the needs of a large area.[151] (Whether these facilities would be private, public, or a combination of the two was not specified.) Whether the production facility is capable of producing only one or several insect viruses, research on both optimization of rearing and virus production should be ongoing. For example, even though *Baculovirus heliothis* has been produced for over a decade, improvements still need to be made, especially in the area of automation,[1] which would lead to greater production of host insects and virus.

IV. THE FUTURE

For unknown or unstated reasons, many entomologists have considered that insect rearing and virus production are low priority areas, not worthy of serious consideration or investigation. It should be remembered that insect rearing is not merely a ''service'', but is the result of research involving many aspects of entomology, e.g., physiology, nutrition, genetics, etc. Virus production is the result of complex research on host-pathogen-environment interrelationships resulting in maximization of both virus quantity and quality. While one may wait for some breakthrough(s) in insect tissue culture, one fact is very clear: ''The development and use of artificial diets and a rearing protocol which provides disease-free insects is at present our best approach to obtain a quality virus-product for identification, standardization, and eventual use.''[119]

In order to produce the most potent or active viruses for control of insect pests, much work needs to be done. The most potent isolates must either be obtained from natural geographical isolates,[93,179] through chemical means,[77,180] or through genetic engineering. Or, more potent viruses could be obtained through passage in heterologous hosts.[54] Once these viruses are available, research would be initiated to determine (for each virus-host system) optimal conditions for virus production. ''As knowledge in microbial control (e.g., application, formulation, persistence, etc.) increases and the successes increase, efforts will not only be directed at increased production and more improvements in production technology, but also for the testing and utilization of other viruses.''[23]

REFERENCES

1. **Ignoffo, C. M. and Anderson, R. F.,** *Bioinsecticides, Microbial Technology,* Vol. 1, 2nd ed., Academic Press, New York, 1979, 1.
2. **Ignoffo, C. M. and Hink, W. F.,** Propagation of arthropod pathogens in living systems, in *Microbial Control of Insects and Mites,* Burges, H. D. and Hussey, N. W., Eds., Academic Press, New York, 1971, 541.
3. World Health Organization, The use of viruses for the control of insect pests and disease vectors, *W.H.O. Tech. Rep. Ser.,* No. 531, 1973.
4. **Yendol, W. G., Hedlund, R. C., and Lewis, F. B.,** Field investigations of a baculovirus of the gypsy moth, *J. Econ. Entomol.,* 70, 598, 1977.
5. **Aloyshina, O. A.,** Study of entomopathogenic viruses in the USSR, in *Proc. 2nd Conf. of Project V, Microbiological Control of Insect Pests of the US/USSR Joint Working Group on the Production of Substances by Microbiological Means,* Ignoffo, C. M., Martignoni, M. E., and Vaughn, J. L., Eds., NSF/RA-800103, PB80-217417, NTIS, Springfield, Va., 1980, 1.
6. **Vanderzant, E. S., Pool, M. C., and Richardson, C. D.,** The role of ascorbic acid in the nutrition of three cotton insects, *J. Insect Physiol.,* 8, 287, 1962.
7. **Vanderzant, E. S., Richardson, C. D., and Fort, S. W., Jr.,** Rearing of the bollworm on artificial diet, *J. Econ. Entomol.,* 55, 140, 1962.
8. **Ignoffo, C. M.,** Bioassay technique and pathogenicity of a nuclear-polyhedrosis virus of the cabbage looper, *Trichoplusia ni* (Hubner), *J. Insect Pathol.,* 6, 237, 1964.
9. **Ignoffo, C. M.,** The nuclear-polyhedrosis virus of *Heliothis zea* (Boddie) and *Heliothis virescens* (Fabricius). I. Virus propagation and its virulence, *J. Invertebr. Pathol.,* 7, 209, 1965.
10. **Ignoffo, C. M.,** The nuclear-polyhedrosis virus of *Heliothis zea* (Boddie) and *Heliothis virescens* (Fabricius). II. Biology and propagation of diet-reared *Heliothis, J. Invertebr. Pathol.,* 7, 279, 1965.
11. **Ignoffo, C. M.,** Insect viruses, in *Insect Colonization and Mass Production,* Smith, C. N., Ed., Academic Press, New York, 1966, 501.
12. **Briggs, J. D.,** Commercial production of insect pathogens, in *Insect Pathology: An Advanced Treatise,* Vol. 2, Steinhaus, E. A., Ed., Academic Press, New York, 1963, 519.
13. **Shapiro, M., Bell, R. A., and Owens, C. D.,** *In vivo* mass production of gypsy moth nucleopolyhedrosis virus, *USDA For. Serv. Tech. Bull.,* No. 1584, 633, 1981.
14. **Thompson, C. G. and Steinhaus, E. A.,** Further tests using a polyhedrosis virus to control the alfalfa caterpillar, *Hilgardia,* 19, 411, 1950.
15. **Chauthani, A. R. and Claussen, D.,** Rearing Douglas-fir tussock moth larvae on synthetic media for the production of nuclear-polyhedrosis virus, *J. Econ. Entomol.,* 61, 101, 1968.
16. **Doane, C. C.,** Infectious sources of nuclear polyhedrosis virus persisting in natural habitats of the gypsy moth, *Environ. Entomol.,* 4, 392, 1975.
17. **Martignoni, M. E.,** Production, activity, and safety of the Douglas-fir tussock moth nucleopolyhedrosis virus, *USDA For. Serv. Tech. Bull.,* No. 1585, 140, 1978.
18. **Doane, C. C.,** Trans-ovum transmission of nuclear polyhedrosis virus in relation to disease in gypsy moth populations, in *Proc. 4th Int. Colloq. Insect Pathol.,* College Park, Md., 1971, 285.
19. **Podgwaite, J. D. and Campbell, R. W.,** Disease in natural gypsy moth populations, in *Proc. 4th Int. Colloq. Insect Pathol.,* College Park, Md., 1971, 279.
20. **Magnoler, A.,** A wheat germ medium for rearing of the gypsy moth, *Lymantria dispar* L, *Entomophaga,* 15, 401, 1970.
21. **Magnoler, A.,** Susceptibility of gypsy moth larvae to *Lymantria* spp. nuclear- and cytoplasmic polyhedrosis viruses, *Entomophaga,* 15, 407, 1970.
22. **Grisdale, D. G.,** A method for reducing incidence of virus infection in insect rearings, *J. Invertebr. Pathol.,* 12, 425, 1968.
23. **Shapiro, M.,** *In vivo* mass production of insect viruses for use as pesticides, in *Microbial and Viral Pesticides,* Kurstak, E., Ed., Marcel Dekker, New York, 1982, 463.
24. **Steinhaus, E. A.,** Diseases of insects reared in the laboratory or insectary, *Calif. Agric. Exp. Sta. Ext. Serv.,* 1, 1953.
25. **Henneberry, T. J. and Kishaba, A. N.,** Pupal size and mortality, longevity, and reproduction of cabbage loopers reared at different densities, *J. Econ. Entomol.,* 59, 1490, 1966.

26. **Sikorowski, P. P.,** Microbiological monitoring in the boll weevil rearing facility, *Miss. Agric. For. Exp. Sta. Tech. Bull.,* 72, 1975.

27. **Stewart, F. F., Bell, M. R., Martinez, A. J., Roberson, J. L., and Lowe, A. M.,** The Surface Sterilization of Pink Bollworm Eggs and Spread of a Cytoplasmic Polyhedrosis Virus in Rearing Containers, USDA-APHIS, 81-27, U.S. Government Printing Office, Washington, D.C., 1976.

28. **Baumhover, A. H., Cantelo, W. W., Hobgood, J. M., Jr., Knott, C. M., and Lam, J. J., Jr.,** An Improved Method for Mass Rearing Tobacco Hornworms, USDA-ARS, S-167, U.S. Government Printing Office, Washington, D.C., 1977.

29. **Shimanuki, H., Lehnert, T., Knox, D. A., and Herbert, E. W., Jr.,** Control of European foulbrood disease of the honey bee, *J. Econ. Entomol.,* 62, 813, 1969.

30. **Tompkins, G. J. and Cantwell, G. E.,** The use of ethylene oxide to inactivate insect viruses in insectaries, *J. Invertebr. Pathol.,* 25, 139, 1975.

31. **Helms, T. J. and Raun, E. S.,** Perennial laboratory culture of disease-free insects, in *Microbial Control of Insects and Mites,* Burges, H. D. and Hussey, N. W., Eds., Academic Press, New York, 1971, 638.

32. **Ignoffo, C. M. and Dutky, S. R.,** The effect of sodium hypochlorite on the viability and infectivity of *Bacillus* and *Beauveria* spores and cabbage looper nuclear-polyhedrosis virus, *J. Insect Pathol.,* 5, 422, 1963.

33. **Getzin, L.,** Mass rearing of virus-free cabbage loopers on an artificial diet, *J. Insect Pathol.,* 4, 486, 1962.

34. **Golanski, K.,** The effectiveness of formalin in controlling jaundice (nuclear polyhedrosis) of the silkworm in Poland, *J. Insect Pathol.,* 3, 11, 1961.

35. **Bell, R. A., Owens, C. D., Shapiro, M., and Tardif, J. G. R.,** Development of mass rearing technology, *USDA For. Ser. Tech. Bull.,* No. 1584, 599, 1981.

36. **Leonard, D. and Doane, C.,** An artificial diet for the gypsy moth, *Porthetria dispar* (Lepidoptera: Lymantriidae), *Ann. Entomol. Soc. Am.,* 59, 462, 1966.

37. **ODell, T. M. and Rollinson, W. D.,** A technique for rearing the gypsy moth, *Porthetria dispara* (L.), on an artificial diet, *J. Econ. Entomol.,* 59, 741, 1966.

38. **Smith, R. P., Wrait, S. P., Tardif, M. F., Hasenstab, M. J., and Simeone, J. B.,** Mass rearing of *Porthetria dispar* (L.) (Lepidoptera: Lymantriidae) for in-host production of nuclear polyhedrosis virus, *J. N. Y. Entomol. Soc.,* 84, 212, 1976.

39. **Lyon, L. R. and Flake, H. W., Jr.,** Rearing Douglas-fir tussock moth larvae on synthetic media, *J. Econ. Entomol.,* 59, 696, 1966.

40. **Spitler, G. H.,** Protection of Indian-meal moth cultures from a granulosis virus, *J. Econ. Entomol.,* 63, 1024, 1970.

41. **David, W. A. L., Ellaby, S., and Taylor, G.,** The effect of reducing the content of certain ingredients in a semisynthetic diet on the incidence of granulosis virus disease in *Pieris brassicae, J. Invertebr. Pathol.,* 20, 332, 1972.

42. **Ignoffo, C. M. and Boening, O. P.,** Compartmented disposable plastic trays for rearing insects, *J. Econ. Entomol.,* 64, 850, 1970.

43. **Howell, J. F.,** Rearing the codling moth on an artificial diet, *J. Econ. Entomol.,* 63, 1148, 1970.

44. **Sheppard, R. F. and Stairs, G. R.,** Effects of dissemination of low dosage levels of a granulosis virus in populations of the codling moth, *J. Econ. Entomol.,* 69, 583, 1976.

45. **Harper, J. D.,** Laboratory production of *Peridroma saucia* and its nuclear polyhedrosis virus, *J. Econ. Entomol.,* 63, 1633, 1970.

46. **Vail, P. V., Anderson, S. J., and Jay, D. L.,** New procedures for rearing cabbage loopers and other lepidopterous larvae for propagation of nuclear polyhedroses viruses, *Environ. Entomol.,* 2, 339, 1973.

47. **Hedlund, R. C. and Yendol, W. G.,** Gypsy moth nuclear-polyhedrosis virus production as related to inoculating time, dosage, and larval weight, *J. Econ. Entomol.,* 67, 61, 1974.

48. **Lewis, F. B.,** Mass propagation of insect viruses with especial reference to forest insects, in *Proc. 4th Int. Colloq. Insect Pathol.,* College Park, Md., 1971, 320.

49. **Simonova, E. Zh.,** Laboratory technique for making preparations of the apple moth virus (transl. from Russian, 1979), in *Pathology of Insects, Ticks, and Mites,* Tsinovskii, Ya. P., Ed., 1972, 154.

50. **Shieh, T. T. and Bohmfalk, G. T.,** Production and efficacy of baculoviruses, *Biotechnol. Bioengin.,* 22, 1357, 1980.

51. **Cunningham, J. C., Howse, G. M., McPhee, J. R., deGroot, P., and White, M. B. E.,** Aerial Application of Spruce Budworm Baculovirus: Replicated Tests with an Aqueous Formulation and a Trial Using an Oil Formulation in 1978, Rep. FPM-X-21, Forest Pest Management Institute, Sault Ste. Marie, Ontario, 1979.

52. **Hostetter, D. L., Pinnell, R. E., Grier, P. A., and Ignoffo, C. M.,** A granulosis virus of *Pieris rapae* as a microbial control agent on cabbage in Missouri, *Environ. Entomol.,* 2, 1109, 1973.

53. **Huber, J. and Dickler, E.,** Codling moth granulosis virus: its efficiency in the field in comparison with organophorus insecticides, *J. Econ. Entomol.,* 70, 557, 1977.

54. **Orlovskaya, E. V.,** Production of preparations based on nuclear polyhedrosis viruses for controlling the cabbage armyworm, *Mamestra brassicae,* and gypsy moth, *Lymantria dispar*, in Proc. 2nd Conf. of Project V, Microbiological Control of Insect Pests of the US/USSR Joint Working Group on the Production of Substances by Microbiological Means, Ignoffo, C. M., Martignoni, M. E., and Vaughn, J. L., Eds., NSF/RA-800103, PB80-217417, NTIS, Springfield, Va., 1980, 54.

55. **Shapiro, M., Martignoni, M. E., Cunningham, J. C., and Goodwin, R. H.,** Potential use of the saltmarsh caterpillar as a production host for nucleopolyhedrosis viruses, *J. Econ. Entomol.,* 75, 69, 1982.

56. **Hamm, J. J.,** Relative susceptibility of several noctuid species to a nuclear polyhedrosis virus from *Heliothis armigera, J. Invertebr. Pathol.,* 39, 255, 1982.

57. **Teakle, R. E.,** Relative pathogenicity of nuclear polyhedrosis viruses from *Heliothis punctigera* and *Heliothis zea* for larvae of *Heliothis armigera* and *Heliothis punctigera, J. Invertebr. Pathol.,* 34, 231, 1979.

58. **Falcon, L. A., Kane, W. R., and Bethell, R. S.,** Preliminary evaluation of a granulosis virus for control of the codling moth, *J. Econ. Entomol.,* 61, 1208, 1968.

59. **Payne, C. C.,** The susceptibility of the pea moth, *Cydia nigricana,* to infection by the granulosis virus of the codling moth, *Cydia pomonella, J. Invertebr. Pathol.,* 38, 71, 1981.

60. **Payne, C. C., Tatchell, G. M., and Williams, C. F.,** The comparative susceptibilities of *Pieris brassicae* and *P. rapae* to a granulosis virus from *P. brassicae, J. Invertebr. Pathol.,* 38, 273, 1981.

61. **Boucias, D. G. and Norton, G. L.,** Susceptibility of *Hyphantria cunea* to a granulosis virus isolated from *Diacrisia virginica, J. Invertebr. Pathol.,* 30, 377, 1977.

62. **Vail, P. V., Jay, D. L., and Hunter, D. K.,** Cross infectivity of a nuclear polyhedrosis virus isolated from *Autographa californica,* in *Proc. 4th Int. Colloq. Insect Pathol.,* College Park, Md., 1971, 297.

63. **Vail, P. V., Sutter, G., Jay, D. L., and Gough, D.,** Reciprocal infectivity of cabbage looper and alfalfa looper nuclear polyhedrosis viruses, *J. Invertebr. Pathol.,* 17, 383, 1971.

64. **Vail, P. V., Jay, D. L., Hunter, D. K., and Staten, R. T.,** A nuclear polyhedrosis virus infective to the pink bollworm, *Pectinophora gossypiella, J. Invertebr. Pathol.,* 20, 124, 1972.

65. **Vail, P. V. and Jay, D. L.,** Pathology of a nuclear polyhedrosis virus of the alfalfa looper in alternate hosts, *J. Invertebr. Pathol.,* 21, 198, 1973.

66. **Harper, J. D.,** Cross-infectivity of six plusiine nuclear polyhedrosis virus isolates to plusiine hosts, *J. Invertebr. Pathol.,* 27, 175, 1976.

67. **Witt, D. J. and Janus, C. A.,** Replication of *Galleria mellonella* nuclear polyhedrosis virus in cultured cells and in larvae of *Trichoplusia ni, J. Invertebr. Pathol.,* 29, 222, 1976.

68. **Kaya, H. K.,** Transmission of a nuclear polyhedrosis virus isolated from *Autographa californica* to *Alsophila pometaria, Hyphantria cunea,* and other forest defoliators, *J. Econ. Entomol.,* 70, 9, 1977.

69. **Lewis, L. C., Lynch, R. E., and Jackson, J. J.,** Pathology of a baculovirus of the alfalfa looper, *Autographa californica,* in the European corn borer, *Ostrinia nubilalis, Environ. Entomol.,* 6, 535, 1977.

70. **Capinera, J. L. and Kanost, M. R.,** Susceptibility of the zebra caterpillar to *Autographa californica* nuclear polyhedrosis virus, *J. Econ. Entomol.,* 72, 570, 1979.

71. **Vail, P. V., Jay, D. L., Stewart, F. D., Martinez, A. J., and Dulmage, H. T.,** Comparative susceptibility of *Heliothis virescens* and *H. zea* to the nuclear polyhedrosis virus isolated from *Autographa californica, J. Econ. Entomol.,* 71, 293, 1978.

72. **Tompkins, G. J., Vaughn, J. L., Adams, J. R., and Reichelderfer, C. F.,** Effects of propagating *Autographa californica* nuclear polyhedrosis virus and its *Trichoplusia ni* variant in different hosts, *Environ. Entomol.,* 10, 801, 1981.

73. **Lynn, D. E. and Hink, W. F.,** Comparison of nuclear polyhedrosis virus replication in five lepidopteran cell lines, *J. Invertebr. Pathol.,* 35, 234, 1980.

74. **Simonova, E. Zh. and Sinitsyna, Z. Q.,** Evaluation of the insecticidal activity of virus preparations, in Proc. 2nd Conf. of Project V, Microbiological Control of Insect Pests of the US/USSR Joint Working Group on the Production of Substances by Microbiological Means, Ignoffo, C. M., Martignoni, M. E., and Vaughn, J. L., Eds., NSF/RA-800103, PB80-217417, NTIS, Springfield, Va., 1980, 154.

75. **Cunningham, J. C., Kaupp, W. J., Howse, G. M., McPhee, J. R., and deGroot, P.,** Aerial Application of Spruce Budworm Virus. Tests of Virus Strains, Dosages, and Formulations in 1977, Rep. FPM-X-3, Forest Pest Management Institute, Sault Ste. Marie, Ontario, 1978.

76. **Stairs, G. R., Fraser, T., and Fraser, M.,** Changes in growth and virulence of a nuclear polyhedrosis virus from *Choristoneura fumiferana* after passage in *Trichoplusia ni* and *Galleria mellonella, J. Invertebr. Pathol.,* 38, 230, 1981.

77. **Wood, H. A., Hughes, P. R., Johnston, L. B., and Langridge, W. H. R.,** Increased virulence of *Autographa californica* nuclear polyhedrosis virus by mutagenesis, *J. Invertebr. Pathol.,* 38, 236, 1981.

78. **Martignoni, M. E., Stelzer, M. J., and Iwai, P. J.,** Baculovirus of *Autographa californica* (Lepidoptera: Noctuidae): a candidate biological control agent for Douglas-fir tussock moth (Lepidoptera: Lymantriidae), *J. Econ. Entomol.,* 75, 1120, 1982.

79. **Stairs, G. R.,** Infection of *Malacosoma disstria* Hubner with nuclear polyhedrosis viruses from other species of *Malacosoma* (Lepidoptera: Lasiocampidae), *J. Insect Pathol.,* 6, 164, 1964.

80. **Stelzer, M., Neisses, J., Cunningham, J. C., and McPhee, J. R.,** Field evaluation of baculovirus stocks against Douglas-fir tussock moth in British Columbia, *J. Econ. Entomol.,* 70, 243, 1977.

81. **Zethner, O., Brown, D. A., and Harrap, K. A.,** Comparative studies on the nuclear polyhedrosis viruses of *Lymantria monacha* and *L. dispar, J. Invertebr. Pathol.,* 34, 178, 1979.

82. **Ponsen, M. B.,** The production of polyhedral viruses from *Barathra brassicae* to control *Adoxophyes orana, Meded. Rijksfac. Landbouwwetensh. Gent.,* 31, 553, 1966.

83. **Jurkovicova, M.,** Activation of latent virus infections in larvae of *Adoxophyes orana* (Lepidoptera: Tortricidae) and *Barathra brassicae* (Lepidoptera: Noctuidae) by foreign polyhedra, *J. Invertebr. Pathol.,* 34, 213, 1979.

84. **Shapiro, M.,** Large scale *in vivo* production of the gypsy moth nucleopolyhedrosis virus at Otis AFB, *USDA For. Serv. Tech. Bull.,* No. 1585, 464, 1981.

85. **Rogoff, M.,** Production responsibilities: private sector and commercial producer of virus, in *Baculoviruses for Insect Control: Safety Considerations,* Summers, M., Engler, R., Falcon, L. A., and Vail, P. V., Eds., American Society of Microbiology, Washington, D.C., 1975, 159.

86. **Martignoni, M. E., Breillat, J. P., and Anderson, N. G.,** Mass purification of polyhedral inclusion bodies by isopycnic banding in zonal rotors, *J. Invertebr. Pathol.,* 11, 507, 1968.

87. **Cline, G. B., Ryel, E., Ignoffo, C. M., Shapiro, M., and Straehle, Z. Q.,** Zonal purification of the nucleopolyhedrosis virus of the cotton bollworm *Heliothis zea* (Boddie), in *Proc. 4th Int. Colloq. Insect Pathol.,* College Park, Md., 1971, 363.

88. **Mazzone, H. M., Breillat, J. P., and Anderson, N. G.,** Zonal rotor purification and properties of a nuclear polyhedrosis virus of the European pine sawfly (*Neodiprion sertifer* Geoffrey), in *Proc. 4th Int. Colloq. Insect Pathol.,* College Park, Md., 1971, 371.

89. **Miller, L. K., Franzblau, Z. Q., Homan, H. W., and Kish, L. P.,** A new variant of *Autographa californica* nuclear polyhedrosis virus, *J. Invertebr. Pathol.,* 36, 159, 1980.

90. **Ossowski, L. L. J.,** Variation in virulence of a wattle bagworm virus, *J. Insect Pathol.,* 2, 35, 1960.

91. **Shapiro, M. and Ignoffo, C. M.,** Nucleopolyhedrosis of *Heliothis:* Activity of isolates from *Heliothis zea, J. Invertebr. Pathol.,* 16, 107, 1970.

92. **Magnoler, A.,** Bioassay of a nucleopolyhedrosis virus of the gypsy moth, *Porthetria dispar* L., *J. Invertebr. Pathol.,* 23, 190, 1974.

93. **Hughes, P. R., Gettig, R. R., and McCarthy, W. J.,** Comparison of the time-mortality responses of *Heliothis zea* to 14 isolates of *Heliothis* nuclear polyhedrosis viruses, *J. Invertebr. Pathol.,* 41, 256, 1983.

94. **Shapiro, M., Robertson, J. L., Injac, M. G., Katagiri, K., and Bell, R. A.,** Comparative infectivities of gypsy moth (Lepidoptera: Lymantriidae) nucleopolyhedrosis virus isolates from North America, Europe, and Asia, *J. Econ. Entomol.,* in press.

95. **Chauthani, A. R., Claussen, D., and Rehnborg, C. S.,** Dosage-mortality data of a nuclear-polyhedrosis virus of the cotton bollworm, *Heliothis zea, J. Invertebr. Pathol.,* 12, 335, 1968.

96. **Shapiro, M., Bell, R. A., and Owens, C. D.,** *In vivo* propagation of the nucleopolyhedrosis virus of *Lymantria dispar,* in Proc. 2nd Conf. of Project V, Microbiological Control of Insect Pests of the US/USSR Joint Working Group on the Production of Substances by Microbiological Means, Ignoffo, C. M., Martignoni, M. E., and Vaughn, J. L., Eds., NSF/RA-800103, PB80-217417, NTIS, Springfield, Va., 1980, 43.

97. **Yadava, R. L.,** Studien über den Einfluss von Wasserglass und Borsäure als chemische Stressoren für die künstlich applizierte Kern- und die latente Cytoplasmapolyedrose der Nonne, mit einer Ammerkung über die genenseitige Beeinflussung beider Krankheiten, *Z. Angew. Entomol.,* 65, 175, 1970.

98. **Thompson, C. G.,** Thermal inhibition of certain virus diseases, *J. Insect Pathol.,* 1, 189, 1959.

99. **Ignoffo, C. M.,** Effects of temperature on mortality of *Heliothis zea* larvae exposed to sublethal doses of a nuclear-polyhedrosis virus, *J. Invertebr. Pathol.,* 9, 290, 1967.

100. **Bullock, H. R.,** Therapeutic effect of high temperature on tobacco budworms to a cytoplasmic-polyhedrosis virus, *J. Invertebr. Pathol.,* 19, 148, 1972.

101. **Boucias, D. G., Johnson, D. W., and Allen, G. E.,** Effects of host age, virus dosage, and temperature on the infectivity of a nucleopolyhedrosis virus against the velvetbean caterpillar, *Anticarsia gemmatalis,* larvae, *Environ. Entomol.,* 9, 59, 1980.

102. **Hunter, D. K. and Hartsell, P. L.,** Influence of temperature on Indian-meal moth larvae infected with a granulosis virus, *J. Invertebr. Pathol.,* 17, 347, 1971.

103. **Pristavko, V. P., Yanishevskaya, L. V., and Rezvatova, O. I.,** Some infectious diseases of the insectary reared codling moth, *Laspeyresia pomonella* L., in *Proc. 4th Int. Colloq. Insect Pathol.,* College Park, Md., 1971, 262.

104. **Kobayashi, M. and Kawase, S.,** Absence of detectable accumulation of cytoplasmic polyhedrosis viral RNA in the silkworm *Bombyx mori,* reared at a supraoptimal temperature, *J. Invertebr. Pathol.,* 35, 96, 1980.

105. **Wallis, R. C.,** Incidence of polyhedrosis of gypsy-moth larvae and the influence of relative humidity, *J. Econ. Entomol.,* 50, 580, 1957.

106. **Orlovskaya, E. V.,** Infection of gypsy moth larvae with the nuclear polyhedrosis virus and its effect of the fertility of the moth and survival of its progeny (trans. from Russian), *Biol. Control Agric. Forest Tests Proc. Symp.,* 1956, 54.

107. **Orlovskaya, E. V.,** Conditions determining the initiation and development of nuclear polyhedrosis epizootics in Lepidoptera, in *Insect Pathology and Microbial Control,* van der Laan, P. A., Ed., North-Holland, Amsterdam, 1967, 282.

108. **Wallis, R. C.,** Environmental factors and epidemics of polyhedrosis in gypsy moth larvae, in *Proc. 11th Int. Congr. Entomol.,* Vienna, 1962, 827.

109. **Owens, C. D.,** Controlled environments for insects and personnel in insect-rearing facilities, in *Advances and Challenges in Insect Rearing,* King, E. G. and Leppla, N. C., Eds., U.S. Department of Agriculture, Washington, D.C., 1984, 58.

110. **Wolf, W. W.,** Controlling respiratory hazards in insectaries, in *Advances and Challenges in Insect Rearing,* King, E. G. and Leppla, N. C., Eds., U.S. Department of Agriculture, Washington, D.C., 1984, 64.

111. **Ridet, J. M.,** Etude des conditions optimales d'elevage et d'alimentation de *Lymantria dispar* L., *Ann. Soc. Entomol. Fr. (N.S.),* 8, 653, 1972.

112. **Steinhaus, E. A. and Thompson, C. G.,** Preliminary field tests using a polyhedral virus to control the alfalfa caterpillar, *J. Econ. Entomol.,* 42, 301, 1949.

113. **McEwen, F. L. and Hervey, G. E. R.,** Control of the cabbage looper with a virus disease, *J. Econ. Entomol.,* 51, 626, 1958.

114. **Cunningham, J. C.,** Polyhedrosis viruses infecting the eastern hemlock looper, *Lambdina fiscellaria,* in *Proc. 4th Int. Colloq. Insect Pathol.,* College Park, Md., 1971, 292.

115. **McIntyre, T. and Dutky, S. R.,** Aerial application of virus for control of a pine sawfly, *Neodiprion pratti pratti, J. Econ. Entomol.,* 54, 809, 1961.

116. **Rollinson, W. D., Hubbard, H. B., and Lewis, F. B.,** Mass rearing of the European pine sawfly for production of the nuclear polyhedrosis virus, *J. Econ. Entomol.,* 63, 343, 1970.

117. **Zarin'sh, I. A.,** Use of entomopathogenic viruses, in Proc. 2nd Conf. of Project V, Microbiological Control of Insect Pests of the US/USSR Joint Working Group on the Production of Substances by Microbiological Means, Ignoffo, C. M., Martignoni, M. E. and Vaughn, J. L., Eds., NSF/RA-800137, PB80-217417, NTIS, Springfield, Va. 1980, 77.

118. The IPM Practitioner, Virus of European pine sawfly registered, 5, 1, 1983.

119. **Ignoffo, C. M.,** Production, identification, and standardization of insect viral pathogens, *Entomophaga,* 10, 29, 1965.

120. **Doane, C. C.,** Trans-ovum transmission of a nuclear-polyhedrosis in the gypsy moth and the inducement of virus susceptibility, *J. Invertebr. Pathol.,* 14, 199, 1969.

121. **Vago, C.,** Predispositions and interrelations in insect disese, in *Insect Pathology: An Advanced Treatise,* Vol. 1, Steinhaus, E. A., Ed., Academic Press, New York, 1963, 339.

122. **David, W. A. L. and Gardiner, B. O. C.,** The incidence of granulosis deaths in susceptible and resistant *Pieris brassicae* (Linnaeus) larvae following changes of population density, food, and temperature, *J. Invertebr. Pathol.,* 7, 347, 1965.

123. **Ignoffo, C. M.,** Possibilities of mass-producing insect pathogens, in *Insect Pathology and Insect Control,* van der Laan, P. A., Ed., North-Holland, Amsterdam, 1967, 91.

124. **Tinsley, T. W.,** The potential of insect pathogenic viruses as pesticidal agents, *Ann. Rev. Entomol.,* 24, 63, 1979.

125. **Tardif, J. G. R.,** Techniques for Large-Scale Production of Gypsy Moth Diet, USDA-APHIS, 81-23, U.S. Government Printing Office, Washington, D.C., 1975.

126. **Shorey, H. H. and Hale, R. L.,** Mass rearing of the larvae of nine noctuid species on a simple artificial medium, *J. Econ. Entomol.,* 58, 522, 1965.

127. **Yamamoto, R.,** Mass rearing of the tobacco hornworm. II. Larval rearing and pupation, *J. Econ. Entomol.,* 63, 1427, 1969.

128. **Shapiro, M., Bell, R. A., and Owens, C. D.,** Evaluation of various artificial diets for in vivo production of the gypsy moth nucleopolyhedrosis virus, *J. Econ. Entomol.,* 74, 110, 1981.

129. **Burton, R. L.,** Mass rearing the corn earworm in the laboratory, USDA-ARS (Ser.), 33-134, 1969.

130. **Burton, R. L.,** A low-cost artificial diet for the corn earworm, *J. Econ. Entomol.,* 63, 1969, 1970.

131. **Shapiro, M.,** unpublished data.

132. **Biever, K. D. and Wilkinson, J. D.,** A stress-induced granulosis virus of *Pieris rapae, Environ. Entomol.,* 7, 572, 1978.

133. **Baugher, D. G. and Yendol, W. G.,** Virulence of *Autographa californica* baculovirus preparations fed with different food sources to cabbage loopers, *J. Econ. Entomol.,* 74, 309, 1981.

134. **Shapiro, M. and Bell, R. A.,** Enhanced effectiveness of *Lymantria dispar* (Lepidoptera: Lymantriidae) nucleopolyhedrosis virus formulated with boric acid, *Ann. Entomol. Soc. Am.,* 75, 346, 1982.

135. **Vanderzant, E. S.,** Defined diets for phytophagous insects, in *Insect Colonization and Mass Production,* Smith, C. N., Ed., Academic Press, New York, 1966, 273.

136. **Moore, I. and Navon, A.,** Calcium alginate: a new approach in the culturing of insects applied to *Spodoptera littoralis* (Boisduval), *Experientia,* 25, 221, 1969.

137. **Moore, I. and Navon, A.,** Artificial rearing of the codling moth (*Carpocapsa pomonella* L.) on calcium alginate diets, *J. Econ. Entomol.,* 66, 561, 1973.

138. **Navon, A. and Moore, I.,** Artificial rearing of the codling moth (*Carpocapsa pomonella* L.) on calcium alginate diets, *Entomophaga,* 16, 381, 1971.

139. **Spencer, N. R., Leppla, N. C., and Presser, G. A.,** Sodium alginate as a gelling agent in diets for the cbbage looper, *Trichoplusia ni* (Lep., Noctuidae), *Entomol. Appl.,* 20, 39, 1978.

140. **Patana, R.,** Rearing cotton insects in the laboratory, *U.S. Dep. Agric. Prod. Res. Rep.,* 108, 1969.

141. **Shapiro, M. and Bell, R. A.,** Production of the gypsy moth, *Lymantria dispar* (L.) nucleopolyhedrosis virus, using carrageenans as dietary gelling agents, *Ann. Entomol. Soc. Am.,* 75, 43, 1982.

142. **David, W. A. L. and Gardiner, B. O. C.,** Rearing *Pieris brassicae* L. on semisynthetic diets with and without cabbage, *Bull. Entomol. Res.,* 56, 581, 1966.

143. **Dulmage, H. R., Martinez, A. J., and Correa, J. A.,** Recovery of the nuclear polyhedrosis virus of the cabbage looper, *Trichoplusia ni,* by coprecipitation with lactose, *J. Invertebr. Pathol.,* 16, 80, 1970.

144. **Rollinson, W. R.,** personal communication, 1979.

145. **Stairs, G. R.,** Effects of a wide range of temperatures on the development of *Galleria mellonella* and its specific *Baculovirus, Environ. Entomol.,* 7, 297, 1978.

146. **Ignoffo, C. M. and Shapiro, M.,** Characteristics of baculovirus preparations processed from living and dead larvae, *J. Econ. Entomol.,* 71, 186, 1978.

147. **Shapiro, M. and Bell, R. A.,** Biological activity of *Lymantria dispar* nucleopolyhedrosis virus from living and virus-killed larvae, *Ann. Entomol. Soc. Am.,* 74, 27, 1981.

148. **Harper, J. D.,** Preliminary testing of a nuclear polyhedrosis virus to control the variegated cutworm on peppermint, *J. Econ. Entomol.,* 64, 1573, 1971.

149. **Shapiro, M., Owens, C. D., Bell, R. A., and Wood, H. A.,** Simplified, efficient system for *in vivo* mass production of the gypsy moth nucleopolyhedrosis virus, *J. Econ. Entomol.,* 74, 341, 1981.

150. **Ignoffo, C. M.,** Specificity of insect viruses, *Bull. Entomol. Soc. Am.,* 14, 265, 1968.

151. **Falcon, L. A.,** Problems associated with the use of arthropod viruses in pest control, *Ann. Rev. Entomol.,* 21, 305, 1976.

152. **Martignoni, M. E. and Steinhaus, E. A.,** *Laboratory Exercises in Insect Microbiology and Insect Pathology,* Burgess Publishing, Minneapolis, 1961.

153. **Hughes, P. R. and Wood, H. A.,** A synchronous peroral technique for the bioassay of insect viruses, *J. Invertebr. Pathol.,* 37, 154, 1981.

154. **Finney, D. J.,** *Probit Analysis,* Cambridge University Press, London, 1962.

155. **Berkson, J.,** A statistically precise and relatively simple method of estimating the bioassay with quantal response, based on the logistic function, *J. Am. Stat. Assoc.,* 48, 565, 1953.

156. **Hoskins, W. M. and Craig, R.,** Uses of bioassay in entomology, *Ann. Rev. Entomol.,* 7, 437, 1962.

157. **Fisher, R. A.,** Bioassay of microbial pesticides, in *Analytical Methods for Pesticides, Plant Growth Regulators, and Food Additives,* Zweig, G., Ed., Academic Press, New York, 1963, 425.

158. **Bucher, G. E. and Morse, P. M.,** Precision of estimates of the median lethal dose of insect pathogens, *J. Insect Pathol.,* 5, 289, 1963.

159. **Mechalas, B. J. and Dunn, P. H.,** Bioassay of *Bacillus thuringiensis* Berliner-based microbial insecticides. I. Bioassay procedures, *J. Insect Pathol.,* 6, 214, 1964.

160. **Burges, H. D. and Thomson, E. M.,** Standardization and assay of microbial insecticides, in *Microbial Control of Insects and Mites,* Burges, H. D. and Hussey, N. W., Eds., Academic Press, New York, 1971, 591.

161. **Vail, P. V.,** Standardization and quantification: insect laboratory studies, in *Baculoviruses for Insect Pest Control: Safety Considerations,* Summers, M., Engler, R., Falcon, L. A., and Vail, P. V., Eds., American Society of Microbiology, Washington, D.C., 1975, 44.

162. **Martignoni, M. E. and Ignoffo, C. M.,** Biological activity of baculovirus preparation: *in vivo* assay, in Proc. 2nd Conf. of Project V, Microbiological Control of Insect Pests of the US/USSR Joint Working Group on the Production of Substances by Microbiological Means, Ignoffo, C. M., Martignoni, M. E., and Vaughn, J. L., Eds., NSF/RA-800103, PB80-217417, NTIS, Springfield, Va., 1980, 138.

163. **Lewis, F. B.,** Gypsy moth nucleopolyhedrosis virus, *USDA For. Serv. Tech. Bull.,* No. 1584, 484, 1981.

164. **Lee, H. H. and Miller, L. K.,** Isolates of genotypic variants of *Autographa californica* nuclear polyhedrosis virus, *J. Virol.,* 27, 754, 1978.

165. **Crook, N. E. and Payne, C. C.,** Comparison of three methods for ELISA for baculoviruses, *J. Gen. Virol.,* 46, 29, 1980.

166. **Longworth, J. F. and Carey, G. P.,** The use of an indirect enzyme-linked immunosorbent assay to detect baculovirus in larvae and adults of *Oryctes rhinoceros* from Tonga, *J. Gen. Virol.,* 47, 431, 1980.

167. **Krywienczyk, J. and Bergold, G. H.,** Serological relationships of viruses from some lepidopterous and hymenopterous insects, *Virology,* 10, 308, 1960.
168. **Norton, P. W. and DiCapua, R. A.,** Serological relationship of nuclear polyhedrosis viruses. I. Hemagglutination by polyhedral inclusion protein from the nuclear polyhedrosis virus of *Porthetria (Lymantria) dispar, J. Invertebr. Pathol.,* 25, 185, 1975.
169. **Vail, P. V., Morris, T. J., and Collier, S.,** An RNA virus in *Autographa californica* nuclear polyhedrosis virus preparations: gross pathology and infectivity, *J. Invertebr. Pathol.,* 41, 179, 1983.
170. **Podgwaite, J. D. and Cosenza, B. J.,** Bacteria of living and dead larvae of *Porthetria dispar* (L.), *U.S. For. Serv. Res. Note,* NE-50, 1966.
171. **Lewis, F. B.,** personal communication, 1979.
172. **Podgwaite, J. D. and Bruen, R. B.,** Procedures for the microbiological examination of production batch preparations of the nuclear polyhedrosis virus (baculovirus) of the gypsy moth, *Lymantria dispar* L., *For. Serv. Gen. Tech. Rep.,* NE-38, 1978.
173. **Lewis, F. B. and Podgwaite, J. D.,** Safety evaluations of gypsy moth nucleopolyhedrosis virus, *USDA For. Serv. Tech. Bull.,* No. 1584, 474, 1981.
174. **Scotti, P. D. and Longworth, J. F.,** Purification of polyhedra from insect tissue extracts contaminated with a small isometric virus, *J. Invertebr. Pathol.,* 32, 216, 1978.
175. **Longworth, J. F.,** The replication of a cytoplasmic polyhedrosis virus from *Chrysodeixis eriosoma* (Lepidoptera: Noctuidae) in *Spodoptera frugiperda* cells, *J. Invertebr. Pathol.,* 37, 54, 1981.
176. **Morris, T. J., Vail, P. V., and Collier, S. S.,** An RNA virus in *Autographa californica* nuclear polyhedrosis virus preparations: detection and identification, *J. Invertebr. Pathol.,* 38, 201, 1981.
177. **Vail, P. V., Morris, T. J., Collier, S., and MacKay, B.,** An RNA virus in *Autographa californica* nuclear polyhedrosis virus preparations: incidence and influence on baculovirus activity, *J. Invertebr. Pathol.,* 41, 171, 1983.
178. **Lawson, F. R. and Headstrom, R. L.,** Small plant for production of *Trichoplusia* NPV, *USDA For. Serv. Tech. Bull.,* No. 1576, 37, 1978.
179. **Vasiljevic, L. and Injac, M.,** A study of gypsy moth viruses originating from different geographical regions, *Plant Prot.,* 24 (No. 124-125), 169, 1973.
180. **Reichelderfer, C. F. and Benton, C. V.,** The effect of 3-methyl-cholanthrene treatment on the virulence of a nuclear polyhedrosis virus of *Spodoptera frugiperda, J. Invertebr. Pathol.,* 22, 38, 1973.

Chapter 3

CELL CULTURE METHODS FOR LARGE-SCALE PROPAGATION OF BACULOVIRUSES

S. A. Weiss and J. L. Vaughn

TABLE OF CONTENTS

I. INTRODUCTION

Insects represent about 75% of the known animal species in the world, with about 1300 host/virus records in which over 800 species were infected with one or more of 22 viral diseases.[1] About 300 insect species have been reported to have nuclear polyhedrosis virus (NPV) diseases, the majority (243) of which were isolated from Lepidoptera. Nuclear polyhedrosis viruses from *Heliothis zea, Orgyia pseudotsugata, Lymantria dispar, and Neodiprion sertifer* are registered with the Environmental Protection Agency (EPA) for insect control. All of these viruses are presently produced in insects. This process has several disadvantages including the need for an elaborate facility for growing and maintaining insects, numerous personnel, and a final product that contains many contaminating microorganisms and large amounts of insect proteins and cuticles. The insect proteins, cuticles, and contaminating microorganisms may be allergenic to man and thus may represent a hazard to production workers and others handling the product. Therefore, insect cell culture technology is under evaluation as a method for the production of insect viruses for use as microbial insecticides. This has provided a major impetus for insect tissue culture and virus research.[2,3]

If cell cultures from insects could be optimized to provide adequate virus yields at competitive costs, there would be several advantages to using large, steady-state culture systems to produce viruses for pest control.[4] For example, specific cell strains could be selected, preserved, and stored by cryogenic means for future use. Clones which may be highly susceptible to insect viruses and plaque purified viral isolates could be selected from these lines. The final product would be free of allergenic insect proteins, insect cuticle, and contaminating microorganisms.

It has been recommended that entomopathogenic viruses for pest control should be produced in cell culture. Specifically, "economical *in vitro* production of viruses and other pathogens through tissue culture techniques would be one of the most important breakthroughs, not only in insect pathology, but also in fermentation technology."[5] In this review we will discuss the technologies that have been developed and tested in the continuing attempts to achieve this breakthrough.

II. CONTINUOUS INSECT CELL LINES AND STRAINS

A. Availability

It has been evident from the inception of insect cell culture that, for the purposes of virus production, it would be most desirable to establish continuous cell lines. Primary insect cell cultures would not provide the required amount of cell material for many studies nor would they produce an appreciable quantity of virus. The rapid developments in the field of insect cell culture are exemplified by the fact that, of 171 cell lines reported, more than 121 were described between 1976 and 1982.[6-8] These established cell lines are now being employed in a wide variety of basic and applied research studies on hormones, genetics, insect cell metabolism, and biochemistry, and to study the replication of viruses in vitro. Modern techniques permit the routine culturing of primary cells that subsequently will become established cell lines. These established lines can be used for commercial applications employing methods and substrates used to grow cells in suspension.

The development of continuous insect cell lines was first accomplished by Grace,[9] and since that time this technology has become almost routine for skilled researchers working with species in the family Lepidoptera. Cell lines from the most economically important lepidopterous species have been tested for replication of viruses that are currently registered for control of insect pests.[2] Moreover, additional polyhedrosis viruses that are being considered for viral insecticides can replicate in one or more of the currently available cell lines.[10] Continuous cell lines could be cloned and a cell strain selected that gives a uniform population doubling time (PDT), produces a high titer of virus, and is free of contaminating agents. A variety of established insect lines can be obtained directly from individual investigators[6-8] and some can be purchased at a nominal cost from the American Type Culture Collection.[11]

B. Quality Control

One of the most important requirements for cell culture is that media, reagents, and cell substrates be free of adventitious microorganisms. Presently, conventional insect cell culture media contain fetal bovine serum, which is an uncharacterized growth supplement.[4] Various lots of animal sera used in cell culture were found to have wide differences in chemical and hormonal parameters which were responsible for variability and lack of reproducibility in cell culture studies.[12,13]

It is of paramount importance that continuous cell lines carried routinely in laboratories for research and production purposes do not get cross-contaminated with adventitious agents that may be present in animal sera. To avoid this, reserved lots of serum that have been tested by the manufacturer for the absence of mycoplasma, bacteria, fungi, viral agents, and coliphage should be used. To define the quality of sera further, each lot should be analyzed by a computer-controlled analyzer for identification and quantitation of chemical components. These profiles are supplied by the serum manufacturers on request. Samples of several selected reserved lots are then requested and tested for growth promotion ability employing at least two insect continuous cell lines.

Test cultures should be planted in flasks for monolayer growth and in spinner vessels for suspension growth. Cells are examined daily for formation of cell sheet confluency, absence of toxicity, and morphological changes. At the end of 7 days in culture, substrate-based cells are dissociated enzymatically or by a scraping procedure and a viable cell count is performed microscopically.[4] Cells grown in suspension are sampled directly for the viable cell counts. Cells are subcultured for a second passage and observations are carried out for another 7 days. At the end of 7 days in the second passage, quantitative yields are recorded and population doubling time is established, which should be comparable to that of the cells grown in medium and serum previously shown to be efficacious.

If the preliminary testing is satisfactory, then the plating efficiency is tested to determine the quality of cell culture media, animal sera, and other nutrient components. The application of plating efficiency methods with animal[14] and plant cells[15] is well established. These methods, with some modifications, are routinely applied to invertebrate cell culture as a rigorous quantitation for nutritional studies of animal sera and other related components. However, in quality control the most practical method is the relative plating efficiency which depends upon comparison of the number of colonies that will develop under the imposed experimental conditions with control conditions. For dependable testing in house, absolute plating efficiency must first be established with a known control. Absolute plating efficiency relates to the total number of cells that grow to form clones relative to the number of cells contained in the original inoculum which is determined by viable cell count prior to plating.

Table 1 contains the type of data and the calculations required for the plating efficiency evaluation. In this example the absolute plating efficiency of the control was 12.8% and that of the test serum was 21.8%, indicating that the test serum increased colony formation by 9% relative to control. In relative plating efficiency tests, a tested lot of serum, medium, or nutrient component must have the ability to produce cell colonies of acceptable size, morphology, and quantity in comparison to the control media. Tested lots should be acceptable only when the difference in plating efficiency of test lots is not more than 10% less than that of the control lot.

Once a specific reserved lot of animal serum meets the criteria for plating efficiency, a quick examination for absence of adventitious viral agents is carried out employing an electron

Table 1
**PLATING EFFICIENCY EVALUATION OF FETAL BOVINE
SERA**

Medium lot no: 8004   Type of medium: IPL-41 + 10% FBS
Control serum lot no: 300135   Test serum lot no: 300224
Assay cells IPL-Sf-21AE   Cell inocula: 500 cells per 5 mℓ

Observation and Colony Counts

Control serum		Test serum	
Flask no.	No. colonies	Flask no.	No. colonies
1	46	7	62
2	96	8	58
3	49	9	105
4	86	10	136
5	43	11	184
Mean: 64 ± 25[a]		Mean: 109 ± 53	
PE = 12.8%[b]		PE = 21.8%[b]	

[a]   This represents number of colonies per T-25 flask (mean from five flasks ± SD).
[b]   The plating efficiency is defined as the ratio:

$$\frac{\text{No. of colonies at the termination of the test}}{\text{No. of cells inoculated}} \times 100.$$

microscopy procedure analogous to that reported by Smith et al.[16] Briefly, serum samples are ultracentrifuged in a Spinco, fixed angle 50 Ti rotor at 35,000 rpm for 1 hr. The resulting pellet is resuspended in 0.02 to 0.05 mℓ of either distilled H_2O or phosphate-buffered saline (PBS). One or more drops of the suspension is applied to Formvar-coated copper grids and excess fluids are removed by touching filter paper to grid edges (blotting). The grids are washed with distilled water (one or more drops), blotted, and then one drop of 1% phosphotungstic acid (pH 2, 5, or 7) is placed on all grids and once again blotted. Each specimen is exposed to an ultraviolet emitting lamp and examined with an electron microscope.

Mycoplasma are the smallest of free-living microorganisms and are common contaminants of continuous cell lines. The presence of mycoplasma and the source of contamination are difficult to establish and may cause interference with virus detection, alteration in cell morphology, cell lysis, chromosome damage, interference with cell metabolism, and poor growth. Presently, a number of methods are available for detecting mycoplasma in continuous lines.[17] Two methods, applicable to invertebrate cell cultures, are the standard microbiological assay described by McGarrity et al.[18] and the method of cytochemical detection based on the ability of 4'-6-diamino-2-phenylindole to form a fluorescent complex with deoxyribonucleic acid (DNA)[19] as described in instructions provided with the Bioassay Systems Mycoplasma Detection Kit 425 (Bioassay Systems, 100 Inman Street, Cambridge, Mass.).

The rapid development and expansion in cell culture research and the availability of many types of established cell lines require cell line monitoring to insure species identification. Current reports document that interspecies and intraspecies contamination of cell cultures may be as high as 30%. Monitoring of invertebrate cells by various methods has been well established and documented.[20]

Isoenzyme analysis employing cellulose-acetate electrophoresis is reliable and useful for characterization of invertebrate cell lines.[20] Table 2 demonstrates the relative mobilities by

Table 2
RELATIVE ISOZYME MOBILITY IN MILLIMETERS FROM ORIGIN[a]

Cell line	Isozyme							
	AST	G6PD	LD	MD	ME	MPI	NP	PepB
IMC-Hz-1[b]	15.42	− 4.90	13.70	− 1.52	8.66	16.24	− 1.73	NA[c]
	− 5.76			− 5.69				
IPLB-Hz-1075	15.05	− 3.44	12.81	− 2.34	9.61	17.72	− 2.00	10.00
				− 5.33				12.70
IPL-Sf-21AE[b]	21.71	− 5.41	10.67	− 1.60	9.73	18.50	14.83	18.86
	− 10.06			− 4.60	− 1.75	13.90		14.57
IPL-Sf-21AEM[d]	19.72	− 6.87	9.90	− 1.51	9.70	18.80	NA	18.00
								13.73
IPL-Sf-21AE S[e]	19.71	− 5.87	9.31	− 1.14	9.60	18.73	NA	18.63
								13.32

[a] Isozyme mobilities were calculated relative to *Spodoptera frugiperda* (IPL-Sf-21AE) cell line phenotype which was reported previously.[32] (Data provided by D. L. Knudson, Yale University, New Haven, Conn., personal communication.)
[b] Yale University standards.
[c] NA = no activity. Significance of numbers = must be $> +/- 3.0$ mm.
[d] Cells grown and maintained on substrate.
[e] Cells grown and maintained in suspension.

electrophoretic separation on cellulose acetate of cellular isozymes from cell lines frequently used in baculovirus studies. This simplified cellulose acetate electrophoretic technique presently may be the preferred method for routine screening of invertebrate, continuous cell lines from different species.

C. Cryopreservation

The major practical significance of modern cryopreservation technology is that specific cultured cells and strains may be preserved for months and years, resuming a viable state upon thawing and recovery. The cultured cells and strains survive cryopreservation best if they are frozen in their optimum metabolic state, normally when the cells are growing logarithmically.

Procedures for freezing and cryopreservation of insect cells in sufficient volume to seed roller bottles have been established.[4] Cells were taken for freezing in logarithmic phase growth, dispensed into centrifuge tubes, and allowed to settle for 10 min. Then they were centrifuged at 4°C for 5 min at 400 rpm (50 × g) and suspended into one part of spent medium and two parts fresh medium. Viability was determined and the cell density was adjusted to 2×10^7 cells per milliliter. The cell suspension was kept on wet crushed ice (0°C), preferably under the laminar flow hood. An equal volume of new complete growth medium containing 15% dimethylsulfoxide (DMSO) (distilled in glass from Burdick and Jackson Laboratories, Inc., Muskegon, Mich.), precooled on wet ice or in the refrigerator, was added to the cell suspension. The cell suspension, with DMSO, was dispensed at the rate of 4 mℓ per cryotube and the vials were identified by marking with cryomarkers. Prior to freezing, the cell suspension should be tested for sterility both before and after dispensing. General procedures for sterility testing are outlined in *U.S. Pharmacopoeia*, XVII ed., pp. 852—857. Vials containing the cell suspensions were placed in the freezer and frozen at 1°C per minute using a Linde Programmed Freezer BF4 or BF5 (Biological Freezing Systems, Linde Division, Union Carbide Corp., New York, N.Y.). When the temperature of the specimen in the freezer reached − 40°C, the cooling rate was increased by direct immersion of the vials containing frozen cells into liquid nitrogen or vapor phase until the storage temperature was reached.

To recover the frozen cells, the vials were removed from liquid nitrogen storage and thawed rapidly by immersion in a 28°C water bath. Cells were planted into growth medium in appropriate vessels such as T flasks and/or roller bottles at a cell concentration not exceeding 1×10^5 cells per milliliter. Cells were allowed to attach for 3 to 6 hr or overnight. After microscopic examination of the cells, the old medium containing DMSO was replaced with new medium and the cells were incubated further as required. Suspension adapted cells were planted directly into 100 to 250 mℓ suspension medium in spinner flasks adjusted to 175 to 190 rpm on nonheating magnetic stirrers. If cells were frozen at 1×10^7 cells per milliliter and recovered directly in 100 mℓ of medium, an optimal planting density of 1×10^5 cells per milliliter resulted. After 2 to 3 days in spinner culture, the cells were diluted in equal volumes of fresh spinner medium and the concentration of DMSO was reduced to 0.04%. The viability of cells frozen by this procedure was about 85 to 95% for either substrate-based or suspension-adapted cells.

Cells recovered from cryopreservation and cultured directly in 850 cm^2 plastic roller bottles (PRB) have been shown to have insignificant loss in yield efficiency.[4] After 7 days postplanting, these cultures attained 85% confluency and the cell sheets were dissociated by scraping and quantitated. The cultures recovered from cryopreservation yielded $2.93 \times 10^8 \pm 7.5 \times 10^6$ cells per plastic roller bottle with a PDT of 13.76 ± 0.35 hr. In comparison, unfrozen cells yielded 4.90×10^8 cells $\pm 4.42 \times 10^7$ per plastic roller bottle with PDT of 8.35 ± 0.76 hr.[4] In more recent studies,[21] cryopreserved insect cells adapted to suspension were evaluated for their ability to recover directly in the suspension system. After 3 days of growth in spinner flasks in 100 mℓ of medium, the PDT of cells recovered from cryopreservation was 8.77 ± 2.80 hr, whereas the control spinners had a PDT of 6.08 ± 1.86 hr. Although the PDT of substrate grown cells recovered from cryopreservation was increased, results from these studies showed that some of the advantages of cell culture for baculovirus production, such as availability, consistency, and stability, could be obtained from the use of cryopreserved cells to seed starter cultures directly.

III. METHOD FOR THE PRODUCTION OF VIRUSES IN SUBSTRATE-DEPENDENT SYSTEMS

A. General Culture Methods
1. Media, Reagents, and Other Parameters Affecting Cell Culture Growth

To obtain the maximum yield of cells in culture with the minimum PDT, the cells must be provided with ideal nutritional, biological, and biophysical requirements for growth. One of the important variables is the composition of insect cell culture media, which generally resembles hemolymph. To understand some of the requirements for basic insect cell culture media formulation, it is desirable to have an understanding of the physiology of the insect and the biochemical profile of insect hemolymph.[22] Extensive studies have been carried out to develop culture media for insect cells. The basis of the most commonly employed medium for lepidopteran cells was the medium developed by Wyatt[23] and modified by Grace.[9] This medium resembles hemolymph and consists mainly of chemically pure amino acids, vitamins, organic acids and inorganic salts, and originally was supplemented with insect hemolymph. Later the hemolymph from this formulation was replaced by fetal bovine serum, bovine serum albumin, and whole-egg ultrafiltrate.[24] Research findings on the role and utilization of amino acids, minerals, hormones, mitogens, carbohydrates, organic acids, fatty acids, sugars, lipids, and monovalent and divalent cations in cultured insect cells[25] have led to some experimental application of serum-free media for insect cells.[26,27] However, to date these serum-free media are not completely satisfactory for viral replication, which may be attributed to the absence of undefined proteins normally in serum supplements.

Vaughn[28] described methods for the production of NPV in an insect cell line on medium devoid of insect hemolymph but supplemented with a combination of mammalian and avian sera. Those studies suggested that additional developmental work was needed to improve both cell propagation and virus replication. Subsequently, improved methods for the replication of insect cells were reported by Weiss and his co-investigators.[4] In these studies, a powdered medium was purchased commercially and prepared as a complete medium prior to filtration. Avian sera was deleted from the medium which was supplemented only with heat inactivated fetal bovine serum and tryptose phosphate broth. Before considering the final formulation and preparation of this medium and related reagents for cell culture, consideration must be given to the quality of water, chemicals, and air used in such preparations.

The quality of water and the purification process to produce water acceptable for tissue culture must meet three important criteria: (1) all the contaminants present in feed water must be removed, (2) no additional impurities should be introduced, and (3) water must be of consistent purity without variation in quality due to possible fluctuations in feed water contaminants. Specific guidelines for standardization of water for tissue culture are available and should be followed for particular laboratory needs.[30,31] A system composed of tap water flowing through a filter, then through activated carbon and beds of ion exchange resins prior to glass distillation at the end point is usually adequate. The filters and deionizers should be monitored and changed as required. Using a distiller at the end reduces the level of trace metals to a total level of a few parts per billion with specific resistance of > 0.5 MΩ. The specific resistance of distilled water should be monitored continuously. This is achieved by installing a quality monitor Model QM-3 in the water line, to be used with a No. 4457.10 conductivity cell (Culligan, Northbrook, Ill.).

If distilled water is stored more than 48 hr, it should be autoclaved prior to use in media preparation. Water reservoirs, piping, and faucet fixtures should be nontoxic homopolymers, of the highest quality that can be obtained with specific designs from commercial sources. [(a) Hydro, Ultrapure® Water Systems, Research Triangle Park, N.C. 27709; (b) Barnstead Co., Boston, Mass. 03132; and (c) Osmonics, Inc., Minnetonka, Minn. 58343.]

The role of the gaseous environment on mammalian cell function in culture has been extensively studied,[32] and some of these parameters may be considered as possible practical guidelines for growing invertebrate cells in culture. Weiss et al.[29,33] reported on the effect of the gaseous environment on the growth and maintenance of uninfected and baculovirus-infected continuous insect cell lines. In those studies, it was observed that insect cells were extremely sensitive to the lowest levels of toxic agents present in the environment to which cells were exposed. The cells exposed to possible toxic agents became poisoned and thus were unsuitable for further studies. Standard house compressed air is known to contain carbon monoxide and other toxic contaminants and impurities.[4] Carbon monoxide and other impurities should be removed from compressed air prior to use in CO_2 incubators, media filtration, sparging of cells, pressurizing of the bulk cell suspensions, and flow meters used for transferring cells into growth vessels.

Air purification can be accomplished by installation in series after the compressor (Pall Pneumatic Products Corp., Ocala, Fla.) of a Pall® Heat-Les dryer, Model 25 HA1. The Pall® Heat-Les dryer system is designed for industrial application requiring oil vapor removal and $-40°F$ pressure dewpoint. Reverse prefilter and after filter cartridges were included as accessories in the dryer. The prefilters removed 99.99% of the liquid oil, water, aerosols, and mists from the air. This reverse prefilter prevented liquid from coating the desiccant; such a coating reduced the capacity to adsorb water vapor.

The after filter cartridges insured that no abrasive desiccant particles moved downstream to clog or damage sensitive instruments. After the dryer, a regulator valve with a moisture trap and an air line filter [No. 81857 Mine Safety Appliance (MSA) Co., Pittsburgh, Penn.) containing two organic vapor cartridges (MSA 46727) and one particulate cartridge (MSA 79030)] with 0.3 μm porosity were installed. Finally, a Del-Monox® Model 130B air purifier with a BH-3 replaceable cartridge (Deltech Engineering Inc., Century Park, New Castle,

Del.) was installed. This removed carbon monoxide by oxidation to carbon dioxide through chemisorption and catalysis in Hepcalite catalyzer.

Only chemicals of the highest purity should be used for cell culture media and the concentrations of possible impurities should be known. New components and additives for nutritional studies must be given serious attention, and active, nontoxic concentrations must be determined in insect cell lines according to procedures described previously.[3] In addition, the effect of new components and/or additives on plating efficiencies should be determined according to the procedures described in Section II.B. Standard chemicals, particularly those used for basic salts preparations, should be of "biological grade". If metal salts are used, they should be spectrophotometrically pure with the analysis of each major contaminant available. For consistency and stability, and to avoid unnecessarily extended quality control studies, it is most practical to purchase media from commercial sources, preferably in powder form, and then to prepare them in the laboratory according to particular needs. Should the concentration of a particular component already present in the medium need to be increased in a specific formulation, it is suggested that these ingredients be purchased from commercial media suppliers that already have in-house quality control standards. All items coming into contact with media components must be chemically and, where possible, biologically clean. Borosilicate, surgical stainless steel, or appropriate plastic containers should be used for storage of media reagents.

Weiss et al.[4,21] used a powdered tissue culture medium obtained commercially from K. C. Biologicals (Lenexa, Kan.). The ball mill processing of the chemicals and related components of this medium into powder and the weighing and dispensing of the powder formulations by this manufacturer are performed in facilities with less than 10% relative humidity at 72°F.

Table 3 shows the formulation of the complete medium from K. C. Biologicals. The powdered medium was supplied in two parts. Part A consisted of powdered amino acids and vitamins, and Part B contained bases of inorganic salts, organic acids, sugars, and tryptose phosphate broth. Calcium chloride ($CaCl_2$) and sodium bicarbonate ($NaHCO_3$) were omitted from the powdered formulation and were added during preparation. If this medium was to be used for cells grown in suspension, methyl cellulose at 0.25% (15 cP, Sigma Chemical) and Darvan N2® at 0.02% (R. T. Vanderbilt Co.) were added to the formulation. Prior to filtration and adjustment of final volume, the medium was supplemented with up to 10% fetal bovine serum heat inactivated at 56°C for 30 min.

A total of 1-ℓ lots of the complete medium were prepared in a stainless steel tank containing 500 mℓ of tissue culture grade, deionized, distilled water by the following procedures.

1. Part B dissolved in the 500 mℓ by stirring for 20 min with a magnetic bar on a nonheating stirrer.
2. Part A dissolved as in Step 1, and the following added in order listed.
3. 0.5 g NaCl.
4. 0.35 g $NaHCO_3$.
5. 0.5 g of $CaCl_2$, predissolved in 150 mℓ deionized, distilled water.
6. 2.6 g of tryptose phosphate broth (Oxoid product code CM 283).
7. 7.5 mℓ of 2.5 N NaOH to adjust pH of the medium to 6.2 ± 0.01.
8. 12.5 mℓ of 15% NaCl.
9. 100 mℓ of fetal bovine serum heat inactivated for 30 min at 56°C.
10. If the medium is to be used for growth of the cells in suspension, add 50 mℓ of 5% methyl cellulose (15 cP, Sigma product M-7140) to a final concentration of 0.25%. Methyl cellulose was dissolved into uniform solution by stirring with a magnetic bar on the stirrer at 4°C overnight.

Table 3
IPL-41 COMPLETE MEDIUM

Ingredient	Amount (mg/ℓ)
Inorganic salts	
NaH$_2$PO$_4$ · H$_2$O	1160.00
NaHCO$_3$[a]	350.00
KCl	1200.00
CaCl$_2$[a]	500.00
MgSO$_4$ · 7H$_2$O	1880.00
(NH$_4$)Mo$_7$O$_{24}$ · 4H$_2$O	0.04
CoCl$_2$ · 6H$_2$O	0.05
CuCl$_2$ · 2H$_2$O	0.20
MnCl$_2$ · 4H$_2$O	0.02
ZnCl$_2$	0.04
ZnSO$_4$ · 7H$_2$O	0.07
FeSO$_4$ · 7H$_2$O	0.55
AlCl$_3$	2.10
Sugars	
Sucrose	1650.00
Maltose	1000.00
Glucose	2500.00
Organic acids	
Malic acid	53.60
α-Ketoglutaric acid	29.60
Succinic acid	4.80
Fumaric acid	4.40
Other components	
Choline chloride	20.00
Darvan[b]	200.00
Tryptose phosphate broth	2600.00
Methylcellulose[a,b]	2500.00
Amino acids	
L-Arginine HCl	800.00
L-Aspartic acid	1300.00
L-Asparagine	1300.00
β-Alanine	300.00
L-Cystine (dissolve in 1.0 *N* HCl)	100.00
L-Glutamic acid	1500.00
L-Glutamine	1000.00
L-Glycine	200.00
L-Histidine	200.00
L-Isoleucine	750.00
L-Leucine	250.00
L-Lysine HCl	700.00
L-Methionine	1000.00
L-Proline	500.00
L-Phenylalanine	1000.00
DL-Serine	400.00
L-Tyrosine (dissolve in 1.0 *N* HCl)	250.00
L-Tryptophan	100.00
L-Threonine	200.00
L-Valine	500.00
L-Hydroxyproline	800.00
Vitamins	
Thiamine HCl	0.080
Riboflavin	0.080
Calcium pantothenate	0.008
Pyridoxine HCl	0.400
p-Aminobenzoic acid	0.320
Folic acid	0.080

Table 3 (continued)
IPL-41 COMPLETE MEDIUM

Ingredient	Amount (mg/ℓ)
Niacin	0.160
Isoinositol	0.400
Biotin	0.160
Cyanocobalamin	0.240

[a]   Not included in the dry powder formulation. Must be dissolved separately.
[b]   For suspension cultures only.

11.   The final volume of the medium was adjusted to 1 ℓ with deionized, distilled water and stirred for 30 min. The pH of the medium should be 6.2 $\pm$ 0.01 and osmolarity 360 to 375 mOSM.

The medium was sterilized by membrane filtration with a combination of depth prefilter No. 13430-257 (Sartorius, Hayward, Calif.), a 0.45 μm Ultipore NM® (Pall Trinity Micro Corp., Cortland, N.Y.), an AP 32 dacron woven space holder (Millipore Corp., Bedford, Mass.), and a 0.20 μm Ultipore NM® (Pall) assembled in a stainless steel filter holder. Detailed methods for assembly and sterilization of the filtering apparatus were fully described elsewhere.[12]

Appropriate samples of medium were removed, postfiltration, for quality control by the sterility testing protocol described in Section II.B. For practical manufacturing purposes, the first lot of liquid medium produced from a large powdered lot was tested by following recommended quality control protocols. Subsequent lots were not tested for virus, mycoplasma, or plating efficiency.

Weiss et al.[4] have demonstrated that complete medium can be stored for up to 6 months at 2 to 4°C and still satisfactorily support growth of the cells recovered from cryopreservation. When the cells were propagated in 490 cm^2 plastic roller bottles on the freshly produced medium, the average yield was $2.09 \times 10^8 \pm 8.5 \times 10^6$ viable cells per 490 cm^2 roller bottle and the PDT was 10.14 $\pm$ 0.25 hr calculated as recommended by the Tissue Culture Association.[35] Statistically there were no differences in the yield and PDT among the cells grown on freshly produced medium, on the medium stored at 2 to 4°C for 6 months, and on the medium stored under the same conditions for 12 months. Stability of the medium ingredients, particularly amino acids and glutamine, during long-term storage at 2 to 4°C was attributed to the addition of serum and tryptose phosphate broth to the formulation prior to filtration and storage.[29] Inventories of large lots of standardized complete medium for production of baculoviruses in vitro assures consistency in osmotic pressure and pH, as well as uniformity of serum and other uncharacterized supplements.

This formulation differs from the one reported previously in that it contains zinc sulfate ($ZnSO_4 \cdot 7H_2O$) and aluminum chloride ($AlCl_3$). During large-scale production, the cells were exposed to shear stress and undesirable forces which were difficult to eliminate and it was demonstrated[29] that the addition of $AlCl_3$ and $ZnSO_4$ to the medium prevented shear stress, improved cell attachment, and increased cell yield up to 25%. It also was demonstrated that the addition of $AlCl_3$, $ZnSO_4 \cdot 7H_2O$, or both, to medium of cells infected with *Autographa californica* NPV (AcNPV) would result in an increased yield of virus and an increase in the biological activity.[3,34] The addition of 16 μM $AlCl_3$ resulted in an 87.8% higher yield of polyhedra over the control; 24 μM $ZnSO_4$ gave a 121% increase in yield of polyhedra, and a combination of both resulted in a 63% increase in a polyhedra yield. Plaque assays of nonoccluded viruses (NOV) with $AlCl_3$ gave a 4.3-fold higher titer and a 2.3-fold higher titer with $ZnSO_4 \cdot 7H_2O$; a combination of both resulted in a 4.7-fold increase over the control. Bioassay (LD_{50}) showed that polyhedra produced in the presence of 16 μM $AlCl_3$ had a 6.4-fold increase in activity and a 2.5-fold increase with $ZnSO_4 \cdot 7H_2O$; a combination of both increased biological activity of polyhedra nearly 7-fold.

Electron microscope assessment of differences in virus particles and polyhedra in the presence of metals in the medium revealed enhancement of the number of NOV, the number and size of polyhedra, and the number of virus particles per polyhedra.[3] The mean number (2.7) of virions per polyhedra produced in supplemented medium was greater than that obtained in medium devoid of supplements (1.2), which may account for the increase in biological activity. Addition of the $AlCl_3$ and $ZnSO_4 \cdot 7H_2O$ demonstrated that culture media for the growth of invertebrate cells may be designed to increase the yield of the virus and its biological activity.

2. Selection of Surface Substrate

An important parameter in the culture of anchorage-dependent or substrate-based cells involves the surface on which they are grown. Mobility of the cells depends upon the ratio of adhesive forces of the cells with the substrate surface to the intracellular adhesiveness. Thus, selection of surfaces for cell growth can markedly alter performance. However, an ideal surface substrate for attachment and growth of anchorage-dependent or substrate-based insect cells has not been fully defined since there are no insect cell lines known to be substrate dependent.[2] The majority of established, continuous insect cell lines will grow in monolayer adhered to pyrex glass or plastic tissue culture-treated substrates. Most of the plastic tissue culture ware is made out of high quality polystyrene with good optical clarity which is electronically charged for tissue culture application.

Vaughn[36] employed a method to grow insect cell lines in conventional glass roller bottles in which the surface for cell attachment and growth was preconditioned with complete growth medium prior to cell planting. This was accomplished by adding 25 mℓ of complete growth medium to 670 cm^2 glass roller bottles and rolling them for 18 hr. Then, 75 mℓ of cell suspension containing 2 × 10^7 *Spodoptera frugiperda* cells was added to the roller bottles, bringing the total volume to 100 mℓ per bottle. After 8 days of incubation at 26°C, the PDT of the cells was 19 hr compared with 28 hr in conventional static flask cultures. Weiss et al.[4] developed an improved, semiautomated process for growing insect cell lines in polystyrene plastic roller bottles that resulted in a reduction in the PDT to around 10 hr. Detailed methods will be described in Section III.B.1.

The attachment and growth of insect cells on collagen-coated surfaces, on glass, polystyrene, and polyester surfaces treated with poly-D-lysine, on tissue culture treated polyester film, and on treated polystyrene microcarriers (Cytosphere, Lux) have also been studied. With the exception of Melinex®, which is polyester film treated for tissue culture application and which will be discussed in detail in Section III.C, none of these substrates was as satisfactory as polystyrene treated for tissue culture purposes.[69] Whether or not these results indicate that tissue culture grade polystyrene is the most ideal surface for the attached growth of insect cells is not fully established at this time. Comparative and complete studies on adhesion to cell surfaces for insect cells, similar to the studies of Grinnell and co-workers[37] and McKeehan and Ham[38] for mammalian cells, are desirable to define an ideal cell surface substrate for insect cells in culture.

3. Dissociation of Attached Cells

Anchorage-dependent and substrate-based continuous cell lines must be dissociated for subsequent subculturing by methods that preserve maximum viability of the cells. A variety of methods have been employed to achieve satisfactory dissociation of the cells, including such procedures as scraping of the cells with a scraper equipped with silicone rubber blades or with a disposable scraper made out of plastic,[4] mechanical dislodging by pipetting, trituration, shaking the vessel containing cells, and treatment with various enzymatic and versene preparations.[22] The methods used depend on the objective of specific investigators and, in most cases, were not suitable for large-scale application where cost effectiveness must be considered. In addition, none of the above-described procedures have become general methods for insect cells and this may be because insect cells derived from various species have different adhesive properties. Optimum virus production in vitro requires cell quantitation to attain a desirable multiplicity of infection in anchorage-dependent or substrate-based cells. Therefore, it is necessary to develop an inexpensive dissociation method for the large-scale process that yields a high percentage of viable cell recovery and is nontoxic for both the cells and the virus produced in these cells.

Weiss and co-workers[39] reported on the dissociation of IPL-Sf-21AE cells using pancreatin as one of the active agents. Table 4 shows the formulation of the 0.003% pancreatin-0.002% EDTA solution developed to dissociate IPL-Sf-21AE cells grown on substrate. The basic salts and sugars and the concentrations were the same as in the formulation of IPL-41 complete medium except that $CaCl_2$, $AlCl_3$, and $ZnSO_4 \cdot 7H_2O$ were deleted. The preparation can be purchased commercially in powdered form.

The procedures for dissociating the cell monolayers were similar to those used with other enzymes. The dissociating agent was added to the vessels at a volume previously established for each vessel, e.g., 5 mℓ per 75 or 150 cm^2 flask; 25 mℓ per 490 and 880 cm^2 PRB; 40 mℓ per 1750 cm^2 PRB; and 400 mℓ per 9500 cm^2 Dyna cell vessel. The vessel was rotated several times until the cell sheet became wet. Then the Pan-EDTA was decanted, leaving a small amount of Pan-EDTA in the vessel, which was then incubated at 26 to 28°C for 2 to 5 min until the cells were dissociated. When the cells were loose and detached from the substrate in the vessel, complete growth medium was added to the cells. The cell suspension was gently pipetted to obtain single cells for counts, subculturing, and subsequent experiments. Although cells harvested by scraping with the policeman had a viability of 94 and 96% after 4 and 7 days in culture, respectively, compared with 89 and 90%, respectively, for cells dissociated with the Pan-EDTA, the enzyme method is more practical in a large production system.

B. Roller Bottle Culture

1. Cell Growth

Roller bottles provide a method for conveniently expanding a substrate based culture method to produce large volumes of insect cells. The system provides considerable savings in time, labor, and space compared with an expansion of flask culture methods. Little modification of a functioning stationary culture system, other than determining the optimum rotation speed, is required to change to the roller bottles. Vaughn[28] developed one of the first successful roller bottle culture methods for *S. frugiperda* cell lines using bottles with 670 cm^2 of growth surface. The culture volume was 75 mℓ and the bottles were rotated at a rate of one revolution per 10 min. The cells grew rapidly with a PDT of 19 hr compared with 28 hr in static cultures. Cell yields ranged from 3 to 5 $\times$ 10^8 cells per bottle, an increase of 20 to 30 times over the inoculum in 7 days.

Weiss et al.[4] expanded and improved the roller bottle system using the same cell line, IPL-Sf-21AE. They used roller bottles with total growth surface areas of 490 cm^2, 850 cm^2, and 1750 cm^2 and obtained cell yields equal to 20 times the inocula from all size bottles. The PDT ranged from 8.35 hr in the 850 cm^2 bottles to 10.22 hr in the 490 cm^2 bottles. The labor required to establish large numbers of bottles was reduced by using a semiautomatic system for dispensing the cell inoculum into a large number of bottles. In this system, a stock cell suspension was prepared in fresh growth medium in a Kimex bottle. The cells were maintained in suspension by gentle stirring on a nonheating, magnetic stirrer and dispensed into roller bottles by placing the cell suspension under 0.5 lb/in.^{2} pressure with purified, filtered air.

Table 4
FORMULATION OF 0.003%
PANCREATIN[a] — 0.002% EDTA[b]
DISSOCIATING SOLUTION FOR INSECT
CELLS

Ingredient	Amount (mg/mℓ)
Deionized distilled H$_2$O (800 mℓ)	
NaH$_2$PO$_4$ — H$_2$O	1160.00
NaHCO$_3$	350.00
KCl	1200.00
MgSO$_4$ · 7H$_2$O	1880.00
(NH$_4$)Mo$_7$O$_{24}$ ·4H$_2$O	0.04
CoCl$_2$ · 6H$_2$O	0.05
CuCl$_2$ ·2H$_2$O	0.02
MnCl$_2$ · 4H$_2$O	0.02
ZnCl$_2$	0.04
FeSO$_4$ · 7H$_2$O	0.55
Sucrose	1650.00
Maltose	1000.00
Glucose	2500.00
Ethylenediaminetetracetic acid	20.00
Pancreatin	30.00

Note: Adjust volume (q.s. with deionized distilled water) to
1 ℓ. pH should be adjusted to 6.2 ± 0.1 and osmolarity
to 360 to 275 mOSM.

[a] Scientific Protein Laboratories, Inc., U.S.P. Grade (Waunakee, Wis.)
[b] Fisher E-478, F.W. 229.25.

These roller bottles were inoculated with cell densities ranging from 1.27×10^7 cells in 100 mℓ growth medium for 490 cm^2 bottles to 3.17×10^9 cells in 250 mℓ medium for 1750 cm^2 bottles. The bottles were rotated at one revolution every 8.5 min. This procedure resulted in remarkably reproducible cell growth. Typical growth curves are shown in Figure 1. Reproducible cell growth is an essential requirement of any large-scale system if the cells are to be used for virus production. Accurate knowledge of cell growth is required to obtain maximum virus yields consistently.

2. Propagation of Virus

In a recent review on the propagation of insect viruses in vitro Stockdale and Priston[10] commented that the future of the large-scale in vitro production of viruses for pest management depended on the satisfactory merging of three factors: (1) financial success (either via the marketplace or with government subsidy), (2) acceptance by regulatory agencies, and (3) solution of the remaining technical problems. Only the third of these is a research problem. Among the unsolved technical problems they listed were achieving high oxygen transfer rates, high cell concentrations, and well-designed media.

Roller bottle systems bypass the need for high oxygen transfer and provide a scaled-up system for the study of the latter two problems. Vaughn[28] reported success in using roller bottles for the production of the NPV in cells of the IPL-Sf-21AE line. In this study it was found that optimum virus yields were achieved if the bottles were inoculated when the cells were in log phase growth. If the inoculation was delayed until the cells reached the stationary phase, no polyhedra were formed. The highest yield obtained was 8.5×10^5 polyhedra per milliliter of medium or 6.3×10^7 polyhedra per bottle.

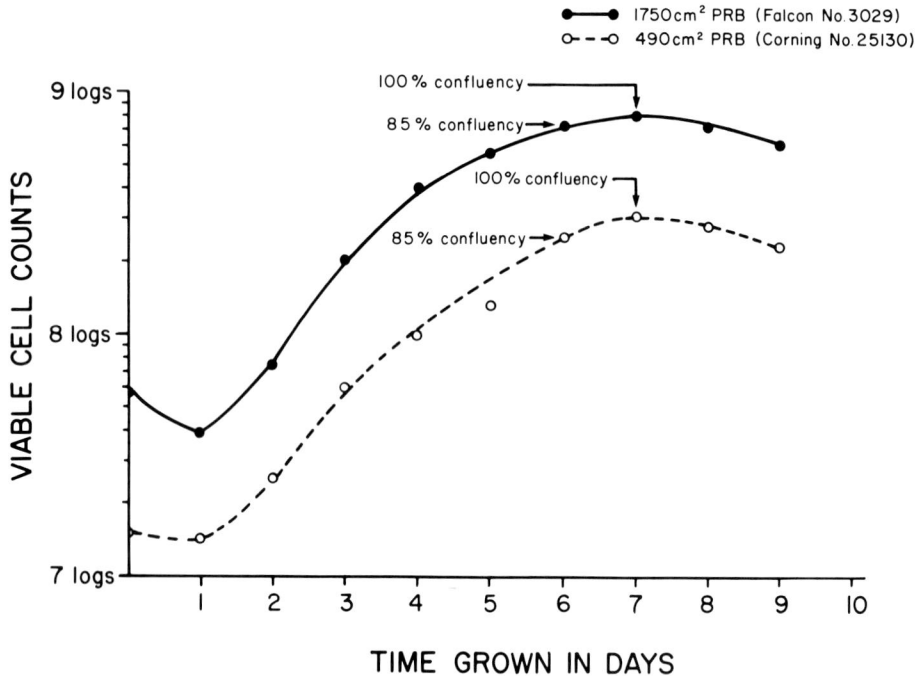

FIGURE 1.   Growth curves of IPL-Sf-21AE cells grown in plastic roller bottles.

Extensive studies on the production of NPV in roller bottle culture systems have been carried out by Weiss and collaborators.[40] IPL-Sf-21AE cells, cultured as described in Section III.B.1, were inoculated at 80 to 85% confluency with the *A. californica* NPV at a multiplicity of infection (MOI) of 0.75 plaque-forming units (PFU) per cell. The virus was added to the culture vessels with the semiautomated pressure system used to establish the cultures, except that a pressure of 1.5 lb/in.^{2} was used instead of 0.5 lb/in.^{2} used for transferring the cells.

Quantitation of the virus yields was accomplished with the electron microscope by a thin-section technique involving the sedimentation of particles onto a Millipore® filter.[41] Modification and application of this method for NOV and polyhedra were described.[40] In addition, morphological analysis of the polyhedra was done with stereological electron microscopic methods.[42,43]

An average maximum polyhedra yield of 6.0×10^7 polyhedra per milliliter was obtained 7 days postinfection. The pattern of visible cytopathic effect was typical of NPV-cell infections in vitro and was reproducible in roller bottles of various sizes. The average size of the polyhedra was 1.9 ± 1 µm and these polyhedra had between 0 and 15 virions per polyhedra per section with a mean of 2.1 ± 2.9 virions per polyhedra per section. It was calculated from this data that the average polyhedra contained 23 virions. Bioassay of the polyhedra against neonate cabbage looper produced a mean LD_{50} of 0.508 ± 0.10 polyhedra per square millimeter of insect diet.

Culture supernatants contained an average of $4.6 \pm 3.1 \times 10^7$ PFU per milliliter. The virion per PFU ratio was estimated to be 223 ± 80. This is similar to the particle per PFU ratio of 128 reported for this virus by Volkman et al.[44]

3. Oxygen Consumption

Available oxygen is a critical parameter and it is necessary to monitor oxygen consumption during large-scale cultivation of insect cells for successful propagation of baculoviruses. Stockdale and Gardiner[45] demonstrated that the demand for O_2 in TN-368 cells ranged from

15 to 45 μg/mg of cell dry weight per hour. When O_2 was not added from external sources, a culture with 1×10^6 cells per milliliter utilized all available O_2 within 17 to 50 min. However, most investigators did not notice this effect in small monolayer cultures or in suspension vessels where sufficient surface area to volume ratio permitted diffusion of O_2 into the medium.

Weiss et al.[33] reported a quantitative method for measuring dissolved oxygen (DO) in uninfected and baculovirus infected IPL-Sf-21AE cells grown and maintained in 490 cm² disposable plastic roller bottles. Dissolved oxygen consumption was measured with a Model IL-531 industrial O_2 measuring system (Instrumentation Laboratory Inc., Lexington, Mass.). After the values were recorded, a computerized program converted the DO from millimeters of Hg to microliters. The data indicated that a significant drop in DO content did not occur in uninfected cells until 6 days postplanting. At this time, these cells had reached 85% confluency with a yield of 2×10^8 cells per roller bottle and a PDT of 9 hr. Considering cell yield per 490 cm² roller bottle at 6 days postplanting and the DO consumption, and assuming that this was an ideal condition for these cells, they calculated that 9.05×10^{-4} $\mu\ell$ of DO was required per viable cell or 1 $\mu\ell$ of DO for 1.10×10^3 cells. It appeared that the ratio of surface area of the plastic roller bottle (490 cm²) to the volume of the vessel (1360 mℓ), provided for sufficient diffusion of O_2 into the growth medium (100 mℓ) to supply the amount of DO (1.81×10^5 $\mu\ell$) consumed by the cell suspension. A steady, but slower, drop in DO was observed between 6 and 9 days in uninfected cells, after which the consumption of DO began to level off, probably due to slowing down of cell growth.

In the perfusion system used in the Dyna cell propagators described in Section III.C, a perfusion rate of 25 mℓ/hr appeared to provide all the DO required by the growing cells. At day 0, the DO level in the medium from the IPL-Sf-21AE cells was 7.7×10^6 $\mu\ell$ per Dyna cell unit and remained within this range for 4 days. On the 5th day, DO level decreased to 6.2×10^6 $\mu\ell$ per bulk cell vessel and remained at this level until 9 days in culture.[69]

Weiss et al.[33] have also shown that cells infected with AcNPV had a rapid consumption of DO from the first day postinfection. The rapid DO consumption continued until 4 days postinfection and then leveled off. By the 7th day postinfection, virus-infected cells had consumed 48% more DO than uninfected cells. This rapid increase in O_2 consumption from the 1st to the 4th day postinfection suggests that the viral-induced metabolic activity required for the synthesis of viral proteins and nucleic acid occurs before or during the early cycle of viral replication.

C. Bioengineering of the Dyna Cell Propagators
1. Design of a Perfusion System

Employing conventional roller bottles for the large-scale growth of insect cells for NPV production also presents some disadvantages. One of the main disadvantages is the limited surface area to which cells attach and grow. Several vessels have been designed to increase the available growth surface. One such design is a commercially available, sterile, disposable vessel with a high surface area-to-volume ratio, the bulk culture vessel. It was developed by House and co-workers[46] for mammalian cell culture and subsequently evaluated by Nicklin and House[47] for polyoma virus production. The vessel is a 1.7-ℓ clear polystyrene bottle molded in halves to allow the insertion of a Melinex® spiral core and, when assembled, the joints are sealed.[48] The internal surface has 9500 cm² of spiral core film, 125 μm thick with 4 mm spacing. Cells grow on both sides. The vessel has three screw-capped inlets/outlets. The small central inlet is equipped with a tube passing through the center of the spiral core to the bottom of the unit. Of the remaining two outlets, the smaller one serves for outlet of gases and the large peripheral opening serves for addition of cell suspension.

In bulk culture vessels, the cells rapidly produce and secrete excessive lactic acid into the spent medium, the pH drops dramatically, and glucose is consumed more rapidly.[49] The

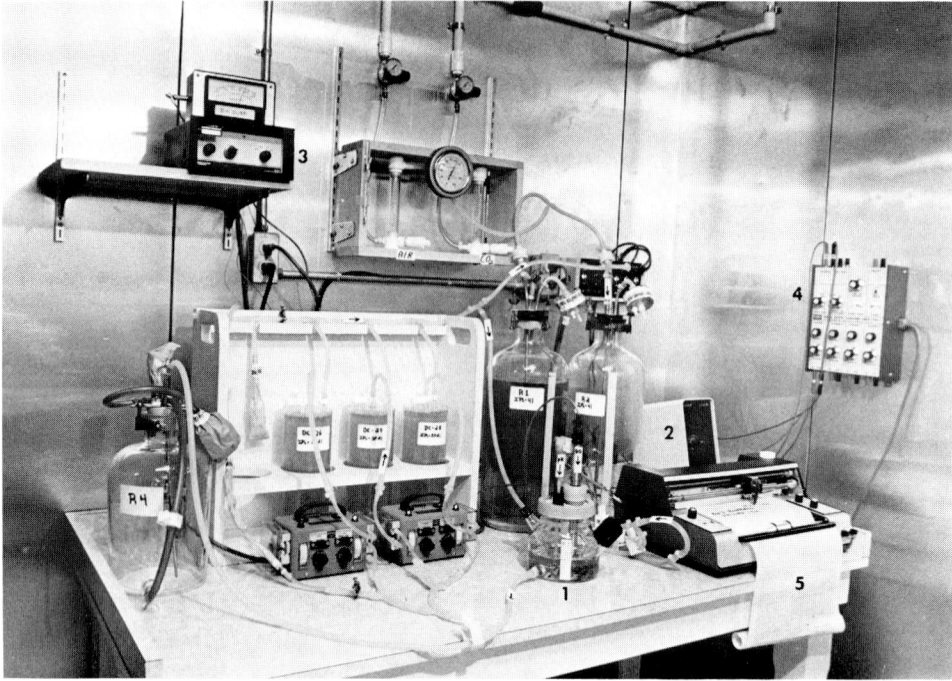

FIGURE 2.   Dyna cell propagators system with three bulk cell vessels in operation. Three bulk cell vessels are marked as DC 26, DC 27, and DC 28. (1) pH-DO-Stat; (2) pH meter; (3) DO meter and amplifier; (4) channelyzer; and (5) recorder for pH and DO values.

bulk culture vessels required further bioengineering modifications to overcome these problems. Figure 2 illustrates the experimental design of the Dyna cell propagators system in a 26.5°C walk-in incubator. There were three bulk cell culture vessels, with 28,500 cm^2 surface area, connected to a steady state, continuous perfusion system leading to a pH-dissolved oxygen-stat which in turn was connected to reservoir R_1 for fresh medium, R_2 for spent medium, and R_4 for waste and sampling. Medium was perfused into the Dyna cell propagators with a medical infusion pump (Model No. 921, Criticon Inc., Miami, Fla.).

The dissolved oxygen monitoring system, IL-531, and a pH meter were connected through a solid state channelyzer (Bailey Instr., Model CHL-1) to an Omni Scribe recorder (Industrial Scientific Inc., Model B500), and the values were constantly recorded on a chart.

2. Cell Growth

The IPL-Sf-21AE cells could be maintained in this system for 30 days without cell degeneration or loss of viability. The culture was established by inoculating each unit with 5×10^8 cells in 1600 mℓ of growth medium and rotating the units at 3.4 r/hr for 48 hr to allow for attachment to the growth surface. After cell attachment, the units were connected to the perfusion system and perfused with fresh medium at the rate of 25 mℓ/hr.

This perfusion rate was sufficient to maintain a constant pH and DO level. After an initial decrease in the 48 hr before perfusion, the levels of glucose also remained constant (Figure 3). In conventional roller bottles without perfusion or medium change, glucose levels decreased rapidly after 24 hr in culture until by the 9th day the glucose was nearly exhausted (broken line on Figure 3). These data indicate that perhaps a lower perfusion rate or a simpler medium formulation could be employed to reduce the cost of the final process.

General procedures for dissociation of insect cells grown on substrate in tissue culture vessels and the formulation of Pan-EDTA solution were described earlier in detail in this

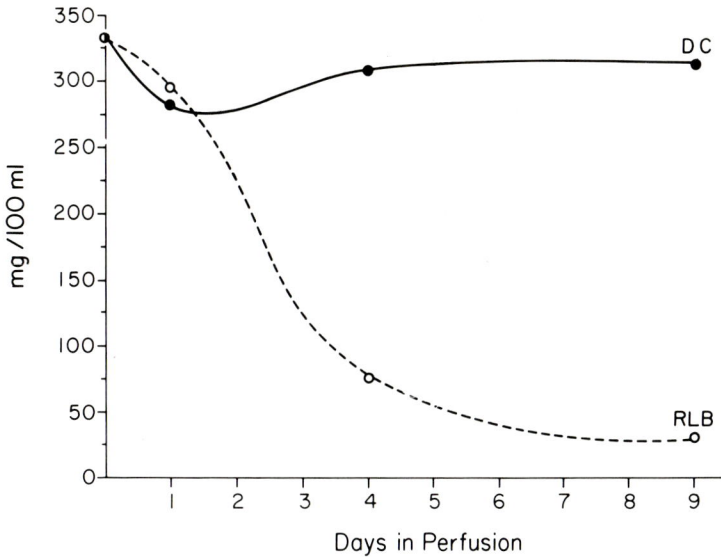

FIGURE 3. Glucose level in the spent media from insect cells in culture. (DC = Dyna cell bulk vessels, perfused; RLB = roller bottle, stationary).

chapter (see Section III.A.3). Initially, some difficulties were experienced with establishing the cell yields. Recovery was, in some experiments, smaller than the initial cell inocula. Similar problems were also reported by Taylor et al.[49] with quantitation of mammalian cells in these vessels. After screening pancreatin from various suppliers, it was found that pancreatin USP grade (Scientific Protein Inc., Waunakee, Wis.) was the most suitable when it was formulated as outlined in Section III.A.3.

The cells in Dyna cell propagators reached 85% confluency at 4 days postplanting and had 90% confluent cell sheets on both sides of unlined film after 11 days in culture. Cells dissociated with Pan-EDTA, and 4 days postplanting, had 89% viability with a fivefold increase in cell numbers and a PDT of 20 hr. DNA content in IPL-Sf-21AE cells grown in Dyna cell propagators at 4 days postplanting, as determined by the Ceriotti method,[50] was 74.6-fold higher per vessel than in cells grown in the conventional plastic roller bottles. Electron microscopy studies indicated that cells released from bulk cell culture vessels by Pan-EDTA were rounded and possessed short bleb-like surface projections that were spread evenly over the cell surface, and were similar in appearance to cells from roller bottles. No significant morphological changes were observed with light microscopy between the cells dissociated from roller bottles and cells released from the bulk cell vessels.

3. Propagation of Virus

When cell growth reached 80 to 85% confluency, as in standard roller bottles, the spent medium was aspirated from the bulk vessels and the units were inoculated with AcNPV diluted in maintenance medium to an MOI of 0.07. The virus was allowed to adsorb to the cells on polyester sheeting by rotating bulk cell culture vessels on the roller apparatus for 24 hr at 3 to 4 r/hr. After adsorption, culture vessels were connected to the perfusion system and perfused with 25 mℓ/hr of fresh medium from the medium reservoir which was sparged with air at rate of 5 $\mu\ell$/min. The cell free virus was continuously pumped into a collection reservoir which was placed directly into a small portable refrigerator to prevent thermal inactivation of NOV. Starting on the 7th day postinfection, polyhedra were noted in effluent from the manifold. When this occurred, units were disassembled from the manifold and the cells dislodged by mechanical circulation of medium, which was then removed. Any remaining cells on the polyester film were removed with Pan-EDTA treatment, combined with the supernatant, and processed for AcNPV yield.

A total of 6.4 ℓ of medium was used for a 9-day culture. Yields of 1.5×10^{11} polyhedra per vessel were obtained. This calculated to a yield of 2.34×10^7 polyhedra per milliliter of medium used. The NOV present in the medium can be concentrated by hollow-fiber ultrafiltration[51] for use in further cell infection if the in vitro passage level has not yet exceeded six to ten passages.

A second method of producing AcNPV in Dyna cell propagators, in which the cells were infected in suspension prior to attachment, was also tested. The cell inoculum was increased from 5×10^8 to 1.5×10^9 cells per vessel, inoculated with virus, and allowed to stir at room temperature for 1 hr for virus adsorption. Then the infected cell suspension was dispensed into the culture vessels and allowed to attach for 48 hr on the roller apparatus before connecting the Dyna cell propagators to the perfusion system. This method saved time and eliminated considerable handling of the units, which decreased the chances of contamination, without a noticeable decrease in the amount of NOV or polyhedra produced.

IV. PRODUCTION OF VIRUSES IN SUSPENSION SYSTEMS

A. General Culture Conditions

Deep tank fermentation has been extensively developed and exploited as a means of producing a variety of microbial products, e.g., antibiotics, enzymes, and growth factors, to list but a few. Cell culturists have endeavored to exploit this large fund of available knowledge and experience whenever possible to culture mammalian cells for a variety of purposes. As a result, a large body of literature has developed on the use of suspension systems for the culture of vertebrate cells. This literature has been recently reviewed by McLimans[52] and interested readers are referred to this publication.

Because no insect cells have been shown to be substrate dependent, suspension culture has been an attractive method for culturing large volumes of these cells. Attempts to culture insect cells in suspension for virus production have generally been with those vessels that were successfully used to culture vertebrate cells. These have included standard spinner flasks,[53-55] spin filter flasks,[21,56] and modified fermentors.[57,58] Roller bottles rotated at 6 to 7 rpm to keep the cells from attaching have also been used as a means of obtaining large numbers of cells in a short time.[59] Two problems appear to have delayed the full utilization of suspension systems for the large volume culture of insect cells: the fragility of insect cells[29] and second the high oxygen demand, particularly for virus-infected cells.[54]

The addition of methyl cellulose to the standard medium used for culturing insect cells has provided some protection for the fragile cells. Vaughn[53] added 0.15% of 400 cP methyl cellulose to protect cells of the Grace *Antheraea* cell line maintained in suspension by a magnetic stirrer rotated at 70 rpm. Hink and Strauss[54] added 0.1% 50 cP methyl cellulose to the TNM-FH medium used to culture the TN-368 cells in spinner flasks and in a New Brunswick MF-205® fermentor. Cell damage was reported in the MF-205 fermentor unless the concentration of methyl cellulose was increased to 0.3%. Weiss et al.[21] added 0.25% 15 cP methyl cellulose and 0.02% Darvan N2® to their medium when the IPL-Sf-21AE cells were cultured in either spinner flasks or spin filter vessels (Virtis Co. Biospin® filter). The spinner bars were rotated at 190 rpm and, in the 1-ℓ flasks, the impeller system was modified with the addition of a second magnetic bar perpendicular to the first to provide additional agitation without increasing the rotation speed. These modifications to the various media and flasks used in suspension culture appear to provide the necessary protection for insect cells as several cell lines have been successfully grown in suspension culture with volumes of up to 9.5 ℓ achieved with cells of the IZD-MB-0503 line from *Mamestra brassicae*.[58]

Meeting the oxygen requirements has proven to be somewhat more difficult. Hink and Strauss[54] reported that growth of TN-368 cells in spin flasks was improved by aeration.

Passing air above the surface of the medium increased growth somewhat, but more than a twofold increase in maximum cell yield was obtained if air was slowly bubbled through the medium. Aeration also improved the growth of IPL-S-21AE cells in spinner flasks when the culture volume exceeded 500 mℓ.[56] Bubbling air through the culture medium extended the growth phase from 8 to 15 days. The cell density increased to 3.36 $\times$ 10[6] cells per milliliter compared to 1.38 $\times$ 10[6] cells per milliliter in unaerated cultures. In similar spinner cultures of 100 and 250 mℓ without aeration, the yield of cells per milliliter was equal to that of the aerated 500 mℓ cultures.

In 2- to 3-ℓ cultures in fermentors, the sparging required to maintain the level of DO at 100% caused excessive foaming, vacuolation of cells, and formation of a precipitate.[57] Silicone DC antifoam emulsion (Dow Chemical Co.) was used to control the foaming and tests demonstrated that a reduced sparging rate that would maintain 50% DO levels provided for adequate cell growth. In the absence of any sparging, the DO level was reduced from 100 to 2% within 24 hr. Miltenburger and David[58] were able to avoid the problems associated with sparging by modifying a 12-ℓ fermentor to enable the introduction of air into the culture medium through an immersed silicone rubber tubing. Oxygen diffused through the tube into the medium without damaging the cells. Reproducible, logarithmic growth rates were achieved in culture volumes up to 9.5 ℓ with cells of the IZD-MB-0503 line from *M. brassicae*.

Weiss et al.[21] achieved growth of the IPL-Sf-21AE cells in volumes up to 3 ℓ in spinner flasks by sparging the cultures with purified air. The air was introduced under the impeller which broke up the bubbles as they entered the medium. Weiss and colleagues[21] determined that the required level of DO was 2.6 $\times$ 10[2] $\mu\ell$/mℓ for a 1-ℓ spinner and 5.3 $\times$ 10[3] $\mu\ell$/mℓ for a 3-ℓ fermentor. A maximum cell density was reached in 7 days. Cultures of less than 250 mℓ in spinner flasks did not require aeration, confirming the results of earlier workers.

B. Conditions for Optimum Virus Yield in Suspension Systems

Among the many parameters studied in suspension cultures, oxygen levels, cell density, and stage of growth have been shown to influence yield of polyhedra. Streett and Hink[60] demonstrated that the oxygen uptake of infected cells was nearly twice that of uninfected cells, with the peak demand occurring at about 14 hr postinfection. No satisfactory method for supplying enough oxygen to meet this high demand has been developed as yet. Instead, Hink and colleagues developed a semicontinuous system for producing the *A. californica* NPV.[57,61] In this system, cells were produced continuously in a 3-ℓ fermentor culture in which the DO was maintained at 50% by sparging with air. Each day, 1500 mℓ of the cell suspension was transferred to a spinner flask and the volume reduced to 500 mℓ to concentrate the cells. After infection with the virus, the culture was sparged with O$_2$ to maintain the DO level at 100%. This system produced four consecutive batches with yields of 4 $\times$ 10[7] to 1 $\times$ 10[8] polyhedra per milliliter. The next two batches had reduced yield.

Weiss et al.[21] also have found that aeration is essential for good polyhedra production in suspension cultures over 500 mℓ in volume. In the 500 mℓ culture, the virus yield 7 days postinfection was 1.5 $\times$ 10[8] polyhedra per milliliter. It was estimated that there were an average of 49 polyhedra per cell and that 98% of the cells were infected. In 1-ℓ cultures without aeration, the virus yield was 9.3 $\times$ 10[7] polyhedra per milliliter. The estimated number of polyhedra per cell was only 24 and only 89% of the cells were infected. When the 1-ℓ cultures were sparged with air, the yields were comparable to the yield from small cultures; 1.2 $\times$ 10[8] polyhedra per milliliter or an estimated 97 polyhedra per cell. Nearly all of the cells (98%) were infected.

The effects of cell density and growth stage were first demonstrated by Vaughn who found that cultures in roller bottles inoculated after the cells reached confluency did not produce polyhedra.[28] Studies by Stockdale and Gardiner[62] reported that this could be overcome in stationary batch cultures if the cells in late-logarithmic or early-stationary phase were refed with fresh medium and diluted to a density of less than 4×10^5 cells per milliliter. Later studies by Lynn and Hink,[63] using cells synchronized by a double thymidine block demonstrated that the yield of polyhedra per cell was not influenced by the growth stage the cells were in when infected. However, the percentage of cells infected was significantly different when infected in S phase (99.03%) then when infected in G_2 (93.68%). Wood et al.[64] have recently shown that, when the cells in a static culture reach confluency, viral synthesis is shut off by an unidentified mechanism associated with cell to cell contact.

It is not yet known whether or not such a shut-off mechanism exists when cells are grown in suspension. However, Hink et al.[65] have reported that yields of *A. californica* polyhedra per milliliter from TN-368 cells grown in suspension decreased if the cell density exceeded 3.8×10^6 cells per milliliter. However, the average number of polyhedra per cell was maximum at only 7.65×10^5 cells per milliliter. Weiss et al.[21] found that infection of IPL-Sf-21AE cells was successful only when the density did not exceed 2×10^5 cells per milliliter in fresh medium. Cultures were examined for number of polyhedra formed, percentage of cells possessing polyhedra, number of NOV particles, number of polyhedra per milliliter, and infectivity of NOV in TN-368 cells at 3, 6, and 7 days postinfection. Cultures established with 5×10^4 and 1×10^5 cells per milliliter in 50-, 100-, 250-, and 500-mℓ spinner flasks (with the same order in volume) and infected with 0.01 and 0.005 MOI were 98% infected by 6 days postinfection. Up to 60 polyhedra per cell were produced in these cultures. The titer of NOV was in the range of 7.9×10^7 PFU per milliliter to 1.6×10^8 PFU per milliliter. When the cell inoculum was increased to 2.5×10^5 cells per milliliter, and the remaining conditions were the same, production of polyhedra per cell decreased by 75%.

C. Perfusion of Suspended Cell Cultures

Earlier in this chapter (see Section III.C.2) the application of a perfusion system for IPL-Sf-21AE cells grown in Dyna cell propagators on polyester sheeting was described. A similar perfusion system was tested in modified Moduculture® vessels equipped with a 1-μm Biospin® stainless steel filter element for perfusion of insect cells in suspension.[21] The principal feature of the Moduculture® is the Biospin® filter that, as part of the impeller assembly, is used to remove spent medium and cell free virus without accumulation of the cells and polyhedra on the surface or inside the filter. With this system it might be possible to prevent or delay the cytopathogenic effect of NOV on suspended cultures by removing the spent medium containing the NOV and perfusing fresh medium into the infected and the dividing uninfected cells, and thus obtain maximum production of polyhedra. Figure 4 illustrates and describes the perfusion of the suspended cell cultures in the Moduculture®-Biospin® filter system. The Moduculture® flask was planted with 3 ℓ of cell suspension at 1×10^5 cells per milliliter and infected with AcNPV at an MOI of 0.005 PFU per milliliter from the seventh passage in tissue culture. The impeller assembly was rotated at 200 rpm. Dissolved oxygen was maintained at 5.3×10^3 $\mu\ell/m\ell$ by aerating with purified air at 10 to 15 mℓ/min and the perfusion rate of the media was adjusted to 25 mℓ/hr or 600 mℓ/day.

Complete virus production was accomplished 6 days postplanting with 98% of the cells containing polyhedra. There were an average of 44 polyhedra per cell and 1.8×10^8 polyhedra per milliliter or 5.4×10^{11} polyhedra per 3000 mℓ of cell suspension. During 6 days of the virus production cycle with perfusion at 25 mℓ/hr, a total 3.6 ℓ of fresh medium was utilized and a total of 6.6 ℓ of spent medium containing 2.4×10^9 NOV per milliliter was produced.

The significant factor in the production of AcNPV in IPL-Sf-21AE by this suspension method was the increase in virions per polyhedra. In previous studies employing a conventional method for producing the virus in plastic roller bottles,[40] there were 2.10 ± 2.90

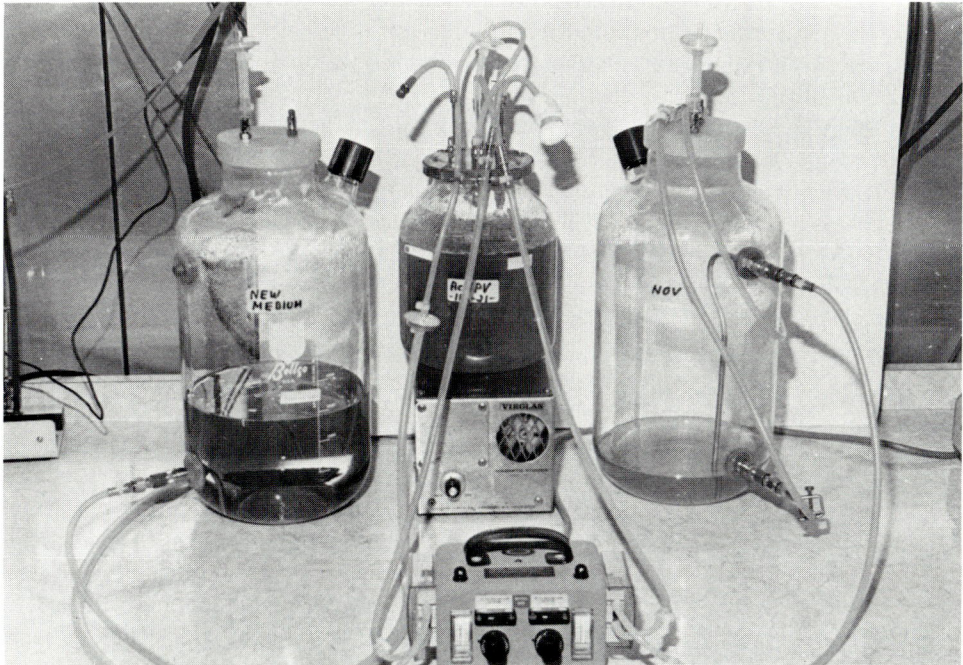

FIGURE 4. Production of AcNPV in perfused IPL-Sf-21AE cells in suspension. At center is a Moduculture® vessel containing 3 ℓ of AcNPV infected cell suspension. On the left is a 9-ℓ vessel with the new medium which is pumped into the Moduculture® vessel with the medical infusion pump (in front). On the right is the NOV collection reservoir which is generally placed in a portable refrigerator. The spent, cell-free medium is continuously pumped into this reservoir from the Moduculture® vessel via a port connected to 1 μm Biospin filter which is assembled between two Teflon® impellers inside the Moduculture® vessel. The cell suspension in the unit is continuously sparged with 10 mℓ of purified air. The infected cells were perfused with 25 mℓ/hr of medium, or 3.6 ℓ during 6 days for complete virus production. During operation, the complete system is connected to a pH-DO monitor and recorder (not shown in the figure).

virions per polyhedra per section. This was improved to 2.70 ± 0.12 virions per polyhedra per section by supplementing the medium with aluminum chloride and zinc sulfate.[3] The polyhedra produced in 3-ℓ perfused Moduculture® flasks with Biospin® filters contained 6.94 ± 4.82 virions per polyhedra per section and 5.65 ± 5.23 virions per polyhedra per section. It could be postulated that the increase in virions per polyhedra per section may be correlated with the continuous nutrient supplied to the cells in suspension culture, controlled with a steady-state system, and more readily available to the cells in the submerged culture than in the substrate-based culture.

V. CONCLUSIONS

In this chapter, we have described the most recent technology for the large-scale propagation of insect cells and cultivation of baculovirus. This technology has resulted from research efforts by many investigators who contributed to the current "state of the art" status of insect cell culture. Successful use of invertebrate cell culture could have an impact on the agricultural economy and human and veterinary medicine by providing quantities of biologicals at greatly reduced cost. Many techniques for insect cell culture evolved through modification of animal cell culture procedures. Because insect cells in culture appear much more sensitive to shear stress[29] and chemical toxicity,[3] modification of equipment and care in the selection of supplies are necessary.[4] By modifying quality control procedures used in propagation of mammalian cells to fit the requirements of invertebrate cells as outlined in this chapter, invertebrate cell culturists and virologists can standardize their reagents to improve overall cell and virus yields.

The application of rigorous quality control standards, the improvement of cell nutrition, and the increased understanding of the effects of biophysical factors on the growth of insect cells have resulted in a steady decrease in the reported population doubling times of insect cells in culture. As an example, the PDT of the IPL-Sf-21AE cells reported in the several papers reviewed here decreased from 24 to 25 hr to less than 10 hr. This decrease represents a considerable reduction in the time and money required to obtain the quantity of cells needed for virus production. Further improvements in media formulation are certain to occur more rapidly in the future than they have in the past as a result of the development of chemically defined media that will permit quantitative studies of cell nutrition without the interference of undefined nutrient sources.[26] Prior efforts have been directed toward increasing either the number of cells or the number of polyhedra obtained. However, it has been demonstrated, first by Tompkins et al.[66] in insects, and later by Weiss et al.,[3,21,40] that certain factors related to the efficacy of the polyhedra are influenced by the host insect or cell. The incorporation of virions into polyhedra can be increased by changing the conditions under which the virus is replicated. Very few specific factors affecting the virion incorporation into polyhedra have been identified but this is a promising area that needs further study.

Oxygen supply probably represents one of the most critical biophysical factors limiting the large-volume production of baculoviruses. It has been clearly demonstrated that baculoviruses will not replicate successfully in cells grown in suspension cultures exceeding 500 mℓ in volume due to depletion of the dissolved oxygen. Therefore, it appears that definition of the required level of dissolved oxygen for large-scale production of baculoviruses in insect cells is mandatory. Methods for supplying oxygen, however, must take into consideration the fragile nature of insect cells. The heavy sparging required to provide the cells' requirements for oxygen has been shown to damage the cells and reduce yields. However, recent studies have shown that new methods, such as the use of silicone tubing to introduce oxygen into the medium by diffusion or improved design of impellers and their placement in conjunction with the air inlets, can avoid or overcome the damage resulting from sparging.

Recent research in the development and testing of large-volume systems has resulted in laboratory models of both the substrate based cultures and suspension cultures. These model systems have given improved cell yields and increased numbers of polyhedra both per cell and per milliliter of media. New types of vessel design such as the Dyna cell propagators and new surface conditioners such as Primaria[67] may further improve the efficiency of the substrate based systems.

The relatively long "in vessel" time to grow the insect cells and produce the virus appears to require some continuous control of such conditions as dissolved oxygen and pH. Recent studies with perfusion cultures show considerable promise in this regard. Perfusion systems may also provide the means for recycling medium or for providing specific nutrients only at the times in the production when they are needed. A unique type of perfusion, suspension culture system was designed by Pollard and Khosrovi[68] as a theoretical fermentor for producing baculoviruses in vitro. They proposed a continuous flow, vertical, tubular fermentor with indirect oxygen supply. The cells would be maintained in suspension by the flow of the medium through the fermentor. Oxygen would be provided by diffusion through silicone tubing.

This and other perfusion systems need to be built and tested on a pilot plant scale before adequate costs can be determined. However, Pollard and Khosrovi[68] estimated that their fermentor design would produce enough virus for one million acre treatments per year at a cost of $3/acre-treatment for the formulated product. Such costs would be competitive with other insect control materials. Additional research with this system or other systems described in this chapter should provide continued improvement in the virus yield and eventually lead to economically feasible methods for the production of viral insecticides.

ACKNOWLEDGMENTS

This program was supported by Grant PCM-8021956 from the National Science Foundation, Washington, D.C., and by cooperative agreement No. 58-3204-8-5 from the U.S. Department of Agriculture Science and Education Administration, Beltsville, Md.

REFERENCES

1. **Tinsley, T. W.,** Properties and replication of insect baculoviruses, in *Beltsville Symposia in Agricultural Research,* Vol. 1, *Virology in Agriculture,* Romberger, J. A., Ed., (Landmark Studies) Allanheld, Osmun & Co., Montclair, N.J., 1977, 117.
2. **Vaughn, J. L.,** Insect cells for insect virus production, *Adv. Cell Cult.,* 1, 281, 1981.
3. **Weiss, S. A., Smith, G. C., Vaughn, J. L., Dougherty, E. M., and Tompkins, G. J.,** Effect of aluminum chloride and zinc sulfate on *Autographa californica* nuclear polyhedrosis virus (ACNPV) replication in cell culture, *In Vitro,* 18, 937, 1982.
4. **Weiss, S. A., Smith, G. C., Kalter, S. S., and Vaughn, J. L.,** Improved method for the production of insect cell cultures in large volume, *In Vitro,* 17, 495, 1981.
5. Recommendations. Section IV. Overcoming current technological and conceptual impediments to wide scale use of microbial control agents. A. In vitro production, in *Workshop on Insect Pest Management with Microbial Agents: Recent Achievements, Deficiencies, and Innovations, Ithaca, N.Y., May 1980, Proc.,* Insect Pathology Resource Center, Cornell University, Ithaca, N.Y., 1980, 62.
6. **Hink, W. F.,** A compilation of invertebrate cell lines and culture media, in *Invertebrate Tissue Culture: Research Applications,* Maramorosch, K., Ed., Academic Press, New York, 1976, 319.
7. **Hink, W. F.,** The 1979 Compilation of Invertebrate Cell Lines and Culture Media, in *Invertebrate Systems In Vitro,* Kurstak, E., Maramorosch, K., and Dubendorfer, A., Eds., Elsevier/North Holland, Amsterdam, 1980, 553.
8. **Hink, W. F.,** *The 1983 Compilation of Invertebrate Cell Lines,* Available on request from Dr. W. F. Hink, Dept. of Entomology, Ohio State University, 1735 Neil Avenue, Columbus, Ohio, 43210.
9. **Grace, T. D. C.,** Establishment of four strains of cells from insect tissues grown *in vitro, Nature (London),* 195, 788, 1962.
10. **Stockdale, H. and Priston, R. A. J.,** Production of insect viruses in cell culture, in *Microbial Control of Pests and Plant Diseases 1970—1980,* Burges, H. D., Ed., Academic Press, New York, 1981, 313.
11. American Type Culture Collection, Catalogue of strains, II, in *Cell Lines, Viruses, Antisera,* 4th ed., American Type Culture Collection, Rockville, Md., 1983.
12. **Weiss, S. A., Lester, T. L., Kalter, S. S., and Heberling, R. L.,** Chemically defined serum-free media for the cultivation of primary cells and their susceptibility to viruses, *In Vitro,* 16, 616, 1980.
13. **Price, P. J. and Gregory, E. A.,** Relationship between *in vitro* growth promotion and biophysical and biochemical properties of the serum supplement, *In Vitro,* 18, 576, 1982.
14. **McKeehan, W. L., McKeehan, K. A., Hammond, S. L., and Ham, R. G.,** Improved medium for clonal growth of human diploid fibroblasts at low concentrations of serum protein, *In Vitro,* 13, 399, 1977.
15. **Dougall, D. K.,** Dilution plating and nutritional considerations. B. Plant cells, in *Tissue Culture: Methods and Applications,* Kruse, P. F., Jr. and Patterson, M. K., Jr., Eds., Academic Press, New York, 1973, 261.
16. **Smith, G. C., Lester, T. L., Heberling, R. L., and Kalter, S. S.,** Coronavirus-like particles in nonhuman primate feces, *Arch. Virol.,* 72, 105, 1982.
17. **Dent, P. B., Cleland, G. B., and Liao, S.-K.,** Detection and control of occult mycoplasma contamination in human tumor cell lines, *Cancer Immunol. Immunother.,* 8, 27, 1980.
18. **McGarrity, G. J., Sarama, J., and Vanaman, V.,** Factors influencing microbiological assay of cell-culture mycoplasmas, *In Vitro,* 15, 73, 1979.
19. **Russell, W. C., Newman, C., and Williamson, D. H.,** A simple cytochemical technique for demonstration of DNA in cells infected with mycoplasmas and viruses, *Nature (London),* 253, 461, 1975.

20. **Brown, S. E. and Knudson, D. L.,** Characterization of invertebrate cell lines. IV. Isozyme analyses of dipteran and acarine cell lines, *In Vitro*, 18, 347, 1982.
21. **Weiss, S. A., Peplow, D., Smith, G. C., Vaughn, J. L., and Dougherty, E.,** Biotechnological aspects of a large scale process for insect cells and baculovirus, in *Techniques in Setting Up and Maintenance of Tissue and Cell Cultures*, Vol. C1, Kurstak, C., Ed., Elsevier Biomedical, Ireland, 1986, C 110/1.
22. **Hink, W. F.,** Insect tissue culture, *Adv. Appl. Microbiol.*, 15, 157, 1972.
23. **Wyatt, S. S.,** Culture *in vitro* of tissue from the silkworm, *Bombyx mori* L., *J. Gen. Physiol.*, 39, 841, 1956.
24. **Yunker, C. E., Vaughn, J. L., and Cory, J.,** Adaptation of an insect cell line (Grace's Antheraea cells) to medium free of insect hemolymph, *Science*, 155, 1565, 1967.
25. **Mitsuhashi, J.,** Media for insect cell cultures, *Adv. Cell Cult.*, 2, 133, 1982.
26. **Wilkie, G. E. I., Stockdale, H., and Pirt, S. V.,** Chemically-defined media for production of insect cells and virus *in vitro*, *Dev. Biol. Stand.*, 46, 29, 1980.
27. **Goodwin, R. H. and Adams, J. R.,** Nutrient factors influencing viral replication in serum-free insect cell line culture, in *Invertebrate Systems In Vitro*, Kurstak, E., Maramorosch, K., and Dubendorfer, A., Eds., Elsevier/North-Holland, Amsterdam, 1980, 493.
28. **Vaughn, J. L.,** The production of nuclear polyhedrosis viruses in large-volume cell cultures, *J. Invertebr. Pathol.*, 28, 233, 1976.
29. **Weiss, S. A., Kalter, S. S., Vaughn, J. L., and Dougherty, E.,** Effect of nutritional, biological and biophysical parameters on insect cell culture of large scale production, *In Vitro*, 16, 222, 1980.
30. **Pumper, R. W.,** Purification and standardization of water for tissue culture, in *Tissue Culture: Methods and Applications*, Kruse, P. F., Jr. and Patterson, M. K., Jr., Eds., Academic Press, New York, 1973, 674.
31. **Basalik, T. J.,** Reagent grade water for laboratories, *Clin. Lab Proced.*, 5, 18, 1976.
32. **McLimans, W. F.,** The gaseous environment of the mammalian cell in culture, in *Growth, Nutrition and Metabolism of Cells in Culture*, Vol. 1, Rothblat, G. H. and Cristofalo, V. J., Eds., Academic Press, New York, 1972, 137.
33. **Weiss, S. A., Orr, T., Smith, G. C., Kalter, S. S., Vaughn, J. L., and Dougherty, E. M.,** Quantitative measurement of oxygen consumption in insect cell culture infected with polyhedrosis virus, *Biotechnol. Bioeng.*, 24, 1145, 1982.
34. **Weiss, S. A., Smith, G. C., Kalter, S. S., and Vaughn, J. L.,** Chemical enhancement of *Autographa californica* nuclear polyhedrosis virus (ACNPV) replication in cell culture, *In Vitro*, 17, 208, 1981.
35. **Schaeffer, W. I.,** Proposed usage of animal tissue culture terms (revised 1978). Usage of vertebrate cell, tissue and organ culture terminology, *In Vitro*, 15, 649, 1979.
36. **Vaughn, J. L.,** The production of viruses for insect control in large scale cultures of insect cells, in *Invertebrate Tissue Culture: Research Applications*, Maramorosch, K., Ed., Academic Press, New York, 1976, 295.
37. **Grinnell, F., Milam, M., and Srere, P. A.,** Studies on cell adhesion. II. Adhesion of cells to surfaces of diverse chemical composition and inhibition of adhesion by sulfhydryl binding reagents, *Arch. Biochem. Biophys.*, 153, 193, 1972.
38. **McKeehan, W. L. and Ham, R. G.,** Stimulation of clonal growth of normal fibroblasts with substrata coated with basic polymers, *J. Cell Biol.*, 71, 727, 1976.
39. **Weiss, S. A., Peplow, D., Kalter, S. S., and Vaughn, J. L.,** Dissociation of insect cell cultures by pancreatin, *In Vitro*, 18, 298, 1982.
40. **Weiss, S. A., Smith, G. C., Kalter, S. S., Vaughn, J. L., and Dougherty, E.,** Improved replication of *Autographa californica* nuclear polyhedrosis virus in roller bottles: characterization of the progeny virus, *Intervirology*, 15, 213, 1981.
41. **Miller, M. F., Allen, P. T., and Dmochowski, L.,** Quantitative studies of oncornaviruses in thin sections, *J. Gen. Virol.*, 21, 57, 1973.
42. **Baudhuin, P. and Berthet, J.,** Electron microscopic examination of subcellular fractions. II. Quantitative analysis of mitochondrial population isolated from rat liver, *J. Cell Biol.*, 35, 631, 1967.
43. **Wibo, M., Amar-Costesec, A., Berthet, J., and Beaufay, H.,** Electron microscope examination of subcellular fractions. III. Quantitative analysis of the microsomal fraction isolated from rat liver, *J. Cell Biol.*, 51, 52, 1971.
44. **Volkman, L. E., Summers, M. D., and Hsieh, C.-H.,** Occluded and nonoccluded nuclear polyhedrosis virus grown in *Trichoplusia ni*: comparative neutralization, comparative infectivity, and *in vitro* growth studies, *J. Virol.*, 19, 820, 1976.
45. **Stockdale, H. and Gardiner, G. R.,** Utilization of some sugars by a line of *Trichoplusia ni* cells, in *Invertebrate Tissue Culture: Applications in Medicine, Biology and Agriculture*, Kurstak, E. and Maramorosch, K., Eds., Academic Press, New York, 1976, 267.
46. **House, W., Shearer, M., and Maroudas, N. G.,** Method for bulk culture of animal cells on plastic film, *Exp. Cell Res.*, 71, 293, 1972.

47. **Nicklin, P. M. and House, W.,** Large scale production of virus (communication to the editor), *Biotechnol. Bioeng.*, 18, 723, 1976.
48. **Firket, H.,** Polyester sheeting, in *Tissue Culture: Methods and Applications,* Kruse, P. F., Jr. and Patterson, M. K., Jr., Eds., Academic Press, New York, 1973, 378.
49. **Taylor, W. G., Evans, V. J., Fox, C. H., Camalier, R. F., and Sanford, K.,** Evaluation of the dyna-cell vessel for production of surface-substrate dependent cells, *Biotechnol. Bioeng.*, 17, 1847, 1975.
50. **Ceriotti, G.,** Determination of nucleic acids in animal tissues, *J. Biol. Chem.*, 214, 59, 1955.
51. **Weiss, S. A.,** Concentration of baboon endogenous virus in large-scale production by use of hollow-fiber ultrafiltration technology, *Biotechnol. Bioeng.*, 22, 19, 1980.
52. **McLimans, W. F.,** Mass culture of mammalian cells, *Methods Enzymol.*, 58, 194, 1979.
53. **Vaughn, J. L.,** Growth of insect cell lines in suspension culture, in *International Colloquium on Invertebrate Tissue Culture,* 2nd, Tremezzo, 1967, Proceedings: Barigozzi, C., Ed., Istituto Lombardo, Accademia di Scienze e Lettere, Milan, 1968, 119.
54. **Hink, W. F. and Strauss, E. M.,** Growth of the *Trichoplusia ni* (TN-368) cell line in suspension culture, in *Invertebrate Tissue Culture: Applications in Medicine, Biology, and Agriculture,* Kurstak, E. and Maramorosch, K., Eds., Academic Press, New York, 1976, 297.
55. **Gardiner, G. R. and Stockdale, H.,** Two tissue culture media for production of lepidopteran cells and nuclear polyhedrosis viruses, *J. Invertebr. Pathol.*, 25, 363, 1975.
56. **Vaughn, J. L. and Goodwin, R. H.,** Large-scale culture of insect cells for virus production, in *Beltsville Symposia in Agricultural Research,* Vol. 1, *Virology in Agriculture,* Romberger, J. A., Ed., Allanheld, Osmun & Co., Montclair, N.J., 1977, 109.
57. **Hink, W. F. and Strauss, E. M.,** Semi-continuous culture of TN-368 cell line in fermentors with virus production in harvested cells, in *Invertebrate Systems In Vitro,* Kurstak, E., Maramorosch, K., and Dubendorfer, A., Eds., Elsevier/North-Holland, Amsterdam, 1980, 27.
58. **Miltenburger, H. G. and David, P.,** Mass production of insect cells in suspension, *Dev. Biol. Stand.*, 46, 183, 1980.
59. **Hilwig, I. and Alapatt, F.,** Insect cell lines in suspension, cultivated in roller bottles, *Z. Angew. Entomol.*, 91, 1, 1981.
60. **Streett, D. A. and Hink, W. F.,** Oxygen consumption of *Trichoplusia ni* (TN-368) insect cell line infected with *Autographa californica* nuclear polyhedrosis virus, *J. Invertebr. Pathol.*, 32, 112, 1978.
61. **Hink, W. F.,** Production of *Autographa californica* nuclear polyhedrosis virus in cells from large-scale suspension cultures, in *Microbial and Viral Pesticides,* Kurstak, E., Ed., Marcel Dekker, New York, 1982, 493.
62. **Stockdale, H. and Gardiner, G. R.,** The influence of the condition of cells and medium on production of polyhedra of *Autographa californica* nuclear polyhedrosis virus *in vitro, J. Invertebr. Pathol.*, 30, 330, 1977.
63. **Lynn, D. E. and Hink, W. F.,** Infection of synchronized TN-368 cell cultures with alfalfa looper nuclear polyhedrosis virus, *J. Invertebr. Pathol.*, 32, 1, 1978.
64. **Wood, H. A., Johnston, L. B., and Burand, J. P.,** Inhibition of *Autographa californica* nuclear polyhedrosis virus replication in high-density *Trichoplusia ni* cell cultures, *Virology,* 119, 245, 1982.
65. **Hink, W. F., Strauss, E. M., and Ramoska, W. A.,** Propagation of *Autographa californica* nuclear polyhedrosis virus in cell culture: methods for infecting cells, *J. Invertebr. Pathol.*, 30, 185, 1977.
66. **Tompkins, G. J., Vaughn, J. L., Adams, J. R., and Reichelderfer, C. F.,** Effects of propagating *Autographa californica* nuclear polyhedrosis virus and its *Trichoplusia ni* nuclear polyhedrosis virus variant in different hosts, *Environ. Entomol.*, 10, 801, 1981.
67. **Guhl, T. A. and Umstatter, D. M.,** Laminin and fibronectin adsorption to a new surface for cell culture: effect on cell attachment and spreading, *In Vitro,* 19, 290, 1983.
68. **Pollard, R. and Khosrovi, B.,** Reactor design for fermentation of fragile tissue cells, *Process Biochem.*, 13, 31, 1978.

Chapter 4

ECOLOGY AND EPIZOOTIOLOGY OF BACULOVIRUSES

Hugh F. Evans

TABLE OF CONTENTS

I. INTRODUCTION

Although not recognized as viral infections, insect diseases have been known from their effects on insect populations probably at least from the first rearing of silkworms *(Bombyx mori)*, reputedly dating from 4000 B.C.[1] In fact most early circumstantial evidence from accounts of possible baculovirus infections accrue from observations of silkworm diseases. Steinhaus[1] detailed many anecdotal references from sericultural literature and regarded the first accounts of probable baculovirus infection as dating from at least the 16th century. Since that time the role of disease in sericulture has received considerable attention; notable advances in describing the transmission of infectious diseases were made by Bassi[2] and later Pasteur.[3]

However, the earliest references to recognizable baculovirus diseases in natural populations in the field probably relate to the nuclear polyhedrosis virus (NPV) of *Lymantria monacha*. Virus diseased larvae of this species exhibit a tendency to migrate to, and eventually die in, the tops of trees, and this condition is traditionally known as "wipfelkrankheit" (tree top disease) in Germany. It is not known when the term was first coined although epizootics of wipfelkrankheit were described by Hofmann in 1891.[4] Recognition of the disease as viral in origin came in 1909,[5] although polyhedra had been described from diseased silkworm larvae much earlier.[6,7] Final confirmation that inclusion bodies contained the infective entity came in 1924 when Komárek and Breindl[8] dissolved polyhedra in weak alkali and observed the liberated virions under the light microscope. Bergold[9] later showed, by electron microscopy, that the virions were embedded in the matrix of the inclusion bodies.

Observations on wipfelkrankheit in *L. monacha* and later the "wilt" disease (NPV) of gypsy moth *(L. dispar)*[10,11] represent the first data on the ecology of baculoviruses. By 1946 when Steinhaus[12] produced his treatise on insect microbiology the presence of virus diseases in 47 species of insect had been described. Natural epizootics of baculoviruses were also reported in populations of several forest and rangeland pests in the U.S. and Europe. Steinhaus noted that "polyhedral disease is present wherever there is a large number of insects and these natural epidemics seem to be just as effective as those which are supposedly assisted by the virus artificially introduced."[12] These words will be recognized by insect ecologists as an early description of the density-dependence of baculovirus diseases. In combination with Steinhaus' other observations on the conditions that initiate infection, they provide a remarkable insight into the far-reaching conclusions of workers involved in early observational baculovirus ecology.

Approaches from the 1940s onward became more analytical and included many detailed studies on the relationships between host and virus. Notable among these accounts, Bird,[13-15] Bird and Elgee,[16] and Bird and Burk[17] presented comprehensive descriptions of field and laboratory studies of the NPV of *Gilpinia hercyniae*. Quantitative data on the role of natural and artificially introduced NPV were presented and were sufficiently detailed to permit comparison with a later outbreak of NPV in *G. hercyniae* populations in mid-Wales.[18]

Progress in this area of research has centered on improved links between study of host population dynamics and virus ecology as a single integrated discipline. Tanada[19-22] has produced several excellent reviews on epizootiology of insect pathogens, especially viruses. Each review has described the state of the art and reading them in sequence serves to illustrate how ideas and quality of data have changed over the years. The opening phrase in his 1963 review is as relevant now as when it was first written — "An important goal in insect pathology is to determine the fundamental principles governing the disease dynamics in groups or populations of insects."[19] The concept of dynamics with reference to disease growth was particularly significant. It is only during the last 15 years that real progress has been made on the quantitative description of epizootics. This has resulted from improved techniques of virus detection, enumeration, and characterization with the parallel develop-

Table 1
THE MAJOR COMPONENTS OF ECOLOGICAL
STUDY

Biosphere
Ecosystems }   classical   }
Communities }   ecology   }
Populations }   }
Organisms   }
Organ systems   }   baculovirus ecology
Organs   }
Tissues   }
Cells   }
Subcellular organelles   }
Molecules   }

Modified from Krebs, C. J., *Ecology, The Experimental Analysis of Distribution and Abundance*, Harper & Row, New York, 1978. With permission.

ment of simple predictive models for baculovirus epizootiology. In this respect we lag behind
our counterparts in vertebrate or plant pathology and often have to adapt techniques of
analysis from their work.

The aim of this chapter is to illustrate how basic ecological parameters can be quantified
and used to describe the epizootiology of baculoviruses in the field. In this respect it represents
a synopsis of current selected literature and also encompasses personal ideas of how various
areas of research should be approached. It is hoped that it may provide stimuli for those
established in, or new to, the field, with the aim being always to produce data that can be
used to generate new trains of thought and new principles to advance our understanding of
baculovirus epizootiology.

II. BASIC ECOLOGY OF BACULOVIRUSES

In discussing the ecology of baculoviruses it is convenient to begin by providing a definition
of the term in the context of baculovirus epizootiology. Although baculoviruses can not be
classified in the same way as other biotic agents in the environment, they nevertheless share
many of the attributes of populations of plants or animals. By regarding them in this way
they can be included in a suitable general definition of ecology first coined by Andrewartha
namely, "ecology is the scientific study of the distribution and abundance of organisms."[23]

Under this definition we are concerned with two fundamental parameters, distribution and
abundance, but as Andrewartha and Birch[24] pointed out, the two components are intimately
related so that distribution can be defined in terms of abundance. Taking the concept further
and broadening it to include the general epizootiology of baculoviruses the aim of the present
section is to define and discuss those components of baculovirus ecology that delimit the
progress of disease in time and space. It must be stressed that, although considered separately,
each variable is part of a broader continuum of ecological factors that interact to drive the
epizootic. This approach is, of necessity, highly selective and some attributes may be
neglected or indeed omitted. This underlines one of the major problems for ecologists,
namely defining the level at which interrelationships should be studied. There is a sequence
of ecological strata covering every level of system organization which was well illustrated
by Krebs[25] and which is reproduced, with modifications, in Table 1.

Krebs considered that classical ecological studies primarily encompassed ecosystems,
communities, and populations and that in these areas each had its own unique characteristics.
In baculovirus ecology we have the paradox of whether we are studying the ecology of the
virus or of its host insect. It is clear that all the levels of organization in Table 1 apply to

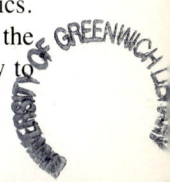

baculovirus ecology and that, in some ways, the host insect is equivalent to the biosphere from the standpoint of the virus. However, it is more realistic to view baculovirus ecology as a dynamic interaction between the virus and its host so that the ecology of the host is considered equally with the ecology of the virus itself.

Although there is a place for observational baculovirus ecology it is increasingly apparent that a more rigorous quantitative approach is necessary if predictive modeling of virus-host interactions is to be a practical proposition.

Ultimately, the assumptions developed through fundamental ecological studies have to be tested in the field, and it is often the case that laboratory studies form the basis for these tests. Unpredictable variations in the field may make such assumptions poor or invalid and great care has to be exercised in this approach. Environmental fluctuations, particularly meteorological changes, provide the greatest host independent variations, and it is in this area that modeling of ecological parameters is most difficult. This is reflected in the lack of broad ranging ecological studies that actually include a wide range of epizootiological parameters. Those data that are available provide, at best, a partial description of the nature of baculovirus disease in the host population.

An additional limiting factor in furthering our understanding of baculovirus ecology is that fundamental data on virus and host responses are often obtained as a result of unrelated studies such as quality control in bioassay or mass propagation of virus. In these cases it is likely that important data relating to virus ecology and epizootiology may be overlooked. Despite these limitations, when several examples supporting a single ecological principle exist in the literature, a trend may become apparent and this can be applied ultimately to description of field relationships. The following discussion, therefore, includes the ways in which both laboratory and field data influence baculovirus ecology. The aim is to develop a common theme between the various characteristics so that one parameter can be viewed as part of the overall virus-host continuum.

A. Infectivity

There are many data on the ways in which insect hosts respond to challenge by baculoviruses. These have been discussed by Burges and Thomson,[26] Evans,[27,28] and by Hughes and Wood in this volume. Here it is proposed to deal only with the host responses that influence the overall ecology of baculoviruses in the field.

In insect hosts, the major pathway for virus ingress is per os in the larval stage,[29] a period when metabolic activity and growth are extremely rapid. The sequence of events from ingestion to infection and death have been documented,[29] although the complete pathway is not fully understood.[30] However, from an ecological standpoint the primary criterion is response of the host to a given quantity of inoculum.

Description of this parameter begins by regarding the host population as a series of individuals, of which a proportion will be susceptible to a defined quantity of virus. Two parameters define this level of susceptibility in populations of insects.

1. Intrinsic Variation at the Individual Level

At a constant age, susceptibility is a function of intrinsic variation in host response and this will be reflected in the range of responses in larvae to defined inocula. For example, Hughes et al.[31] infected newly emerged *Trichoplusia ni* larvae with *Autographa californica* NPV and computed an LD_{50} of 1.2 polyhedral inclusion bodies (PIBs) with 95% fiducial limits of 0.6 and 1.7 PIBs. Since all the larvae tested were treated identically and were of virtually constant size and from the same insectary stock, the observed variation can be regarded as the least that is likely to occur in similar populations of the host in the field. There are other examples where careful experimental technique appears to have minimized variation in response as reflected in the slopes and limits of dosage-mortality re-

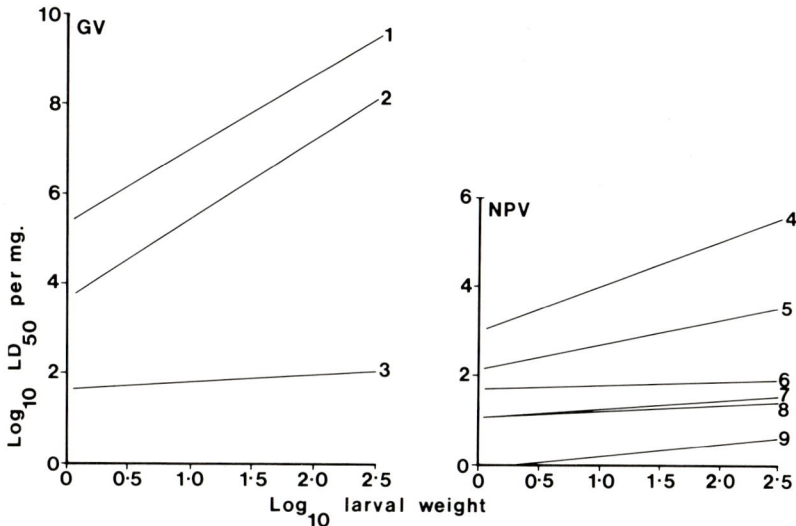

FIGURE 1.   The relationships between $\log_{10}$, LD_{50} and $\log_{10}$ body weight for selected hosts and baculoviruses. The value b = slope of the regression line. Granulosis virus (GV), (1) *Hypantria cunea*, b = 1.68;[36] (2) *Pieris brassicae*, b = 1.77;[37] (3) *P. rapae*, b = 0.17;[37] Nuclear polyhedrosis virus (NPV), (4) *H. cunea*, b = 1.02;[36] (5) *Mamestra brassicae*, b = 0.54;[28] (6) *Operophtera brumata*, b = 0.08;[32] (7) *Gilpinia hercyniae*, b = 0.20;[38] (8) *Heliothis zea*, b = 0.14,[34] (9) *Trichoplusia ni*, b = 0.26.[35]

sponses.[27,28,32,33] No matter how careful the technique, it is impossible to eliminate the intrinsic heterogeneity of the host population with respect to virus.

Of greatest significance ecologically is the now well-recognized fact that susceptibility of larvae to baculovirus infection decreases with increasing larval age. This occurs over and above the inherent variability at a given age. It therefore provides a basis on which to quantify the dosages necessary to produce a required level of mortality for larvae of different ages. This stems from the finding that, in the majority of cases, age-related changes in susceptibility can be linked directly to increasing body weight.[27,34,35]

The relationship between larval susceptibility and weight tends to be specific to the host and the virus being studied, and therefore no unifying quantitative link between the two parameters can be produced. This is illustrated in Figure 1 where log LD_{50} per milligram has been plotted against log weight of larvae for a number of host/virus relationships. There are clearly large differences in both slope and position reflecting the greater extent of resistance to infection in some species. As a general rule it is those species showing the greatest range of susceptibility that have the largest slope values. Such a finding is intuitively obvious since there is no upper limit to the number of inclusion bodies that might be required to initiate infection so that some mature larvae may be regarded as virtually resistant to infection. For example, late sixth-instar *Mamestra brassicae* larvae showed little response to dosages in excess of 5×10^7 PIBs.[28] On the other hand there is a finite limit to body weight and this, for each species, is the main parameter determining the slope of the LD_{50} per milligram relationship.

2. Variation at the Population Level

Notwithstanding the interaction of individuals at different population densities (see host density effects) the main variation in observed responses at the population level lies in the relative proportions of individuals that are susceptible or infected in a given population. Specifically, a population can be regarded as a series of susceptible or nonsusceptible individuals. This truism forms the basis for many models in plant pathology,[39,40] and the

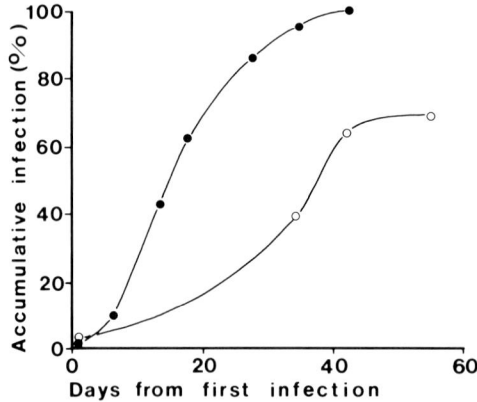

FIGURE 2.   Accumulative increase in nuclear polyhedrosis virus infection in larvae with time for *Neodiprion sertifer* ● (reproduced from Kaupp[42]) and
Spodoptera frugiperda[43] ○ (extrapolated from disease
incidence and initial host populations).

same theoretical assumptions can be applied to description of the buildup of baculovirus
infection in an insect population.[41]

Before any infections have arisen the entire population can be regarded as susceptible;
the level of susceptibility depends on larval age as described above, with a small proportion
of larvae near maturity showing full maturation resistance. As individuals become infected
the probability of infection in the remaining population decreases, i.e., at any one time the
proportion of the population remaining susceptible is given by $(1 - x)$ where x is the
proportion of larvae already infected. Thus the greater the value of x, the lower the probability
of virus encountering a susceptible host. This law of diminishing returns is typically described
by the accumulative logistic growth of infection in a population with the proportion of
infected individuals increasing asymptotically (Figure 2).

Curves that are sigmoid are difficult to compare directly and it is therefore necessary to
transform the data using logarithmic scales and expressing infection as a proportion of the
remaining population and not of the entire original population, i.e., $\log_{10} (x/1 - x)$. Such
a transformation normalizes the data so that the buildup of infection in a population can be
expressed as a daily infection rate described by the slope of the regression line.

Examples of this approach have been plotted in Figure 3 where the transformed relationships from Figure 2 are shown. The infection rate (y) varied considerably between the
examples possibly reflecting fundamental differences between the insect orders. In addition,
when conditions were known to be similar, as was the case for the NPV of *G. hercyniae*
in both Canada and Wales, the value of y was remarkably constant. However, there is also
the complication of host density which tends to influence y in low host populations by
decreasing the rate of infection in the larval population.

The curves just described were accumulative and expressed the total incidence of infection
with time. If the instantaneous infection level at various time points is examined there is an
interesting difference between the patterns of response to NPV infection for different orders
of insects, particularly between the Hymenoptera and Lepidoptera. Although there are limited
data on the natural incidence of infection during an entire host generation, the examples
shown in Figure 4 illustrate the fundamental difference between population responses in the
two insect orders. Infection typically increases exponentially in both cases. In Hymenoptera
this ends asymptotically, whereas in Lepidoptera it decreases again to a low level even
though many individuals may still be present. Such responses may be observed only when

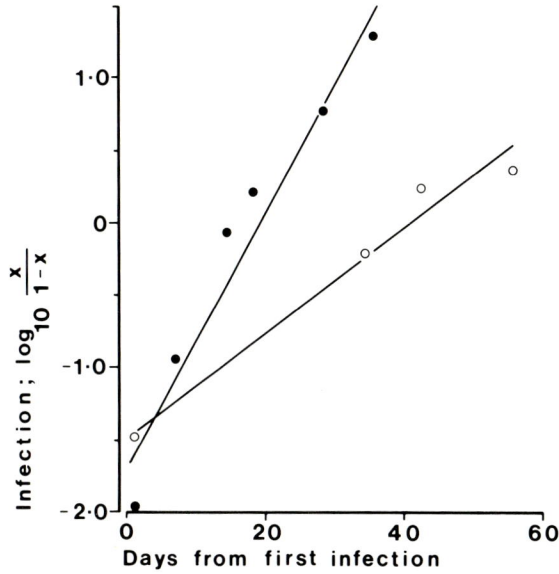

FIGURE 3.   The relationships between transformed nuclear polyhedrosis virus infection (as $\log_{10} x/1 - x$) and time for *Neodiprion sertifer* ● ($y = -1.7 + 0.091x$; reproduced from Kaupp[42]) and *Spodoptera frugiperda*[43] ○ ($y = -1.47 + 0.04x$; data from Figure 2).

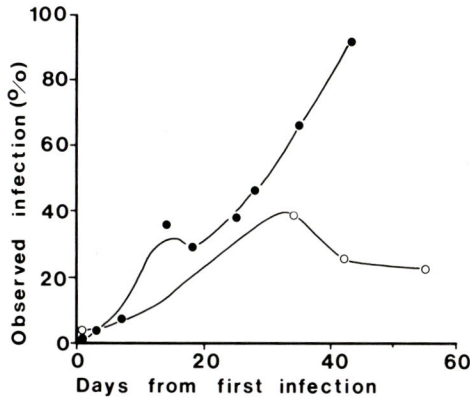

FIGURE 4.   The incidences of nuclear polyhedrosis virus infection in larvae at the times of sampling for *Neodiprion sertifer* ● (reproduced from Kaupp[42]) and *Spodoptera frugiperda*[43] ○.

host generations are discrete. Overlapping host generations will tend to obscure any pattern to the infection curve.

An explanation of the differences between the insect orders may be attributable to the ways in which susceptibility changes with age. As illustrated in Figure 1, the Lepidoptera, with few exceptions, show extremely large variation between young and mature larvae. It follows that decreasing numbers of individuals, combined with sharply decreasing susceptibility as the larvae age, result in greatly reduced probability of a finite virus source being sufficiently large to induce even moderate mortality. The infection rate thus declines rapidly

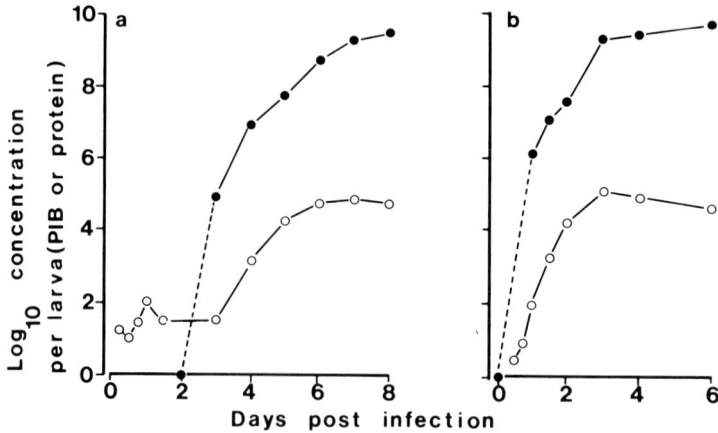

FIGURE 5.   Growth of nuclear polyhedrosis virus in fifth-instar larvae of (a) *Mamestra brassicae* (reproduced from Evans et al.[44]); (b) *Heliothis armigera* (reproduced from Kelly et al.[45]). ● $\log_{10}$ concentration of polyhedra, ○ $\log_{10}$ concentration of protein (ng).

towards the end of the host generation. In the Hymenoptera there is also declining probability of infection as more of the population is affected, but the remaining individuals remain highly susceptible, and the infection rate is therefore a function of host abundance and distribution only.

B. Virus Availability

It follows from the above discussion that infection in a host population will be related quantitatively to the amount of virus available for larval ingestion. Data on virus productivity in host larvae are therefore necessary as a basis for furthering our understanding of infection processes in the field. The mode of replication has been discussed by Granados and Williams (Volume I, Chapter 4), whereas virus growth as a component of mass propagation systems was detailed by Shapiro (Volume II, Chapter 2). Both these discussions provide data on sites and quantities of virus production in different virus-host systems. Here it is proposed to outline briefly the roles of site of replication and age of the host in the buildup of infection in field populations.

The units used for quantifying NPV and granulosis virus (GV) populations are usually concentrations of inclusion bodies, i.e., polyhedra (NPVs) and granules (GVs), respectively. We therefore concentrate normally on those entities that can be quantified using relatively simple optical equipment. When virus growth is measured in this way the rate at which the number of inclusion bodies increases is generally logistic with time. This is illustrated for NPV growth in *M. brassicae*[44] and *Heliothis armigera*[45] in Figure 5. The earliest identifiable appearance of PIBs was generally about 2 to 4 days postinoculation at 22°C. However, this was a limit imposed by the accuracy of light microscope counts of PIBs which prevents estimation below about 1×10^6 PIBs per milliliter. It is, of course, possible to extrapolate back to estimate early growth, but a more accurate method is to estimate production of virus particles and to relate this to PIB production. This has been done for the above examples using enzyme-linked immunosorbent assay (ELISA) to estimate virus particle protein concentration. These data are also shown in Figure 5 where it is clear that ELISA was able to detect replication much earlier and at a higher level of sensitivity than PIB counts. However, the ELISA curve, although logistic in its early phase, declined towards the end of replication. This is assumed to reflect masking of the antigenic properties of virus particles as they became occluded in PIBs. It is therefore not possible to correlate virus particle growth with

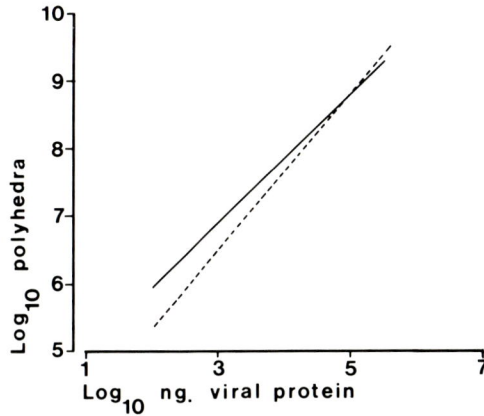

FIGURE 6. The relationship between $\log_{10}$ ng viral protein (ELISA) and $\log_{10}$ number of polyhedra (microscopy) for the early phase of nuclear polyhedrosis virus growth in *Mamestra brassicae* (— log y = 3.13 + 1.13 log x, reproduced from Evans et al.[44]) and *Heliothis armigera* (--- log y = 4.08 + 0.94 log x[45]).

Table 2
PEAK PRODUCTIVITIES OF BACULOVIRUS INCLUSION BODIES IN SELECTED INSECT HOSTS

Virus[a]	Host	Peak productivity[b]	Productivity per mg body wt [b]	Ref.
NPV	*Heliothis zea*	1.5×10^{10}	3.08×10^7	46
	Lymantria dispar	4.9×10^9	1.0×10^7	46, 47
	Mamestra brassicae	3.4×10^9	9.2×10^6	44
	Operophtera brumata	1.0×10^9	1.0×10^7	32
	Trichoplusia ni	1.3×10^{10}	4.3×10^7	46
	Gilpinia hercyniae	2.0×10^8	5.0×10^6	17
	Neodiprion sertifer	1.0×10^8	2.5×10^6	46
	N. swainei	1.3×10^7	2.0×10^5	48
GV	*Cydia pomonella*	2.0×10^{10}	2.0×10^7	46, 49
	Phthorimaea operculella	4.5×10^9	—	50
	Pieris rapae	6.2×10^9	—	51

[a] NPV = nuclear polyhedrosis virus; GV = granulosis virus.
[b] Inclusion bodies.

PIB growth for this phase of replication. However, during the exponential phase of growth there was a linear relationship between the two variables (Figure 6).[44,45] As the use of immunological techniques in the field becomes more refined it will be possible to use methods like ELISA to detect, and perhaps more importantly, identify conclusively the presence of baculoviruses quantitatively.

Where data are available the logistic form of virus growth has been shown to occur in both lepidopterous and hymenopterous (sawfly) larvae, and is similar for all larval stages. Examples of peak productivities of inclusion bodies are given in Table 2. In all cases where body weight was stated there was a direct correlation between virus production per milligram and host body weight at death. In Lepidoptera this represented up to 40% of body weight,[52] while in Hymenoptera (sawflies) the proportion was of the order of 2 to 5%,[38] presumably reflecting the restriction of virus replication to the midgut in the latter group.

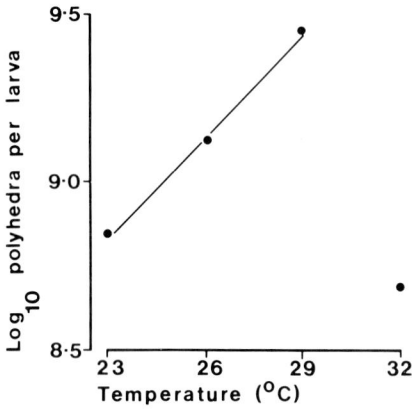

FIGURE 7.   The relationship between nuclear polyhedrosis virus productivity ($\log_{10}$ polyhedra per larva) and temperature for *Lymantria dispar* larvae ($\log_{10} y = 6.55 + 0.099 \log x^{54}$).

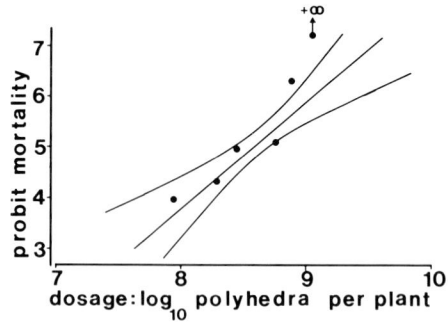

FIGURE 8.   The relationship between probit mortality of neonate *Mamestra brassicae* larvae against $\log_{10}$ concentration of polyhedra on the cabbage plants as a result of previous nuclear polyhedrosis virus mortality. Probit $y = -12.69 + 2.06 \log_{10}$ polyhedra (x). The 95% fiducial limits are shown. (Reproduced from Evans, H. F. and Allaway, G. P., *Appl. Environ. Microbiol.*, 45, 493, 1983. With permission.)

The rate of virus growth is temperature dependent and seems to be well correlated with host metabolic rate. Stairs[53] showed that virus growth was suspended in ligatured larvae which were themselves in an arrested state of development. This occurred at a range of temperatures thus showing the overriding influence of host metabolism on virus replication. Temperature related larval growth and baculovirus replication are therefore closely related as shown in Figure 7. Here PIB concentration per larva is plotted against temperature for NPV growth in *L. dispar* larvae. The linear relationship between the two variables confirms the dependence of virus growth and host metabolism on temperature up to 29° C., but thereafter host metabolism and hence virus production was disrupted.[54]

Extrapolation of the general principles of virus growth and host development allow quantification of virus productivity in natural populations of the host. For example, laboratory data on PIB growth in *M. brassicae* larvae were used to estimate, from host age and level of virus infection, the total quantities of NPV produced during primary and secondary stages of infection in populations of the insect in the field. It was thus possible to link virus production on individual cabbage plants with the subsequent infection levels in later populations of larvae feeding on those plants (Figure 8).[56] Similarly Wigley,[32] working on *Operophtera brumata* populations, used laboratory virus productivity data to quantify virus production and utilization in field populations of the host. ELISA was used by Kaupp[42] in his studies of the epizootiology of *Neodiprion sertifer* NPV, while solid-phase radioimmunoassay was used by Kalmakoff and Crawford[57] to detect NPV antigen in droppings from birds that had been feeding on infected *Wiseana* sp. in the field. Thompson and Scott[58] quantified the total amount of NPV PIBs produced per ha of forest in their studies of the epizootiology of this disease in populations of *Orgyia pseudotsugata*. This awareness of the potential uses of laboratory derived virus growth curve data has wide implications for the development of predictive models in baculovirus epizootiology, and this point will be expanded in later sections.

C. Host Density Effects

The discussion above has provided a simplified introduction to host response at the individual and population levels. Such simplification inevitably means that data have been drawn only from examples of hosts with discrete generations. These are characterized by synchrony of egg hatch and larval development making it relatively simple to describe the population age structure at a given time. This theme can be taken further as an introduction to the effects of host density on the buildup of baculovirus infection in a population.

Infection rates and the declining proportions of susceptible individuals with time have been discussed. We now consider the role host density plays in defining the quantities of inoculum required to initiate and maintain infection in a host population. Two sources of inoculum, primary and secondary, may be distinguished with the second stemming from the success of the first.

1. Primary Inoculum

The quantity of primary inoculum in the environment at the start of a host generation is a function of host infection, density, and age in the previous generation and of attrition of this inoculum between host generations. Thus, when young larvae of the next generation commence feeding, the quantity and distribution of virus present is that which survived the period of nonreplication when no larval stages were present. The following sequence of events then takes place.

Initial infections occur in a few individuals, as shown by the early parts of the untransformed curves in Figure 2. This results from young larvae feeding and ingesting virus, the amount of virus ingested reflecting the quantity and distribution of primary inoculum and also the density, distribution, and feeding rates of the host larvae. Responses depend on whether the quantity of virus ingested exceeds the lethal dosage required for a larva of a given weight, and this is defined by dosage-mortality responses as described previously. Replication follows and eventually the larva stops feeding and dies releasing secondary inoculum.

Primary inoculum is constantly declining as a result of environmental factors such as ultraviolet light, rainfall, and chemical conditions on the leaf surface (see Section II.F). This results in a proportionate decline in probability of a host population of decreasing susceptibility encountering sufficient primary inoculum, which is itself declining, to initiate infection. Indeed most data on virus stability in the field suggest that primary inoculum will tend to decline exponentially when exposed on leaf surfaces (Figure 9). A further factor acting against successful infection is the extremely low feeding rates of young larvae. For example, Tatchell[63] estimated the area of leaf surface consumed by *Pieris rapae* larvae of different ages. There was a 118-fold difference in leaf area consumed between the first and fifth instars. Similarly, Harper[64] showed that *T. ni* larvae consumed more than 90% of their food in the final instar. Virus on leaf surfaces may well be localized, and it is therefore likely that low food consumption in young larvae would result in their failing to eat an area of leaf sufficient to encounter a diffuse source of virus. This, together with the limited dispersal capacity of young larvae, mitigates against high percentage infection in sparsely distributed host populations.

As larvae age, the direct influence of primary inoculum will decline further, both as a result of attrition of that inoculum and of decreasing host susceptibility. These factors are compensated for, at least in part, by increased feeding of older larvae so that, for example, fifth-instar *P. rapae* larvae eat about 71% of their total food consumption.[63] However, the net result of these compensatory factors is a sharp decline in response to primary inoculum, the rate of decline being influenced to a great extent by host density.

It is obvious that high host densities, possibly sufficient to consume the entire available food source, will tend to exploit primary inoculum to its fullest possible extent. In these cases, initial infection in the host population will be at its maximum reflecting the total amount of primary inoculum present. Examples to illustrate specifically this intuitive relationship do not exist, probably because young larvae rarely consume the entire food source even at very high population densities. The effects of host density were demonstrated for the NPV infection rates in populations of *G. hercyniae* in mid-Wales.[65] Here the transformed

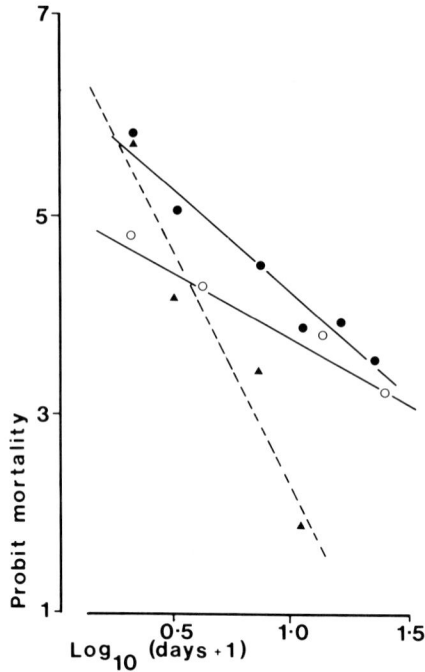

FIGURE 9.   Decay of baculovirus deposits exposed on leaves to solar ultraviolet light (UV). The relationships between probit mortality of test larvae and $\log_{10}$ time of exposure to UV. ● *Trichoplusia ni* nuclear polyhedrosis virus (NPV) on cabbage in Canada,[59] ○ *Heliothis* sp. NPV on corn in U.S.,[60] ▲ *Pieris rapae* granulosis virus on cabbage in Canada.[61] (Reproduced from Entwistle, P. F. and Evans, H. F., *Comprehensive Insect Physiology, Biochemistry and Pharmacology*, Vol. 12, Gilbert, L. I. and Kerkut, G. A., Eds., Pergamon Press, Oxford, in press. With permission.)

infection rates (y) for a number of generations were computed thus including population densities ranging from 5 to 700 per tree subsample. Additional data since that time include densities of as low as one larva per sample tree. There was a reasonable correlation between y and host density although the effect of delayed density dependence of disease resulted in a delay between declining host density and decreasing rate of infection. A similar effect was observed by Kaupp[42] for the NPV of *N. sertifer*. He studied infection rates in populations with variable densities and transformed values (y) varied from 0.01 to 0.09 at a range of host colony densities from 1 to about 70 per tree. It is not possible to distinguish primary and secondary inoculum effects in these examples except in so far as the first appearance of infection tended to be earlier relative to larval emergence when high host densities were present. This indicates more efficient exploitation of the primary inoculum by the higher host numbers as well as the potentially greater quantities of inoculum present as a result of higher infection in the previous season.

2. Secondary Inoculum

The most important effects of host density are those related to production and cycling of secondary inoculum, which will have a large influence on events late in a larval generation.

It has been established that the number of polyhedra produced during the infection process is extremely large, even in young larvae. This quantity increases enormously with increasing body weight. Larvae that die as a result of ingesting primary inoculum increase the quantity of virus present locally by factors as high as 86,000.[44] These localized sources of inoculum provide dosages well in excess of those required for infection in virtually all larval stages. This can result in an exponential increase in infection like those illustrated previously in Figure 2. It is during this phase that host density becomes the key factor in the spread of the disease through the two parameters stressed by Andrewartha and Birch[24] — distribution and abundance.

a. Distribution

Greater host density inevitably results in wider distribution. Increased numbers of host larvae are available to exploit the increasing quantity of secondary inoculum being produced, thereby increasing the effective "searching efficiency" of the virus. This acts as a density-dependent mortality factor so that the proportion of hosts infected increases as the density of hosts increases.

Another important factor is the way in which the host feeds. Colonial feeders, such as *N. sertifer* are known to transmit infection very rapidly through a population as a result of secondary cycling of inoculum. Thus Kaupp[42] demonstrated that a single diseased larva introduced into a feeding pine sawfly colony eventually resulted in the spread of infection throughout the colony. The rate of spread of infection between colonies on the same tree was also density-dependent above an apparent threshold of about two colonies on a 2-m high tree. Efficient spread of infection between larvae has also been shown in other species whose feeding habits tend to bring the larvae into close proximity to each other. Examples include the NPVs of *Neodiprion lecontei*,[38] *Malacosoma fragile*,[66] *Dasychira basiflava*,[67] and *Euproctis chrysorrhoea*[68] all of which feed gregariously at some time during their life cycles.

b. Abundance

Interaction between larvae is dependent upon host distribution and abundance. The mechanism of interaction is through secondary cycling of inoculum as larvae become infected and eventually die. The release and availability of this inoculum to other larvae is a critical component of the epizootic, and the mechanism of release and distribution of virus differs considerably between insect orders depending on the organs infected.

When infection is localized to the midgut the release of inoculum is gradual as the epithelial cells break down. Excellent examples are provided by larval sawflies and adult rhinoceros beetles. In the case of sawflies, NPV PIBs are released from the gut cells within a relatively short period following infection and are then voided continuously in the feces. In addition, degeneration of gut cells is accompanied by appearance of PIBs in the fluid emitted in defensive regurgitation.[69] Similarly, infected *Oryctes rhinoceros* adults void large quantities of the nonoccluded baculovirus in their feces as a result of gut breakdown.[70] Secondary inoculum produced in this manner is capable of being disseminated rapidly through the host population well before the peak release of virus that results when larvae die from the disease.

When most of the body tissues are infected, but the gut is a site of initial replication only, as in larval Lepidoptera, virus release can be described as inundative. Virtually all the virus produced in a larva becomes available over a short period following rupture of the fragile body wall after larval death. Secondary inoculum of this type produces very large, but localized, reservoirs of virus that can become readily available for consumption by a large host population. In some situations the effect may be increased by the attraction of larvae to the virus-killed cadaver as is the case in *Mamestra brassicae*.[71] Conversely, Capinera et al.[72] showed that insect remains acted as a phagodeterrent in lyophilized preparations of *L.*

dispar NPV. In this example the effect was not specifically related to *L. dispar* insect remains since a similar response was observed when *Hyphantria cunea* larval remains were tested against *L. dispar* larvae. These contradictions may reflect the host species studied or even whether the attraction to newly killed larvae changes to deterrency when putrefaction sets in.

It is generally agreed that baculoviruses act in a delayed density-dependent manner thereby inducing maximum mortality some time after the host population has reached a peak.[22,41,73] This will be discussed more fully later, but the concept is clearly central to the present consideration of host density effects. The appearance of epizootic virus growth in a population will therefore reflect the numbers of susceptible hosts and the quantities of primary and secondary inoculum present. Populations with discrete generations have periods when no hosts are present, and the production of secondary inoculum ceases. Infection growth is then cyclic with respect to the individual host generations.

When populations are multivoltine and have overlapping generations, the situation becomes more complex. In these cases there is an ongoing addition of new susceptible larvae to the population. Infection is governed by larval numbers and age distribution, which in turn define the quantity of secondary inoculum produced. Under such conditions the period over which released inoculum is required to be effective is reduced, and the probability of further infection is increased.

The tendency in overlapping host populations will therefore be for infection to buildup with time to a peak level which, unlike hosts with discrete generations, will tend to be maintained at this high level until host larval recruitment is completed and the host population declines. Such conditions pertain in warmer climates where host voltinism is not limited and it is possible to have as many as 12 generations per year with considerable overlap between them. This is the case for *Heliothis* sp. and *Spodoptera* sp. in Thailand[74] and at a less serious level for these species in Indonesia.[75]

Unfortunately, quantitative data on the role of baculoviruses in hosts with overlapping generations is scant. Ehler[76] showed that when *T. ni* had three generations on cotton, although they did not overlap widely, NPV infection generally increased between one generation and the next. Thus, despite adverse conditions for virus retention on cotton, sufficient virus was produced in the first two generations to induce increased mortality in the third. Samples taken when populations were very large in the third generation yielded up to 91% NPV mortality in some plots.

D. Specificity

Baculoviruses that can replicate in more than one host species have a considerable advantage over those that are monospecific. These advantages lie in the larger numbers of hosts available for virus replication and in the increase in inoculum that can result. Other factors such as wider spatial or temporal contiguity of host and virus will also have significant effects on the population dynamics of the virus.

Specificity of baculoviruses has not been studied in sufficient detail to state unequivocally that certain viruses have particular host ranges. Nevertheless, as Ignoffo[77,78] pointed out, viruses that produce inclusion bodies, and the occluded baculoviruses in particular, tend to be more host specific than viruses that do not produce inclusion bodies. Within the baculoviruses, the GVs are regarded as being most specific while the nonoccluded baculovirus of *O. rhinoceros* is also relatively restricted in its host range. The NPVs are less specific but range from those infecting single hosts such as *N. sertifer* NPV[38] through those that infect a few species in the same genus, such as *Heliothis* sp. NPV[79] to those such as *A. californica* NPV that have a relatively wide host range.[62]

Not all cross transmission can be regarded as significant since the dosages required to infect secondary hosts may be unrealistically high. Thus although *A. californica* NPV is known to infect at least 34 species of the order Lepidoptera, the dosages required to infect the majority of these are very high. For example, in excess of 5×10^5 PIBs were required to infect *Galleria mellonella* larvae.[80] When this is compared with the LD_{50} for fifth-instar *T. ni* larvae of as low as ten polyhedra,[81] the dosages required for some secondary hosts would require massive reservoirs of virus in the field.

There also remains the question of identity of progeny viruses. Few data are available on cross infection where the input and output viruses have been characterized fully. Indeed it is only the advent of refined techniques such as restriction endonuclease analysis (REN)[82] that has allowed meaningful conclusions to be drawn from cross-infection studies. True cross infections do occur and have been established beyond doubt. For example, *T. ni* NPV will replicate in *Spodoptera exigua* without modification of polyacrylamide gel electrophoresis profiles or serological reactions.[83] This was expanded by Shapiro[84] in discussing possible use of alternate hosts for virus production. This is an important area of research with regard to optimizing virus productivity for field control trials but also provides valuable data on potential ecological relationships between baculoviruses and host range.

Obviously the laboratory demonstration of a wide host range in a virus can have ecological significance only if the alternative hosts are within the same environment. The fact that *A. californica* NPV can infect *S. littoralis* or *M. brassicae* has no epizootiological meaning since the insect hosts never occur in the same locality. We must, therefore, consider only examples from ecosystems where hosts and viruses coexist so that they come into contact for at least some of the susceptible period of the alternative hosts.

Evans and Harrap[85] have discussed the role of secondary hosts in virus persistence. Synchrony of host and virus is a central theme in the ecology of baculoviruses, and an expanded host range acts to increase the period of synchrony and hence of virus replication and production. It was shown above that greater replication results in acceleration of infection growth by more efficient secondary inoculum production. When virus is restricted to a single host species, infection will be limited by voltinism of the host and thus ultimately by host density. The presence of secondary hosts increases both the spatial and temporal range of the virus.

Dispersal of virus is likely to occur when the primary and secondary hosts are in different localities and some agent acts to spread virus between the two (see next section for discussion of dispersal). Infectivity in the secondary host might differ considerably from that in the primary host, but this acts only on the probability of a particular level of infection being initiated. Once the disease is established, it proceeds under the limits "driving" any epizootic. Surprisingly, there are no unequivocal examples of infection passing to spatially separated secondary hosts, presumably as a result of the difficulty of establishing beyond doubt that the event had occurred and was not simply normal development of homologous virus in the secondary host.

Secondary hosts may also be separated in time from the primary host. This has mixed implications for synchrony of host and virus since the virus has to persist or disperse in the environment before the secondary host becomes available for infection. If this period is prolonged, then the quantity of virus present may decline to levels too low to initiate infection in the secondary host, especially if that host is intrinsically less susceptible. However, the period could be shorter than that between generations of the primary host. Replication in the secondary host would, therefore, increase the general level of virus persisting in the environment. This would, in turn, increase the probability of maintenance of infection in the primary host. Examples of this occurring in the field are not available although circumstantial evidence for cross transmission of NPV in nymphalid butterflies[86] and pests of crucifers[21] exists and the probable temporal spread over generations of the various species would fulfill the conditions for secondary host transmission through time.

E. Transmission

Transmission of baculoviruses is related to all facets of their ecology. The mechanisms of transmission will be considered under virus persistence, whereas the outcome and ecological implications of the transmission process are discussed in this section.

Baculovirus transmission is generally divided into spatial and temporal components, although the two interact in a single continuum. It is, however, convenient to consider each mode of transmission individually before attempting to bring them together in a unified concept.

1. Horizontal Transmission

In essence, horizontal transmission effectively means virus dispersal. The ability of a virus to disperse, or more appropriately to be dispersed, is an important characteristic if it is to remain in contact with a changing host population or spatially separated subpopulations of the host. In addition, the parameter defining dispersal is generally infection in the host population, and this will obviously depend entirely on the presence of hosts. Virus itself may be present when no hosts are available to act as a means of dispersal, and hence the extent of horizontal transmission may be underestimated. Three phases of dispersal, primary, secondary, and tertiary may be distinguished regardless of the scale involved for a particular host and virus. Thus it is possible to compare dispersal as a mechanism in the laboratory, in small field plots, or over large areas on the same conceptual basis. Entwistle et al.[18] have published a detailed comparative analysis of these three phases in their description of *G. hercyniae* NPV dispersal. These data are used extensively in the following discussion, with the addition of further information where appropriate.

a. Primary Dispersal

Dispersal in its primary phase can be observed most clearly from isolated epicenters of disease in initially healthy host populations. This limitation means that few data are available on primary dispersal, and even these tend to be the result of artificial initiation of infection. Entwistle et al.[18] quoted three examples of virus-host interactions that could provide quantitative primary dispersal data while extrapolation of data from Reed[87] provides a fourth. The examples are the NPV of *G. hercyniae* in Wales[18] and Canada,[17] the NPV of *Malacosoma disstria* in Canada,[88] the nonoccluded baculovirus of *O. rhinoceros* in the Tonga Islands,[89] and the GV of *Phthorimaea operculella* in Australia.[87]

Despite the disparate nature of these examples, the pattern of primary dispersal was similar in them all (Figure 10). Here it is possible to identify a rapid decline in disease incidence with distance from the epicenter, the scale depending on the virus and host studied. During the following host generation, disease incidence expanded into a wave-like form with a tendency for a decline in mortality at the disease epicenter. This was a prelude to the secondary phase which will be considered later.

The curves generated in Figure 10 are difficult to compare directly, a problem familiar to plant pathologists studying primary dispersal of fungal diseases.[39] However, using techniques of analysis developed by them, it is possible to normalize the infection curves using a logarithmic transformation of the data.[90] The procedure is, therefore, to plot $\log_{10}$ (proportionate disease incidence + 1) against $\log_{10}$ (distance from the epicenter + 1). The linear relationships between the transformed variables are then described by

$$\log_{10} y = a - b(\log_{10} x) \qquad (1)$$

The slopes of such relationships are negative and are described as the gradients of dispersal. The steeper the gradient, the shorter the distance covered per unit time.

Transformations of the data in Figure 10 are plotted in Figure 11 where the dispersal

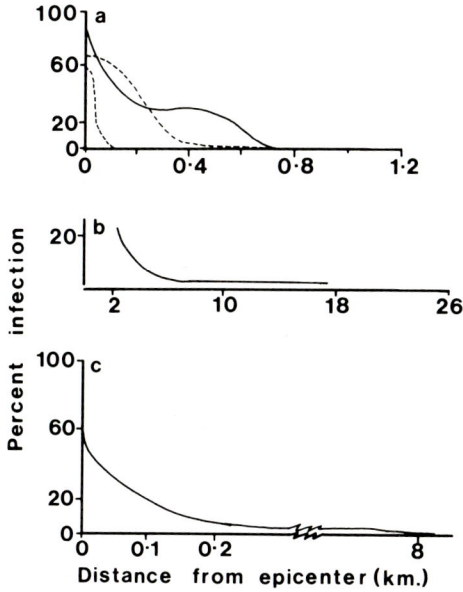

FIGURE 10.   Patterns of primary dispersal of baculovirus disease through various host populations. (a) Nuclear polyhedrosis virus of *Gilpinia hercyniae* in Wales (—)[18] and Canada (---)[17]; (b) nonoccluded baculovirus of *Oryctes rhinoceros* in Tongatapu, Tonga Islands;[89] (c) granulosis virus of *Phthorimaea operculella*[87] (note split x-axis). (a and b reproduced from Entwistle, P. F., Adams, P. H. W., Evans, H. F., and Rivers, C. F., *J. Appl. Ecol.*, 20, 473, 1983. With permission.)

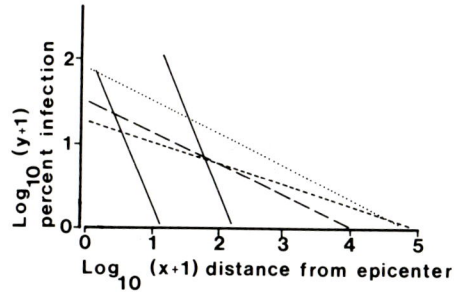

FIGURE 11.   Logarithmic transformations of primary dispersal patterns for baculovirus diseases. — Nuclear polyhedrosis virus (NPV) of *Gilpinia hercyniae*, b = -1.9;[18] ---- NPV of *Malacosoma disstria*, b = -0.37;[88] ..... granulosis virus of *Phthorimaea operculella*, b = -0.39;[87] --- nonoccluded baculovirus of *Oryctes rhinoceros*, b = -0.25.[89] (All lines except Reference 87 reproduced from Entwistle, P. F., Adams, P. H. W., Evans, H. F., and Rivers, C. F., *J. Appl. Ecol.*, 20, 473, 1983. With permission.)

gradient values are given for each curve. Information from a study of the dispersal of *N. sertifer* NPV in Scotland indicated a gradient of -0.87[91] compared with -1.98 for the NPV of *G. hercyniae* in Wales and Canada. The reason for the greater dispersal of *N. sertifer* NPV is unclear although the gregarious feeding habits of the larvae may tend to result in relatively larger numbers of infection-bearing adults.

The two examples of primary dispersal of baculoviruses in lepidopterous hosts indicated close similarity in gradient values even though the ecology of the insects themselves were so dissimilar that no strong inference can be drawn from this. It was postulated by Stairs[88] that the main mechanism of dispersal of *M. disstria* NPV was adult moths, and the slope of the gradient would therefore have been defined to a considerable extent by their flight capacities. Other mechanisms such as wind dispersal of contaminated dust were postulated for the GV of *P. operculella*.[92]

The extremely low gradient of dispersal for *O. rhinoceros* baculovirus relects the strong capacity for flight in the adult stage confirming that it is through this mechanism that both primary and secondary dispersal takes place. Indeed, release of infected adults into the environment was the initial step in the spread of the baculovirus through the beetle populations on Tongatapu shown in Figure 10.[70,89]

Gradients of primary dispersal, therefore, provide quantitative descriptions of the capacity of a baculovirus to disperse through populations of its host. In general, the gradient will depend on the mechanism driving the observed dispersal but will also be limited by the density of the host. Success of virus dispersal must be dependent on the presence of sufficient susceptible host stages to support replication of virus, the incidence of infection then being dependent on the parameters discussed in the sections above. No clear relationship with host

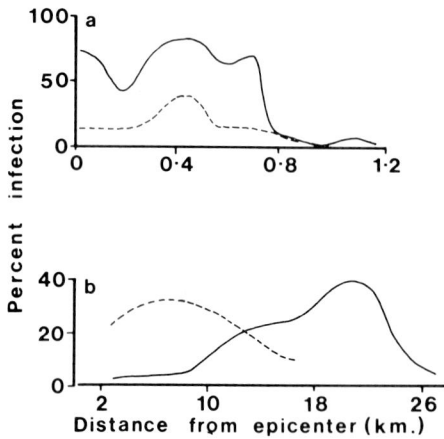

FIGURE 12. Patterns of secondary dispersal of baculovirus disease through host populations. (a) Nuclear polyhedrosis virus of *Gilpinia hercyniae*, --- Canada,[17] — Wales;[18] (b) nonoccluded baculovirus of *Oryctes rhinoceros* --- 250 days from initiation, — 450 days from initiation.[89] (Reproduced from Entwistle, P. F., Adams, P. H. W., Evans, H. F., and Rivers, C. F., *J. Appl. Ecol.*, 20, 473, 1983. With permission.)

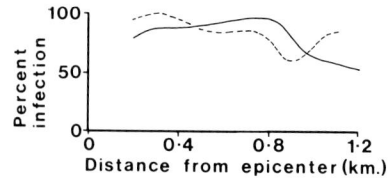

FIGURE 13. Patterns of tertiary dispersal of the nuclear polyhedrosis virus of *Gilpinia hercyniae* in Wales. --- NE transect, — SW transect. (Reproduced from Entwistle, P. F., Adams, P. H. W., Evans, H. F., and Rivers, C. F., *J. Appl. Ecol.*, 20, 473, 1983. With permission.)

density was observed for dispersal of the NPV of *G. hercyniae* in mid-Wales, although primary dispersal in very low populations was not studied.[18] However, in all cases the clear pattern of primary dispersal is lost as soon as the host population responds to the wider distribution of virus and a smooth transition to secondary dispersal takes place.

b. Secondary Dispersal

This phase is characterized by the development of a "wave-like" form in the untransformed data. This is normally accompanied by a decrease in disease incidence near the epicenter, probably reflecting density-dependent response to the decreasing host population. Progress of the wave with time is similar for the NPV of *G. hercyniae* and the nonoccluded baculovirus of *O. rhinoceros*, the only examples where data were available (Figure 12). The scale of dispersal is the main difference between the two examples with disease incidence in *O. rhinoceros* populations peaking 450 days from initiation at 21 km from the epicenter. This reflects both adult flight capacity and overlapping generations of the host.

Secondary dispersal waves are dependent on inoculum production in the primary phase and on the continuing presence of susceptible hosts in the areas in front of the wave. Delayed density-dependence of disease incidence may be partially compensated for by large bodies of inoculum persisting to provide sufficient virus to induce substantial mortality in the reduced populations near the epicenter.

c. Tertiary Phase

Entwistle et al.[18] defined a tertiary phase of dispersal known as an "interference phase" typified by the lack of clear pattern to the spread of disease which was eventually prevalent throughout the host environment. Such a pattern of disease incidence is illustrated for *G. hercyniae* NPV in Figure 13. Here, there was a generally high level of disease throughout the dispersal study area. Virus was, in fact, present over a large area in sawfly populations in several forests.[93]

It is reasonable to assume that the poorly defined tertiary phase reflected disease spread

Table 3
VERTICAL TRANSMISSION OF BACULOVIRUSES

Virus[a]	Host	Mechanism	Ref.
NPV	*Gilpinia hercyniae*	Infected adult gut	38
	Neodiprion sertifer	Infected adult gut	38
	Colias philodice eurytheme	Paste applied to female genitalia	96
	Heliothis zea	Virus fed to adults	97
		Virus dust in light traps	98
	Trichoplusia ni	Adults sprayed with virus suspension or fed virus in sugar solution	99
GV	*Sesamia nonagrioides*	Larvae sublethally infected virus passed as external contaminant	100
	Pieris brassicae	Virus applied to female genitalia	101
NOBV	*Oryctes rhinoceros*	Infected adult gut	102

[a] NPV = nuclear polyhedrosis virus; GV = granulosis virus; and NOBV = nonoccluded baculovirus.

from a multiplicity of disease centers that were generated by discontinuous dispersal of virus by adult sawflies[15,94] or birds.[95] Spread of this nature would eventually overlap through primary and secondary dispersal from the discontinuous secondary epicenters, resulting in the confused picture characteristic of the tertiary "interference phase".

Horizontal transmission of baculoviruses, therefore, provides valuable contrasts with the dispersal characteristics of other pathogens. It is possible, as shown above, to use the methods of analysis of the other disciplines to quantify the patterns of dispersal in baculoviruses, although there are few data available on which to use these approaches.

2. Temporal Transmission

This is a broad category that includes all aspects of virus transmission between generations of the host insect. It therefore encompasses vertical transmission in the classical sense of passage of virus via the adult stage to its progeny[96] as well as virus persistence as a link between temporally separated susceptible hosts.[85] Both categories of temporal transmission provide a source of primary inoculum for the new host generation. Mechanisms of virus persistence will be considered in the next section, but some facets relating to the quantitative transmission of virus will be included here.

Vertical transmission is arguably the most efficient means of passage of virus to newly hatched host larvae. In the field, this stems from the direct link between infection in the late larval stages passing via the adults to the progeny thus avoiding the losses likely to occur through persistence in the external environment. Examples of successful virus transmission via the adult stage are included in Table 3, although none of these cases can be regarded as transovarian transmission by incorporation in the egg.[96]

Quantitative links between virus present in the adults and infection in progeny are few. Melamed-Madjar and Raccah[100] reported that 62% of *Sesamia nonagrioides* females fed NPV PIBs transmitted infection to their progeny. Similarly *Heliothis zea* females fed with various concentrations of NPV transmitted virus in proportion to these dosages with the rate of transmission declining with time.[97] However, it has not been demonstrated conclusively that sublethally infected lepidopterous larvae can give rise to infection bearing adults. Indeed, the effects of late larval infections tend to be deleterious in the adult stage. *S. littoralis* adults appeared normal but fecundity and fertility were reduced relative to normal females as a result of sublethal larval NPV infection.[33] The NPV of *Epiphyas postvittana* produced a similar effect with an additional sex ratio bias towards males.[103] Even where vertical transmission has been demonstrated, the adults involved tend to be reproductively inferior to healthy adults as in *O. rhinoceros.*[104] True transovum transmission appears to be confined to those species having gut infections in the adult stage, thereby limiting examples to sawflies, in the cases of occluded baculoviruses, and rhinoceros beetles, in the case of nonoccluded baculoviruses. It is probable that vertical transmission by lepidopterous adults in the field results from external contamination of newly emerged females.

The ecological implications of vertical transmission relate particularly to threshold host densities for virus infection. Although density-dependence of baculovirus infection has been demonstrated[73,105] it is normally regarded as being operative above a threshold host density characteristic of the virus and host being studied. For example, Kaupp[42] demonstrated that development of NPV infection in *N. sertifer* populations required densities in excess of two feeding colonies per tree. Vertical transmission operates independently of host threshold density since virus is passed directly to progeny. It is, therefore, potentially an important mechanism for maintenance of virus in low-density host populations.

In sawflies the conditions required for passage of virus through the adult tend to occur only at relatively high host densities. The probability of NPV infection in the adult stage of *G. hercyniae* appears to be highest following epizootic buildup of infection while the host population is still relatively large.[38] This reflects the large quantities of secondary inoculum produced under these conditions which provide the dosages necessary (10^5 to 10^6 PIBs per larva) to infect late stage larvae.[106] At lower host densities the probability of such conditions being met is minimal and infection in the adult stage is unlikely. Natural passage of NPV via sublethal larval infection through adult Lepidoptera has not been documented. For example, Elmore and Howland[99] were unable to demonstrate NPV transmission via adults surviving epizootics in populations of *T. ni* larvae, despite the large quantities of virus present as a consequence of high infection rates.

Persistence of virus outside the host insect is the most common means of transmission from one period to another, and thus from one generation to another. This subject has been reviewed by Evans and Harrap[85] who emphasized the quantitative role of persistent inoculum in the development of epizootics. Inoculum may persist between host generations by a variety of means, but the key parameters defining the success of persistence are the quantities of virus remaining viable and the probability of inoculum being encountered by susceptible hosts.

Study of an epizootic of NPV in *G. hercyniae* populations in Wales has provided quantitative data on the amounts of virus persisting between host generations. Evans and Entwistle[107] demonstrated a direct link between the amount of inoculum produced in 1 year and primary infection in the following year (Figure 14). Virus retention was relatively poorer at higher initial inoculum levels possibly reflecting a finite carrying capacity on foliage of different tree species. The fate of virus on foliage was measured during the winter months when it was noted that, following an initial buildup as larval corpses disintegrated, the level of infection declined rapidly towards the beginning of the new larval season (Figure 15). During this period the half-life of impure virus deposits was 55 days,[107] confirming the remarkable capacity of spruce foliage for virus retention during the winter months when no larvae were available for virus replication. Decline of virus was exponential over the 3 months immediately preceding the new larval season. It therefore follows that large quantities of potential inoculum are lost at this stage which will clearly influence the buildup of primary infection in the new larval generation. Evans and Harrap[85] carried out approximate calculations of

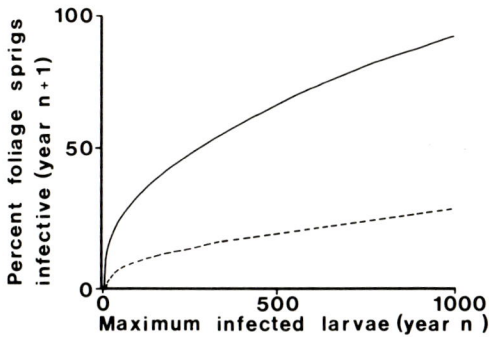

FIGURE 14. The relationship between maximum numbers of nuclear polyhedrosis virus infected *Gilpinia hercyniae* larvae in 1 year and the percent of foliage sprigs infective by bioassay to neonate larvae in the following year. — Norway spruce, --- Sitka spruce. (Reproduced from Evans, H. F. and Entwistle, P. F., *Microbial and Viral Pesticides*, Kurstak, E., Ed., Marcel Dekker, New York, 1982, 449. With permission.)

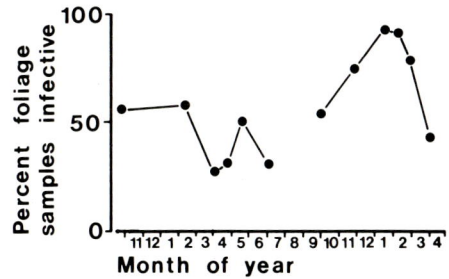

FIGURE 15. The levels of virus on Norway spruce foliage detected by bioassay of foliage sprigs to neonate larvae. No samples were taken during the larval season. (Reproduced from Evans, H. F. and Entwistle, P. F., *Microbial and Viral Pesticides*, Kurstak, E., Ed., Marcel Dekker, New York, 1982, 449. With permission.)

this loss and estimated that there was up to a 10,000-fold decrease in PIBs when measured as numbers utilized effectively in the process of primary infection. Similarly, the NPV of *Orgyia leucostigma* was retained on balsam fir foliage in an infective state over the winter months, although no measure was given of the rate of decline during that period.[108]

Absence of foliage on deciduous trees during the winter does not necessarily imply that sites for virus retention are absent. NPVs of various lepidopterous species have been shown to persist well on the nonfoliage sites of deciduous trees, but quantitative data on the numbers of PIBs retained are lacking. However, Thompson and Scott[58] calculated the total quantities of *O. pseudotsugata* NPV PIBs produced per ha, during a virus epizootic in populations of the moth on Douglas fir. They concluded that although up to approximately 3.8×10^{15} PIBs were produced per ha very little of this inoculum source remained on the tree, the rest being incorporated in soil and duff below. Even at these high initial productivities, only 0.16% of the PIBs in the soil layer remained active after 1 year, a finding similar to that for *M. brassicae* NPV in agricultural soil.[109]

Soil is, in fact, regarded as the most stable long-term site of virus retention, and examples have been quoted in agriculture *(T. ni* NPV under cole crops),[110] forestry *(O. pseudotsugata* NPV),[58,111] and grassland *(Wiseana* sp. NPV).[57] Quantitative assessment in terms of viable PIBs produced variable conclusions. For instance Jaques[112] reported a loss of 85% in viable virus in soil over a 318-week period in studies of *T. ni* NPV. In contrast, Payne[113] and Evans[109] studying *P. rapae* GV and *M. brassicae* NPV, respectively, showed that viable soil-born virus declined exponentially over a 1-year period to about 2% of the original concentration. Retention of virus is therefore variable, and it is also pertinent to question the ecological role of soil as a reservoir. Links between virus in soil and infection in populations of the hosts on the plants above must be established before the ecological significance of soil virus can be assessed fully.

Direct contact between plants and soil offers the highest probability of movement of virus from soil to foliage. Jaques[110] showed that cabbages grown in soil containing *T. ni* NPV were sufficiently contaminated to induce epizootics in larval populations. Direct application of NPV to soil below soybean plants controlled *Pseudoplusia includens* and the virus was shown to persist between host generations.[114] However, Baugher and Yendol[115] and Payne[113] carrying out similar experiments on *T. ni* NPV and *P. rapae* GV, respectively, were unable to establish a quantitative link between soil virus and infection of larvae on the plants above.

It is probable that variation in these results is attributable as much to host density as it is to presence of virus in soil. Indeed the role of host density becomes extremely important if soil-borne virus is to be effective in situations where there is no direct contact between foliage and soil. The demonstrated long-term retention of *O. psuedotsugata* NPV in soil below Douglas fir trees was postulated to be an effective source of inoculum for initiation of infection in expanding host populations.[58] It must be concluded that temporal transmission through the medium of virus retention in soil is a long-term mechanism which is significant when host density is sufficiently high to encounter the limited quantity of virus cycled back to the host plants.

Horizontal and temporal transmission have, so far, been considered separately but, as mentioned earlier, are part of the same continuum. Development of disease in a host population will obviously progress with time but is also subject to dispersal as virus can come into contact with the remaining susceptible hosts in the population. Such a three-dimensional framework has been envisaged by Entwistle for the epizootiology of *G. hercyniae* NPV.[116] Entwistle produced a schematic representation of the likely course of virus development in space and time incorporating a buildup of infection over several host generations followed by eventual collapse of the host population. At the same time, the front of the epizootic wave spread outwards leaving a declining disease incidence at the center. Such a pattern will obviously be specific to the host and virus being studied. Cycles of infection and the scale of dispersal will be governed by many of the ecological factors discussed in this chapter.

F. Baculovirus Persistence

When the ecology and epizootiology of baculoviruses are examined, a single factor, that of virus persistence, dominates the dynamics of virus-host interactions. Most of the observed relationships result from the ability of the virus to persist between periods of host availability. In this section, the mechanisms of virus persistence are discussed and the significance of some of these in relation to virus ecology is considered. In this context two main categories can be recognized, namely, biotic and abiotic.

1. Biotic Persistence

The role of vertical transmission via the host adult has already been discussed in relation to spatial and temporal passage. Other biotic agencies of virus persistence include secondary hosts, predators, and parasitoids.

Effective persistence of virus in secondary hosts depends on replication in that host and, more significantly, transmission back to the original host. The question of specificity has already been considered, but few examples are available to illustrate the potential of secondary hosts as a means of virus persistence in the field. Circumstantial evidence of cross transmissions of an NPV between a number of nymphalid butterflies in the same habitat and on the same host plant was provided by Cunningham.[86] He assumed that the high degree of serological relatedness of the NPVs in the various species reflected infection by baculovirus common to them all. Tanada[21] concluded, from laboratory cross-transmission tests, that several alfalfa pests in California shared GV and NPV diseases, each contributing to a common pool of persistent virus. *P. brassicae* and *P. rapae* can be infected by the same GV, and could therefore act as secondary hosts in the field.[113]

Predators, by virtue of their direct action on the host insect, act as possible regulatory factors in their own right. In this respect, they may tend to reduce the potential virus productivity in a host population through removal of susceptible hosts or by loss of inoculum through feeding on infected or virus-killed hosts. They can, therefore, be regarded as a competitive element having an adverse influence on the replicative success of the virus. On the other hand, predators (and parasitoids to be considered later) can also indirectly increase

Table 4
TRANSMISSION OF BACULOVIRUSES BY PREDATORS

Virus[a]	Predator	Class	Host insect	Ref.
NPV	*Mitopus morio*	Arachnida	*Gilpinia hercyniae*	91
	Ephippiger bitterensis	Insecta	*Lymantria dispar*	117
	Nabis tasmanicus	Insecta	*Heliothis punctigera*	118
	Oechalia schellenbergii	Insecta	*H. punctigera*	119
	Podisus maculiventris	Insecta	*Trichoplusia ni*	20
GV	*Arboela ibis*	Aves	*H. armigera*	121
NPV	*Cyanocitta cristata*	Aves	*L. dispar*	122
	Parus ater	Aves	*G. hercyniae*	123
	Passer domesticus	Aves	*T. ni*	124
	Apodemus sylvaticus	Mammalia	*G. hercyniae*	91
	Peromyscus leucopus	Mammalia	*L. dispar*	122

[a] NPV = nuclear polyhedrosis virus; GV = granulosis virus.

the potential of a virus. This is achieved by increasing virus dispersal and survival capacity through fuller exploitation of available host populations.

It is generally true that predators have been studied from the viewpoint of virus dispersal, although it is also true that they act as a means of virus persistence outside the host itself.[85] The mechanism of predator-aided persistence is generally through passage of infective virus through the gut. Examples of predator transmission of baculoviruses are given in Table 4. It is clear that the phenomenon is widespread in many predators ranging from insects to small mammals. Variability in the length of time and quality of this effect appears to be related to the predator group being studied. However, no systematic investigations of the relationships of predator gut pH and enzyme activity to baculovirus degradation have been carried out. Nevertheless, there is a link between the rate of feeding by predators and the appearance of virus in feces following ingestion of infected hosts. Cage tests of titmice (*Parus* spp.) feeding on NPV-infected *G. hercyniae* larvae indicated that virus first appeared within 30 min of ingestion.[95] Lautenschlager and Podgwaite[125] demonstrated that small mammals in the genera *Apodemus, Blarina,* and *Peromyscus* passed infective virus in their droppings in about 1 hr.

Following a single ingestion of virus, the length of time over which infective virus was passed in the feces varied considerably but extended to 24 hr for small mammals,[91,125] 4 days for the heteropteran *Nabis tasmanicus*,[118] 5 days for *Ephippiger bitterensis* (Orthoptera),[117] 6 days for coal tit, *Parus ater*,[91] and 15 days for the pentatomid *Oechalia schellenbergii*.[119]

The proportion of infective virus passed through the gut intact will also have a strong influence on the success of this means of persistence. In small mammals, the percentage survival can be as low as 0.05% in the flying squirrel *Glaucomys volans*, under 5% for the mouse *Peromyscus leucopus*, and as high as 15% for the raccoon *Procyon lotor*.[125] Survival of virus in the guts of birds appears to be somewhat higher, and up to 34% survival of NPV was reported after passage through the gut of house finch *Carpodacus mexicanus*.[125] Entwistle[91] expressed the opinion that percentage survival of baculoviruses in invertebrates was high, although no comprehensive studies had been carried out to determine this point.

Gut pH in predators is generally neutral to acid, i.e., < pH 7.0,[126,127] and it is therefore a little surprising that survival of virus following gut passage is so poor in mammals and birds. It is possible that pH, enzyme action, and physical damage may act together to produce the observed loss since each factor acting alone would be less likely to do so.

Predator involvement in virus persistence is an implied rather than a proven fact. It has been assumed that since virus is passed in an infective state in the droppings and that

Table 5

TRANSMISSION OF BACULOVIRUSES BY PARASITOIDS

Virus[a]	Parasitoid	Order[b]	Family	Mode of development	Host insect	Ref.
NPV	Apanteles melanos-celus	H	Braconidae	Endo	Lymantria dis-par	128
GV	A. glomeratus	H	Braconidae	Endo	Pieris rapae	129
NPV	Campoletis sonoren-sis	H	Ichneumonidae	Endo	Heliothis vires-cens	130
	Hyposoter exiguae	H	Ichneumonidae	Endo	Trichoplusia ni	131
	Lopyroplectus ob-longopunctatus	H	Ichneumonidae	Endo	Neodiprion sertifer	132
	Parasetigena sylves-tris	D	Tachinidae	Ecto	L. dispar	133
	Sarcophaga aldrichii	D	Calliphoridae	Ecto	Malacosoma disstria	134
	Voria ruralis	D	Tachinidae	Ecto	T. ni	135

[a]   NPV = nuclear polyhedrosis virus; GV = granulosis virus.
[b]   H = Hymenoptera; D = diptera.

droppings containing virus undoubtedly occur in the field, then passage to the susceptible host must occur. Conclusive proof of the final act of transmission to the host does not exist, but it seems quite likely that it must take place when host populations are high and hence more likely to encounter the localized sources of virus. As an example, Entwistle[91] calculated that during an NPV epizootic in *G. hercyniae* populations on spruce, each tree in the forest studied would receive about 95 infective droppings per year representing around 5×10^9 PIBs per tree. This is equivalent to 25 million early instar LD_{50} dosages per tree and represents a significant reservoir of viable inoculum.

Parasitoids act in similar way to predators in increasing virus dispersal and hence improving the probability of virus persistence. In general, the mechanism for this is by external con-tamination rather than gut passage. Examples of this phenomenon are summarized in Table 5 where ecto- and endoparasitic modes of development have been distinguished. In both categories external contamination of the parasitoid is involved, the direct contact of endo-parasitoids being regarded as the most effective. Most of the data have been accrued from the braconid parasitoids in the genus *Apanteles*. The efficiency of transmission in some members of this genus is such that up to eight hosts can be infected following oviposition in a single infected larva.[136]

Three possible mechanisms of transmission by parasitoids were identified by Raimo et al.[128] The first was inoculation of intact inclusion bodies via the ovipositor. However, Entwistle[91] has pointed out that injection of intact inclusion bodies into the hemocoel does not normally result in infection. The second mechanism postulated was injection of virions into the hemocoel, which in reality is probably not distinguishable from the first method. Thirdly, the parasitoid could be externally contaminated by contact with an infected host and in turn would contaminate the immediate environment of the susceptible hosts. Distin-guishing between these possible mechanisms is difficult and Raimo et al.[128] found no dif-ference in rate of transmission between the various mechanisms when *Apanteles melanoscelus* transmitted NPV to *L. dispar* larvae.

Virus productivity can obviously be increased if viable inoculum is transmitted between hosts, especially if there is a time delay spanning host generations. There is also a fine balance between the success of the parasitoid and that of the virus within a single host

Table 6
FACTORS INFLUENCING THE
PROBABILITY OF VIRUS PERSISTENCE
OUTSIDE THE HOST INSECT

Probability of persistence	Contributory factor
Low	Immediate loss of virus at lead fall or crop harvest; deposition in extreme environment, e.g., water, beneath rocks, etc.
Medium	Slow physical degradation on foliage/bark; UV inactivation; ingestion by predators/other hosts with no recycling
High	Protection in larval feeding nests, debris mats, etc.; stored products; soil

individual. Obviously development of one at the expense of the other will have marked consequences on the potential role of either as a mortality factor. Some parasitoids accelerate their development in infected hosts and result in decreased susceptibility of the host to virus infection. A fine example of this is the doubled LD_{50} in *Hyposoter exiguae* parasitized *T. ni* larvae exposed to NPV.[131,137] Endoparasitic larvae generally do not survive host death from baculovirus and as such have no part to play in further virus persistence and spread. Some tachinid parasitoids, on the other hand, seem able to survive host death and transmit virus to other hosts. For example, *Voria ruralis* passed NPV soon after emergence from its host, *T. ni*.[135] Entwistle[91] discussed competition between virus and parasitoids and concluded that, in general, the two compete for hosts to the exclusion of one or the other. In some cases, parasitoids have evolved avoidance mechanisms so that they generally oviposit in younger larvae that are less likely to be virus infected while others, like tachinids, can survive host infection or death.

Unequivocal field demonstration of the role of parasitoids is rare, and generally evidence stems from correlation between parasitoid incidence and virus incidence. For example, transmission of NPV to *L. dispar* via *A. melanoscelus* is assumed to take place since there is a strong correlation between the two variables.[128]

2. Abiotic Persistence

This category includes all examples of virus persistence outside the host insect, excluding biotic entities, especially those involving the physical environment in which the host feeds. Predominant among these is the host plant, which is the site of virus ingestion and hence the critical area for virus survival. Other reservoirs of virus persistence such as soil may also be significant, but secondary mechanisms are required to return this source of virus to the host plant if ingestion by the host is to take place.

Release of virus onto the host plant has already been discussed. Persistence of this inoculum will depend in the first instance on how and where it was released and, later, on environmental variables such as ultraviolet light, rainfall, temperature, and pH. The host plant itself will, therefore, have a major influence on virus survival regardless of whether virus release upon host death is gradual or inundative. Table 6 categorizes the likely probabilities of virus persistence in relation to the basic characteristics of host plants. The three probabilities reflect the degree of permanence of foliage either naturally or artificially through crop harvesting.

The category of high probability of persistence includes special cases where the host plant itself is incidental to the success of persistence. Soil as a reservoir has already been discussed briefly. A period of 41 years retention of activity appears to be the longest recorded survival of virus in soil (NPV of *O. pseudotsugata* beneath white fir trees[111]). The decline of virus in soil is gradual and a wide range of times for retention of viability have been described.[62] Combinations of clay content and pH appear to be the main factors responsible for the success of baculovirus persistence in soil. Adsorption to clay colloids is known to occur,[138] and Evans[109] demonstrated a quantitative link between percentage clay content and virus retention for *M. brassicae* NPV. The role played by soil pH has produced conflicting results with *T. ni* NPV; Thomas et al.[139] reported that this virus was inactivated more rapidly at lower pH values, while Jaques and Harcourt[140] concluded that soil pH from 5.5 to 9.0 had little effect on retention. Despite the lack of knowledge on the mechanisms of virus adsorption in soil, the phenomenon is well established, and soil is probably the most stable long-term reservoir for virus retention.

A high degree of protection is afforded to baculoviruses when larvae die in protected situations. Thus, virus that accumulates in debris mats or silk feeding nests is likely to have a high probability of persistence. Similarly the GV of Indian meal moth, *Plodia interpunctella*, is very stable in stored grain as a result of the high level of protection from adverse environmental fluctuations.[141]

The categories defining low and medium probabilities of persistence differ only in degree of permanence of the host plant surfaces since factors that cause attrition and act on the exposed virus will be the same in both cases. The single most important environmental factor acting on baculoviruses is the ultraviolet light component of sunlight (UV).

Inactivation of purified deposits of baculoviruses exposed to UV at a wavelength of 2500 Å is rapid, with GVs being less stable than NPVs.[142] Virus present on the plant as a consequence of larval mortality receives some protection from the disintegrating cadaver. Thus half-lives for pure and impure *G. hercyniae* NPV on spruce foliage were 38 and 55 days, respectively.[107] Thus, although baculovirus PIBs adhere very strongly to plant surfaces, their infectivity may be lost rapidly as a result of UV degradation.[62]

The presence of salt glands on the surfaces of some plants such as cotton[143] may signify yet another agent for inactivation of baculoviruses on plant surfaces. The effect on cotton illustrates this point. Cotton leaves produce dew having a high pH (>pH 9.0) and salts in high ionic concentrations which act together to inactivate baculoviruses.[144,145] Other plants such as cabbages appear to be relatively inert with respect to baculovirus degradation, and it is possible that cotton leaf surfaces represent a rare extreme as a microenvironment.

Some virus may also be lost through ingestion by other hosts which do not cycle the virus back to the original host. Thus, any competition for a limited food source may lead to reductions in the quantities of virus retained on the plant and hence to poorer persistence. Other possible factors tending to inactivate or remove virus include rainfall and temperature. Viruses on leaf surfaces are remarkably resistant to removal by rain. For example, *P. brassicae* GV was still present in large amounts after 5 hr of artificial rain and brushing with an artists brush.[146] Temperatures likely to be experienced in the field would appear to fall short of those known to inactivate occluded baculoviruses. However, there may be local increases in temperature that could act as a further source of virus loss. Apart from the chemical effects demonstrated on cotton, it is, therefore, probably correct to state that UV light is the most significant factor in baculovirus degradation and it is only through protection from it that viruses survive in the environment.

III. EPIZOOTIOLOGY OF BACULOVIRUSES

The discussion on ecology of baculoviruses presented above emphasizes the complex interactions that delimit disease progress in space and time — the epizootiology of baculoviruses. Although various stages of epizoosis can be distinguished, and will be used to illustrate specific points, they form part of a dynamic interaction that is constantly shifting in relation to the changing influences of the component parts. It is therefore proposed to

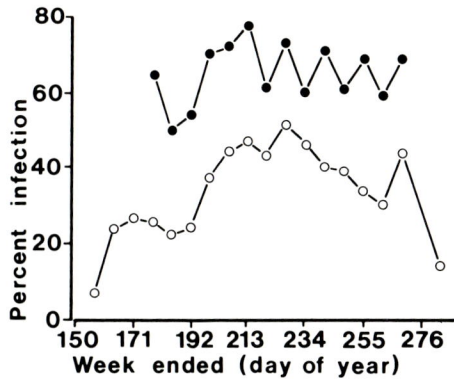

FIGURE 16.   Prevalence of nuclear polyhedrosis virus infection in field populations of *Spodoptera frugiperda* in southeastern Louisiana. ○ Mean percent infection. ● Epizootic threshold defined by 1.64 SD units above expected percent infection. (Reproduced from Fuxa, J. R., *Environ. Entomol.*, 11, 239, 1982. With permission.)

divide the disease process into three components, namely enzoosis, sporadic epizoosis, and cyclic epizoosis, bearing in mind the somewhat artificial divisions being created.

A. The Enzootic State

Baculoviruses can be considered enzootic in two ways reflecting first host population density and secondly virus effectiveness.

Density-dependent virus infection normally occurs above defined host population thresholds so that, conversely, disease remains enzootic below these thresholds. For example, *G. hercyniae* NPV has an extremely low disease incidence when host density is low.[147,148] Fuxa[149] has taken the concept of enzoosis further in his studies of NPV incidence in fall armyworm, *Spodoptera frugiperda*. Based on the definition coined by Steinhaus and Martignoni[150] who classed epizootics as disease with " . . . unusually large number of cases . . . " Fuxa identified a normal level of infection in fall armyworm populations in Louisiana (Figure 16). Fuxa also defined an epizootic threshold as 1.64 Standard Deviation (SD) units above the expected percent infection based on known characteristics of the disease. This is also illustrated in Figure 16. Interestingly, although primary inoculum levels in soil were shown to be similar, NPV infection developed more rapidly in armyworm populations on grass than on corn or sorghum, presumably reflecting more efficient passage of virus back to the host plant. A further example of enzootic disease incidence was provided by Ehler[76] in his comprehensive account of natural enemies of *T. ni* on cotton. He showed that NPV incidence remained below 5% in the majority of cases, although at times of peak host density this rose to about 90% and became epizootic. Ehler and van den Bosch[151] had earlier indicated that disease incidence could be density-dependent and it is therefore likely that a reasonably high threshold density for epizootic disease development was required for the NPV of *T. ni* on cotton.

A more comprehensive quantitative study of enzootic baculovirus disease was carried out by Wigley[32] in investigations of the NPV of winter moth, *O. brumata,* at Wistman's Wood in Dartmoor, U.K. He monitored host and virus (PIB) populations on eight trees over a 3-year period. Larval populations surviving attack by cecidomyid predators exhibited NPV infection ranging from 3.6 to 16.8%, but no clear relationship to the established populations on buds was noted. The disease played an insignificant part in population regulation at these observed host densities and could therefore be regarded as enzootic.

Table 7
EXAMPLES OF SPORADIC BACULOVIRUS
EPIZOOTICS

Virus[a]	Host insect	Host plant	Ref.
NPV	*Autographa californica*	Alfalfa	105
	Heliothis zea	Sorghum	43
GV	*Phthorimaea operculella*	Potato	152
	Pieris rapae	Cabbage	153
NPV/GV	*Scotogramma trifolii*	Russian thistle	154
NPV	*Spodoptera exigua*	Alfalfa	105
	Trichoplusia ni	Cole crops	155, 156

[a] NPV = nuclear polyhedrosis virus; GV = granulosis virus.

These examples of enzoosis are characterized by high virus pathogenicity and respond epizootically to increased host density above a certain threshold. Other viruses may be enzootic as a result of low pathogenicity, either intrinsically or for environmental reasons, so that they are relatively independent of host density and are less likely to become epizootic.

Such an example can be found in Tanada and Omi's[105] study on the epizootiology of NPV diseases of lepidopterous species on alfalfa. *Colias philodice eurytheme, S. exigua,* and *A. californica* were found commonly on alfalfa during the study, with density-dependent NPV epizootics being prevalent in the latter two species. The incidence of virus in *C. philodice eurytheme* populations was, however, consistently low and independent of host density. These characteristics implied low pathogenicity so that primary inoculum was not exploited sufficiently to allow rapid secondary virus production and epizootic growth, even at high host densities. The authors concluded that temperature, humidity, and rainfall had no marked influence on disease incidence and speculated that low efficiency of soil persistence may have been the critical factor under the conditions of the study. Martignoni and Milstead[96] quoted an LD_{50} to NPV in this species (*C. p. eurytheme*) of only 250 PIBs in the fifth instar indicating extremely high intrinsic pathogenicity. It therefore seems that poor temporal transmission or rapid inactivation of secondary inoculum contributed to the maintenance of an enzootic state for this virus.

B. Sporadic Epizootics

Sporadic epizootics are characteristic of those that occur in hosts that exploit unstable habitats such as agricultural crops. The instability of the habitat is a consequence of large environmental changes that tend to remove host feeding sites and with them sources of persistent virus or the host insects themselves. Such modifications to the normal host ecosystem obviously introduce density-independent mortalities that may interrupt natural cycles of host population abundance. Host instability may therefore have a related effect on virus population stability. There are many examples of virus epizootics where the host feeds on crop plants that are removed periodically (Table 7). As Franz[157] pointed out, removal of the crop plant reduces the possibility of the buildup of effective control. This does not preclude the development of epizootics as is evidenced in Table 7. It is more dependent on the ability of the disease to develop fast enough to prevent economic damage, itself a rather arbitrary parameter. Natural epizootics in unstable habitats therefore rely on stable reservoirs of virus as a source of primary inoculum and a high host population for density-dependent disease development.

Increases in the disease rate from generation to generation in the same year has been discussed above. This has particular significance in unstable habitats where "within-season" factors have strong influences on disease buildup since, as a result of crop harvesting, there

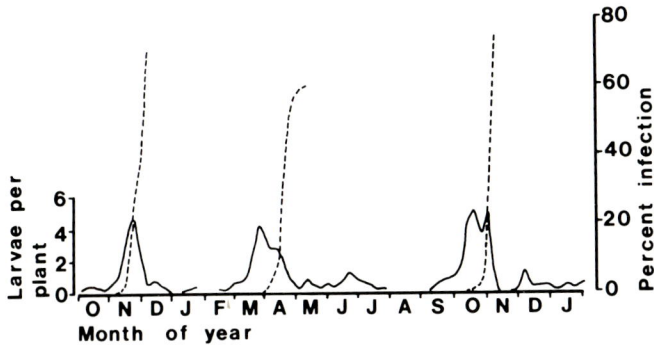

FIGURE 17.   Incidence of granulosis virus (GV) infection in populations of *Pieris rapae* larvae on cabbage plants in Australia. — Larval density per cabbage plant, --- percent GV infection in larvae.[153] (Larval density data reproduced from Hamilton.[153])

is no significant foliar persistence. An excellent example was provided by Hamilton[153] who demonstrated epizootics of GV in populations of *P. rapae* on cabbage in three successive host generations (Figure 17). The increase in disease, indicated by the transformed value (y), generally followed the peak population in each generation leading to rapid decreases in larval numbers per plant. Although the population in the second generation was similar to those in the other generations, GV appeared later, increased more slowly (y = 0.08 compared with 0.11 and 0.16), and reached a lower final incidence. Rainfall during the period of virus increase was considerably lower than normal and may have contributed to the reduced infection rate by decreasing the redistribution of soil-borne virus onto the plants.

High virus pathogenicity favors development of sporadic epizootics. Virus persistent in soil or transmitted vertically through biotic agencies provides the initial inoculum which may be present only locally. Rapid increases in disease incidence and efficient secondary cycling of virus are therefore essential mechanisms if virus disease is to develop epizootically before the host plant is removed. *T. ni* NPV is highly pathogenic in laboratory tests and is one of the few baculoviruses epizootic in unstable agricultural habitats. Jaques,[110] Pimentel,[155] and Hofmaster[156] all described epizootics of this virus on brassicas. Ehler,[76] on the other hand, commented on the rarity of NPV epizootics in *T. ni* populations on cotton. This may reflect the very short-lived persistence of baculoviruses on cotton foliage[158] compounded with the 2-month gap between planting and the appearance of *T. ni* eggs. Most primary inoculum on the leaves, possibly originating from soil, would therefore largely be inactivated by the time susceptible larvae began feeding actively.

Data on the dynamics of NPV dispersal in populations of *M. brassicae* in small plots of cabbages have provided quantitative information on the development of an epizootic in terms of the virus population.[56] Virus accumulating on cabbage plants as a result of initial host mortality provided a large, dispersed, inoculum source for development of a secondary epizootic. The epizootic phase was accompanied by an exponential increase in polyhedral concentration and eventually by a rapid rise in numbers of virus-killed larvae per plant (Figure 18a, b). Most of these died in the fifth and sixth instars when heavy damage to the plants had already occurred, again confirming that natural epizootics tend to act too slowly for pest control when economic thresholds are low. Evans and Allaway[56] suggested that disease incidence in younger larvae would have been higher if larger host densities had been present to exploit the localized sources of primary inoculum.

C. Cyclic Epizootics

Most data on natural epizootics fall into this category and are almost universally associated

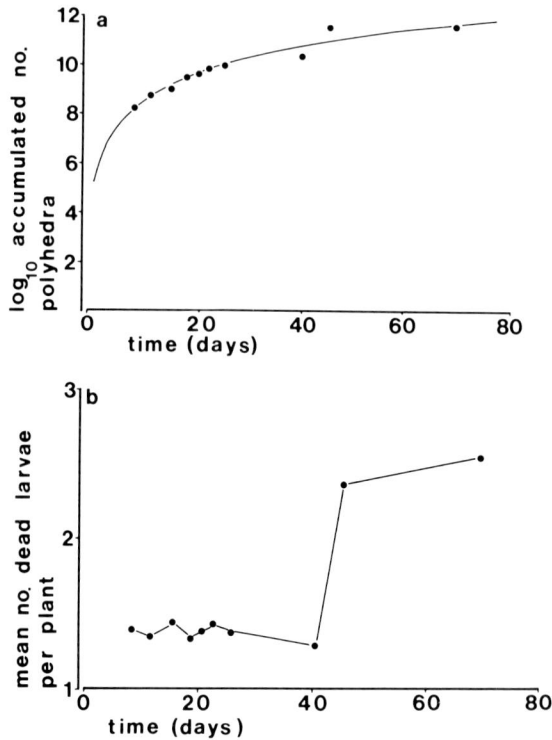

FIGURE 18.   (a) The relationship between accumulated total numbers of *Mamestra brassicae* nuclear polyhedrosis virus (NPV) polyhedra in infected larvae and days since larvae were placed on cabbage plants; (b) mean number of NPV-killed *Mamestra brassicae* larvae per plant and days since larvae were placed on cabbage plants; each point represents larval mortality since the previous sample. (Reproduced from Evans, H. F. and Allaway, G. P., *Appl. Environ. Microbiol.*, 45, 493, 1983. With permission.)

with forest pests. The cyclic nature of pest outbreaks in forests has been noted for many years and, for some species, extensive data on pest population levels go back to the early 1800s.[159,160] Collapse of forest pest outbreaks has often been associated with the appearance of baculovirus epizootics.[161] Table 8 details a few of these, although whether baculoviruses were the primary regulatory factor in all examples is unlikely.

Theoretical considerations as a background to modeling mathematically the role of infectious diseases in the cyclic appearance of pest outbreaks have been presented by Anderson and May.[41,123] The basis of their models was the assumption that the host population can be divided into either susceptible or infected individuals. Further parameters were then introduced to incorporate intrinsic reproductive and death rates for the host population and the rate of death from disease. Refinements such as vertical transmission, latency, and long-term reservoirs of infective virus can be incorporated into the models where appropriate. As an explanation of the population cycles of forest insects, Anderson and May developed a model with the characteristics shown in Figure 19.

Full mathematical details of the model are given in Anderson and May,[41] but the main conclusion in validation tests was that, at least on the theoretical basis of the model, "highly pathogenic microparasites producing very large numbers of long-lived infective stages are likely to lead to non-seasonal cyclic changes in the abundance of their invertebrate hosts and in the prevalence of infection."

Table 8
EXAMPLES OF CYCLIC BACULOVIRUS EPIZOOTICS

Virus[a]	Host insect	Host plant	Period between outbreaks (years)	Ref.
NPV	*Acleris variana*	Hemlock	10 to 15	162, 163
	Bupalus piniarius	Pine	5 to 8	164
	Chloristoneura murinana	Fir	20 to 40	165
	Gilpinia hercyniae	Spruce	>10 to 40	62
	Lymantria dispar	Oak and other hardwoods	5 to 11 in Europe possibly noncyclic in N. America	166, 167
	L. fumida	Japanese fir	5 to 7	168
	L. monacha	Pine	≈70	69
	Neodiprion swaineii	Pine	8	170
	Orgyia pseudotsugata	Douglas fir	7 to 10	58
GV	*Zeiraphera diniana*	Larch	9 to 10	171, 172

[a] NPV = nuclear polyhedrosis virus; GV = granulosis virus.

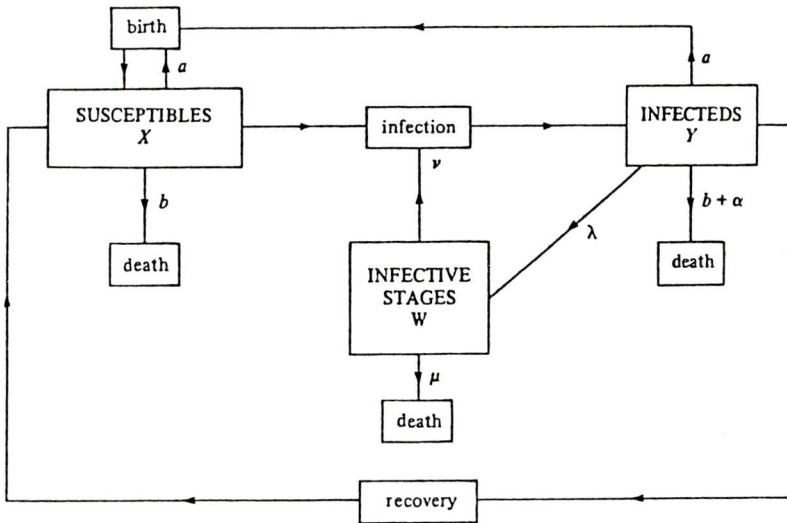

FIGURE 19.   Schematic representation of Anderson and May's model G representing the effects of free-living infective stages of viruses on virus-host interactions: a = host birth rate per individual; b = natural mortality rate of hosts; ν = transmission rate of infective stages; α = disease induced mortality rate; μ = mortality rate of free-living infective stages; λ = rate of production of infective stages per infected host, γ = rate of host recovery from infection. (Reproduced from Anderson, R. M. and May, R. M., *Phil. Trans. R. Soc. Lond. Ser. B.*, 291, 451, 1981. With permission.)

This conclusion was tested against field data for *Zeiraphera diniana* which exhibits cyclic population upsurges in the Engadine area of Switzerland with 9 to 10 years between peaks (Figure 20a).[171] The decline in populations at the end of each cycle was associated with the appearance of GV. Various assumptions about the parameters in Figure 19 were used to test the model. Resultant predictions on the amplitude and length of population cycles depended only on the parameters μ, α, and r (Figure 19) and a composite variable (α + β + γ) representing the net death rate of the host. Figure 20b shows the model output in terms of predicted host population change and virus incidence. The agreement between

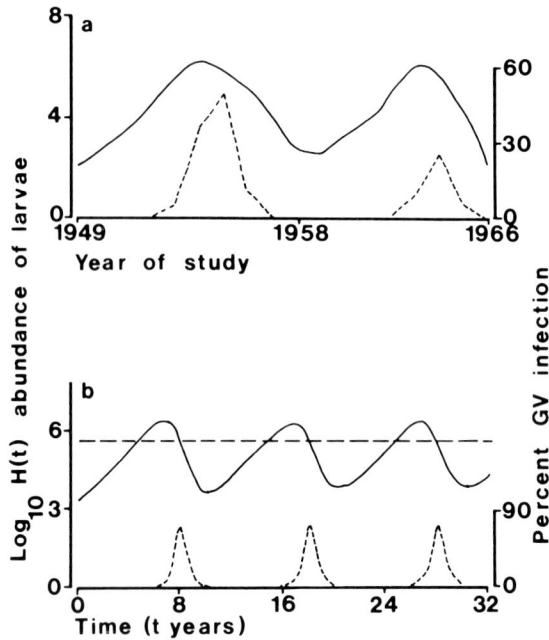

FIGURE 20.   (a) Changes in abundance of *Zeiraphera diniana*
in the European Alps (—) and the percent infection of larvae
with granulosis virus (GV) (---);[171] (b) predicted host abundance, $\log_{10} H(t)$ (—) and percent GV infection, y(t) (---) as
a function of time using Anderson and May's model G with
parameter values approximating those for the *Z. diniana*-GV
interaction in (a); the horizontal line (— — —) indicates the threshold host density for maintenance of the virus within the host
population. (Reproduced from Anderson, R. M. and May, R.
M., *Phil. Trans. R. Soc. Lond. Ser. B.*, 291, 451, 1981. With
permission.)

theory and practice was remarkably good in view of the independence of the model assumptions from the observed data. It is to be hoped that more quantitative observations on
natural epizootics will be produced to test further the assumptions of this and later models.

Although the data on cyclic baculovirus epizootics tend to be qualitative, a number of
features are common to the observed cycles. Reference to Figure 20a serves to illustrate
this point. There is generally a delay in peak virus infection relative to peak host population
as well as a period when host population density falls below a critical threshold. Disease,
therefore, acts in a delayed density-dependent manner above the threshold density. At these
times, disease incidence is measured easily since both infected and healthy hosts are present
in abundance. Data for the NPV of *G. hercyniae* in Canada[174] and Wales[65] support this
finding. In Wales the transformed infection rate (y) varied from 0.13 per day at high host
densities to less than 0.01 per day at low host densities, a measure of both residual inoculum
and host exploitation of that inoculum. Sawfly populations have since declined to extremely
low levels with disease apparently absent. A similar sequence occurred in Canada following
the initial pest outbreak and regulatory epizootic.[174] However, although disease appeared to
be absent in Canada, spraying with DDT to control spruce budworm, *Choristoneura fumiferana,* in New Brunswick led to a resurgence of the sawfly by removal of parasitoid
control of the low level populations.[174,175] This was followed, after the normal delay, by a
decimating NPV epizootic. The possible means by which NPV maintained itself in the

apparent absence of replication in the host has been discussed earlier. The main requirement of a threshold host density for disease growth appears to be validated in this case which also illustrates that baculoviruses are only one among a number of possible regulatory factors.

Doane[73] proposed an extension of the mechanisms of epizootic growth in his résumé of the role of NPV in regulation of *L. dispar* populations. He postulated normal delayed density-dependent growth of disease but suggested that the delay between host population collapse and the eventual decline in infection rate was extended and independent of host density. Very large reservoirs of virus in the environment were assumed to sustain high infection levels even in a declining host population. Such an effect would also tend to keep the host population at a low density until the quantities of inoculum in the environment declined. Data from Campbell's[177] studies of large area effects of NPV in *L. dispar* populations may be explained by this phenomenon. In this way host populations growing rapidly in the middle of a collapsed population will themselves tend to collapse.

This hypothesis is perfectly compatible with Anderson and May's[41] criteria and could be incorporated into the model by varying the parameters, μ, mortality rate of free-living infective stages and, v, the rate at which infective stages successfully infect hosts. Doane[73] has, in fact, provided a description of delayed density-dependent epizoosis differing only in rate from the usual assumptions regarding disease progress.

An alternative hypothesis to explain cyclic epizootics is that proposed by Martignoni and Schmid.[178] They assumed that virus epizootics acted differentially on the more susceptible individuals in a population eventually resulting in small numbers of relatively resistant individuals. Virus was then unable to develop efficiently in this population and hence ceased to act as a selection factor for resistance. Host heterogeneity then increased, inevitably leading to an increasing proportion of susceptible individuals and eventually to further epizootic virus development. As Briese[179] pointed out, the one example of apparent development of resistance in the field quoted by Martignoni,[180] *Eucosma griseana* GV LD_{50} increased 38-fold, could be explained merely by removal of the more susceptible individuals. This does not constitute true resistance since the slope of the dosage-mortality regression line became steeper indicating less heterogeneity in the host population as Martignoni and Schmid[178] postulated. As a factor in determining population response to baculoviruses during the troughs between outbreaks, this relative decrease in susceptibility may act to delay the onset of significant infection, shortening the period between pest outbreaks.

To conclude this section on baculovirus epizootics it must be stressed that, although three states of epizoosis have been discussed, the progress of disease in space and time is a single concept that includes all levels of disease incidence in the host population. Models such as those produced by Anderson and May[41] are valuable in pinpointing the variables that have the greatest influences on disease development. Armed with these concepts, it is now up to field ecologists to gather the necessary data with which to test and refine predictive models. Such data is not generally available in a suitable form at present.

IV. THE USE OF ECOLOGICAL AND EPIZOOTIOLOGICAL DATA TO DESIGN SYSTEM-ORIENTED PEST MANAGEMENT PROGRAMS

Practical field use of baculoviruses is well established and has been reviewed recently by Payne[113] and Entwistle and Evans.[62] Further consideration is given to this aspect by Harper, Young and Yearian, and Huber in this volume. This section will concentrate on ways of using our knowledge of epizoosis to predict likely responses of the host in the field. Two main aspects will be amplified.

A. Modeling of Dosage Rates and Host Mortality

There is an extensive body of data on dosage-mortality responses under laboratory conditions. In most cases these data stand alone and no attempt has been made to use them to predict responses in the field. In part, this has stemmed from the obvious disparity between controlled experiments in the laboratory and the wide ranging variation likely to be experienced by host populations in the field. Nevertheless, it is essential that some effort is made to test laboratory data under field conditions, even if this has to be on the basis of empirical estimates of major parameters. Data on the ecology and population dynamics of the host must also be taken into account.

Pinnock and Brand[181] and Brand and Pinnock[182] outlined a framework for the study of quantitative ecology of pathogens, particularly bacteria, as they apply to the control of pests. A model for the direct effects of *Bacillus thuringiensis* spray against hosts where there was no replication of the pathogen was developed by Pinnock et al.[183] They distinguished a number of important parameters that would also apply when baculoviruses are used as insecticides producing rapid mortality of the pest. These included data on lethal dosage, feeding rates of host larvae, deposition and distribution of the pathogen on the host plant, and the rate of decay of that pathogen population. These data are sufficient to provide working predictions of the expected host responses following pathogen applications. On the same conceptual basis, Payne[113] used data on the GV of *P. rapae* to predict the expected mortalities in the field from a range of applied dosages. Payne's results are summarized in Figure 21 where the linear regression of observation vs. prediction tended to overestimate the expected responses at higher dosages. However, the general form of the relationship is a vindication of the approach used and further refinement of the parameters in the model would probably improve the level of prediction.

When virus growth and secondary cycling of inoculum are included as parameters, modeling the outcome becomes more complex. Indeed, data to enable predictive models of this type to be developed are not often available in a suitable form and inferences about the likely consequences of secondary inoculum growth tend to be qualitative. In our studies on the control of *N. sertifer* by application of NPV, we have quantified the role of host population density on the levels of infection achieved from a range of dosages.[184] Secondary cycling of inoculum increased the incidence of disease in proportion to the density of host feeding colonies per tree. This finding was confirmed by Kaupp[42] who showed that there was rapid transfer of inoculum within and between larval feeding colonies, and this increased with host density.

Decreasing larval susceptibility with increasing age must also be incorporated in predictive models. For this, it is necessary to quantify the changing age structure and density of the host population with time and to incorporate virus productivity and decay rates for both primary and secondary inocula. As an example, the LD_{99} for *M. brassicae* larval instars exposed to NPV in relation to weight is shown in Figure 22.[27,28] This indicates the minimum dosage that each larva of a particular age, and hence weight, must consume in 24 hr or less to become lethally infected. Feeding rates of larvae change dramatically with age,[63,64] so that dosage per unit area must take account of this. On the positive side, greater feeding rates in older larvae will compensate, in part, for their reduced susceptibility. For example, in practical terms *P. rapae* can be controlled by the same dosage of GV capsules per unit area for the first four instars as a result of proportionately greater feeding as larvae age.[113]

Host voltinism and the possible overlap of generations may add further complications since the population age structure has to be assessed with respect to the maximum dosages required to infect all larvae in the population, including the more resistant individuals. In these cases, predictive models must incorporate age structure, dosage, and decay rates, with options for inclusion of multiple virus applications. Control of *Heliothis* sp. on cotton in the U.S. requires a mean of eight to nine applications per season[62] and as a system would benefit from predictive modeling on the basis described above. Analysis of the many studies of *Heliothis* NPV on cotton suggests that control may be improved by increasing the number of applications per season but with fewer polyhedra per application.[62]

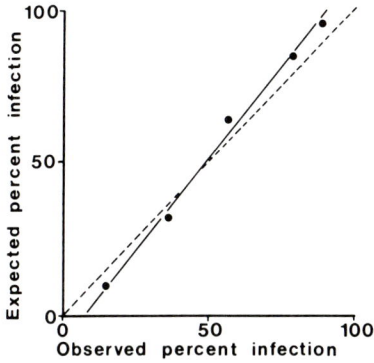

FIGURE 21.   Expected percent infection with granulosis virus (GV) in relation to observed responses (●) following application of GV to field populations of *Pieris rapae*. Expected values were obtained from known dosage-mortality responses, host feeding rates, and virus decay rates. Coincidence of expected and observed values would be described by the dashed line (---). (Data from Payne.[113])

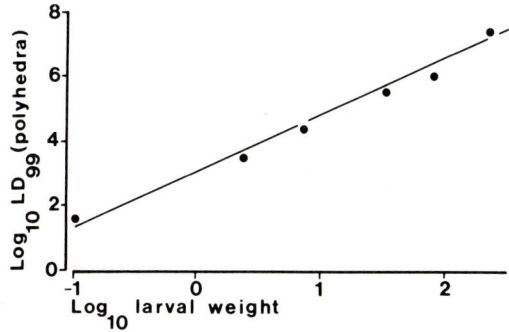

FIGURE 22.   Dosages required to infect 99% of *Mamestra brassicae* larvae with nuclear polyhedrosis virus as larvae age. The relationship between $\log_{10}$ LD$_{99}$ and $\log_{10}$ larval weight was given by $\log_{10} y = 3.03 + 1.65 \log_{10} x$. (Data from Evans.[27,28])

The necessary parameters to model virus application rates are relatively easy to identify but not always possible to estimate quantitatively. This may be a major factor in the dearth of such models in the literature. Anderson and May's[41] models of invertebrate infectious diseases provide for the possibility of incorporating a parameter for artificially introduced pathogens. Basically, control of a pest depends on virus mortality rates (α) exceeding the virus-free growth rate of the host population (r). Thus virus has to be introduced at a rate (A) sufficient to exceed a critical value (Ac) defined by the amount of virus produced naturally in a host population at equilibrium with that virus. The requirement that $\alpha > r$ means that hosts with high reproductive rates (r) will be relatively difficult to control hence requiring larger A, since r is a major component of Ac.

In contrast to the general predictive models of Anderson and May, the role of NPV in *L. dispar* population regulation has been modeled specifically.[185] This model generates predictions on the basis of eight differential equations which are integrated over the larval period. The eight components were derived from known characteristics of the gypsy moth NPV system and include

1.   H(t) = the number of noninfected larvae per hectare
2.   V(t) = the number of infected larvae per hectare
3.   P(t) = the number of PIBs per hectare released on to foliage
4.   P*(t) = the number of PIBs per hectare on tree bark and soil litter
5.   G(t) = dry weight of each larva (mg)
6.   C(t) = accumulated dry weight of foliage consumed per larva (mg)
7.   F(t) = expected dry weight of foliage in absence of pest (kg/ha)
8.   F*(t) = actual dry weight of remaining foliage (kg/ha)

where t = day degrees (°C).

Components of the model incorporated natural and virus-induced host mortality, larval consumption rates, virus production and inactivation rates, NPV transmission rates, and factors limiting infection to predefined limits when PIB populations were extremely high. These parameters fulfill the requirements discussed earlier, and the predictions from the model were said to give realistic output for the observed natural mortality rates in the gypsy moth populations studied. This approach was more system specific than that advocated by Anderson and May,[41] but may provide the stimulus necessary for the development of predictive models for other host-virus systems.

B. Manipulation of the Environment to Induce and Maintain Epizootics

Development of epizootics as a factor in host regulation has been emphasized constantly in this chapter. Most of the data presented has been from naturally occurring epizootics which have not been influenced heavily by human activities. By studying these, a gradual understanding of the components delimiting epizootic development has been accumulated. From these data, it is desirable to identify the key factors in the population dynamics of the virus so that they can be manipulated to increase their effects.

One of the most effective ways of doing this is to increase the level of inoculum in the environment and to maintain it at the higher level for longer periods than normal, i.e., to amplify and extend persistent virus populations. This may be achieved in many ways, some of which have been discussed by Tanada[21] and Evans and Harrap.[85]

Most successes using this approach have been achieved when economic thresholds were high and some level of damage to plants was tolerable. An excellent example was the control of *Wiseana* sp. by NPV in grassland in New Zealand.[57] Studies of a natural epizootic of NPV in *Wiseana* populations led to the interesting finding that natural infection levels were generally higher than those achieved through spraying plots with virus. More detailed studies of virus productivity, persistence, and dispersal indicated that the relative stability of the pasture habitat resulted in accumulation and efficient persistence of NPV in soil. Comparison between ploughed, reseeded, pasture, and undisturbed pasture led to the development of recommendations for a change in pasture management. Ploughing and reseeding diluted persistent inoculum to ineffective levels and resulted in *Wiseana* population resurgence. Kalmakoff and Crawford[57] recommended that damaged areas should be oversown to conserve soil-borne virus and that rotational grazing should be increased to enhance virus spread. Practical tests showed that a population of 60 to 100 larvae per 4 m² was optimal for maintaining enzootic levels of virus infection and consequent persistent virus populations in the soil.

Autodissemination of virus is another attractive approach to pest control that has arisen through basic ecological studies of virus-host dynamics. Ignoffo[186] coined the phrase in his discussion of novel approaches to pest control, although the concept was well established in practice. Contamination of adult genitalia has already been discussed and can be effective in the spread and development of virus infections. However, the best example of this approach, which utilizes epizootiological data to the fullest, is the use of adult beetles to spread nonoccluded baculovirus for the control of *O. rhinoceros*.[102] It was known from laboratory studies that adult *O. rhinoceros* could be infected with the virus which developed in the gut and was voided in large quantities in the feces.[186] This knowledge was used in the development of a program which relied on release of infection-bearing adults in the field. Direct transovum transmission to the progeny resulted in effective control of larvae and rapid dispersal of infection via the strong flying adults.[187] This has proved consistently successful, but Bedford[102] stressed that for continuing control there must be a residual population of beetles present to maintain infection since there is virtually no environmental persistence.

These examples serve to illustrate how basic ecological data can be utilized to enhance natural control using baculoviruses. Since these approaches tend to rely on early initiation of epizootics, there is inevitably a lag period before the required level of control is achieved, and so there has to be a reasonably high tolerance to damage from the pest species. Despite this proviso, manipulation of epizootics has many attractions and has already proved successful.

V. CONCLUSIONS AND FUTURE CONSIDERATIONS

The theme that has been emphasized throughout this chapter is the value of basic ecological data in furthering our understanding of baculovirus epizootiology. Links between laboratory and field provide data on many of the fundamental parameters of virus-host interaction, but these have to be used with care if meaningful extrapolation is to be achieved. It has been demonstrated that quantitative dosage-mortality and host feeding rate studies can be used to predict mortality in field populations of the host.[113,185] These are rare examples of attempts being made to provide a quantitative framework for further development in pest control strategy.

It is in the area of crop protection that epizootiological data have most value since they can indicate novel directions that might otherwise have gone undetected. The *Wiseana* NPV[57] and *Oryctes* nonoccluded baculovirus[102] stories provide excellent examples of this innovative approach. Other strategies such as development of improved virus persistence,[85] combinations of other pathogens, insecticides, and natural enemies (Chapter 5) would all benefit from an integrated ecological approach in assessing their effects.

A number of basic variables must be estimated to provide the framework for predictive models of epizootics. Those that appear, on current knowledge, to have the greatest influence on virus ecology include (1) Host: susceptibility, density, distribution, birth rate, growth rate, feeding rate and (2) Virus: infectivity, productivity, survival rate, specificity.

Many others may be shown eventually to have key influences on baculovirus epizootiology, and it is through descriptive and mathematical modeling of known parameters that gaps in our understanding may be identified.

Future work must therefore aim to improve quantitative data on all aspects of virus ecology with particular attention being paid to study of the ecosystem as a whole. Experimental modification of virus incidence in the field, based on prediction, can provide data for testing and improvement of models, and eventually to utilization in practical pest management programs. It is perhaps appropriate to conclude with the statement made by Waggoner[188] concerning plant diseases, "We shall find the true epidemic muddy and uncomfortable." A similar remark could be made for baculovirus epizootiology, but developments in this field are gradually providing a pathway through the mire.

ACKNOWLEDGMENTS

I thank Mr. J. S. Robertson and Mr. P. F. Entwistle for their constructive comments during the preparation of this manuscript and Mrs. J. Bald for her excellent typing during the final stages of preparation.

REFERENCES

1. **Steinhaus, E. A.,** *Disease in a Minor Chord,* Ohio State University Press, Columbus, 1975, 488.
2. **Bassi, A.,** Maladies des vers à soie: récherches sur la muscardine, *C. R. Seances Hebd. Acad. Sci.,* 2, 434, 1836.
3. **Pasteur, L.,** Étudie sur la maladie des vers à soie, *Gauthiers-Villars, Paris,* 1, 330, 1870.
4. **Hofmann, O.,** Die Schlaffsucht (Flacherie) der Nonne *(Liparis monacha)* nebst einem Anhang, in *Insektentötende Pilze mit besonderer Berücksichtigung der Nonne,* P. Weber, Frankfurt, 1891, 31.

5. **Wahl, B.,** Ueber die polyederkrankheit der Nonne (*Lymantria monacha* L.), *Cent. f. Gesam. For.,* 35, 212, 1909.
6. **Maestri, A.,** *Frammenti anatomici, fiscologici e pathologici sul baco de seta,* Fusi, Pavia, 1856.
7. **Cornalia, E.,** Monografia del bombice del gelso, *Mem. R. Instit. Lombardo Sci. Lett. Arte.,* 6, 3, 1856.
8. **Komárek, J. and Breindl, V.,** Die Wipfelkrankheit der Nonne und der Erreger derselben, *Z. Angew. Entomol.,* 10, 99, 1924.
9. **Bergold, G.,** Die Isolierung des Polyederviren, *Z. Naturforsch.,* 3, 25, 1947.
10. **Reiff, W.,** Einige Flacherie — Experimente mit der "Gypsy moth" *(Liparis dispar), Soc. Entomol.,* 24, 178, 1909.
11. **Jones, H. N.,** Further studies on the nature of the wilt disease of the gypsy moth larvae, in *7th Ann. Rep. State Forester,* Vol. 43, Massachusetts Public Doc., Boston, 1910, 101.
12. **Steinhaus, E. A.,** *Insect Microbiology,* Hafner Press, New York, 1946, 763.
13. **Bird, F. T.,** A Virus (Polyhedral) Disease of the European Spruce Sawfly, *Gilpinia hercyniae,* Ph.D. thesis, McGill University, Montreal, Quebec, 1949, 131.
14. **Bird, F. T.,** Virus diseases of sawflies, *Can. Entomol.,* 87, 24, 1955.
15. **Bird, F. T.,** Transmission of some insect viruses with particular reference to ovarial transmission and its importance in the development of epizootics, *J. Insect Pathol.,* 3, 352, 1961.
16. **Bird, F. T. and Elgee, D. E.,** A virus disease and introduced parasites as factors controlling the European spruce sawfly, *Diprion hercyniae* (Htg) in central New Brunswick, *Can. Entomol.,* 89, 371, 1957.
17. **Bird, F. T. and Burk, J. M.,** Artificially disseminated virus as a factor controlling the European spruce sawfly, *Diprion hercyniae* (Htg) in the absence of introduced parasites, *Can. Entomol.,* 93, 228, 1961.
18. **Entwistle, P. F., Adams, P. H. W., Evans, H. F., and Rivers, C. F.,** Epizootiology of a nuclear polyhedrosis virus (Baculoviridae) in European spruce sawfly (*Gilpinia hercyniae*): spread of disease from small epicentres in comparison with spread of baculovirus diseases in other hosts, *J. Appl. Ecol.,* 20, 473, 1983.
19. **Tanada, Y.,** Epizootiology of infectious diseases, in *Insect Pathology, an Advanced Treatise,* Vol. 2, Steinhaus, E. A., Ed., Academic Press, New York, 1963, 423.
20. **Tanada, Y.,** Epizootiology of insect diseases, in *Biological Control of Insect Pests and Weeds,* DeBach, P., Ed., Van Nostrand-Reinhold, New York, 1964, 548.
21. **Tanada, Y.,** Persistence of entomogeneous viruses in the insect ecosystem, in *Entomological Essays to Commemorate the Retirement of Professor K. Yasumatsu,* Hokuryukan Press, Tokyo, 1971, 367.
22. **Tanada, Y.,** Ecology of insect viruses, in *Perspectives in Forest Entomology,* Anderson, J. F. and Kaya, H. K., Eds., Academic Press, New York, 1976, 265.
23. **Andrewartha, H. G.,** *Introduction to the Study of Animal Populations,* University of Chicago Press, Chicago, 1961.
24. **Andrewartha, H. G. and Birch, L. A.,** *The Distribution and Abundance of Animals,* University of Chicago Press, Chicago, 1954.
25. **Krebs, C. J.,** *Ecology, The Experimental Analysis of Distribution and Abundance,* Harper & Row, New York, 1978.
26. **Burges, H. D. and Thomson, E. M.,** Standardization and assay of microbial insecticides, in *Microbial Control of Insects and Mites,* Burges, H. D. and Hussey, H. D., Eds., Academic Press, New York, 1971, 591.
27. **Evans, H. F.,** Quantitative assessment of the relationships between dosage and response of the nuclear polyhedrosis virus of *Mamestra brassicae, J. Invertebr. Pathol.,* 37, 101, 1981.
28. **Evans, H. F.,** The influence of larval maturation on responses of *Mamestra brassicae* L. (Lepidoptera:Noctuidae) to nuclear polyhedrosis virus infection, *Arch. Virol.,* 73, 163, 1983.
29. **Granados, R. R.,** Infectivity and mode of action of baculoviruses, *Biotechnol. Bioeng.,* 22, 65, 1980.
30. **Kelly, D. C.,** Baculovirus replication, *J. Gen. Virol.,* 63, 1, 1982.
31. **Hughes, P. R., Wood, H. A., Burand, J. P., and Granados, R. R.,** Quantification of the dose-mortality response of *Trichoplusia ni, Heliothis zea,* and *Spodoptera frugiperda* to nuclear polyhedrosis viruses: applicability of an exponential model, *J. Invertebr. Pathol.,* in press.
32. **Wigley, P. J.,** The Epizootiology of a Nuclear Polyhedrosis Virus Disease of the Winter Moth, *Operophtera brumata* L. at Wistman's Wood, Dartmoor, Ph.D. thesis, University of Oxford, Oxford, 1976.
33. **Klein, M. and Podoler, H.,** Studies on the application of a nuclear polyhedrosis virus to control populations of the Egyptian cottonworm, *Spodoptera littoralis, J. Invertebr. Pathol.,* 32, 244, 1978.
34. **Allen, G. E. and Ignoffo, C. M.,** The nucleopolyhedrosis of *Heliothis:* quantitative *in vivo* estimates of virulence, *J. Invertebr. Pathol.,* 13, 378, 1969.
35. **Kunjeku, E.,** Comparative Biological and Biochemical Studies of Three Nuclear Polyhedrosis Viruses Infecting *Trichoplusia ni* (Hüber) (Lepidoptera:Noctuidae), M.S. thesis, Imperial College, University of London, London, 1982, 55.
36. **Boucias, D. G. and Nordin, G. L.,** Interinstar susceptibility of the fall webworm, *Hyphantria cunea,* to its nucleopolyhedrosis and granulosis virus, *J. Invertebr. Pathol.,* 30, 68, 1977.

37. **Payne, C. C., Tatchell, G. M., and Williams, C. F.,** The comparative susceptibilities of *Pieris brassicae* and *Pieris rapae* to a granulosis virus from *Pieris brassicae, J. Invertebr. Pathol.,* 38, 273, 1981.

38. **Cunningham, J. C. and Entwistle, P. F.,** Control of sawflies by baculovirus, in *Microbial Control of Pests and Plant Diseases 1970-1980,* Burges, H. D., Ed., Academic Press, London, 1981, 379.

39. **van der Plank, J. E.,** *Plant Diseases: Epidemics and Control,* Academic Press, New York, 1963.

40. **Kranz, J., Ed.,** *Epidemics of Plant Diseases,* Chapman and Hall, London, 1974.

41. **Anderson, R. M. and May, R. M.,** The population dynamics of microparasites and their invertebrate hosts, *Phil. Trans. R. Soc. Lond. Ser. B.,* 291, 451, 1981.

42. **Kaupp, W. J.,** Studies on the Ecology of the Nuclear Polyhedrosis Virus of the European Pine Sawfly, *Neodiprion sertifer* (Geoff.), Ph.D. thesis, University of Oxford, Oxford, 1981, 363.

43. **Schwehr, R. D. and Gardner, W. A.,** Disease incidence in fall armyworm and corn earworm populations attacking grain sorghum, *J. Ga. Entomol. Soc.,* 17, 38, 1982.

44. **Evans, H. F., Lomer, C. J., and Kelly, D. C.,** Growth of nuclear polyhdrosis virus in larvae of the cabbage moth, *Mamestra brassicae* L., *Arch. Virol.,* 70, 207, 1981.

45. **Kelly, D. C., Edwards, M. L., Evans, H. F., and Robertson, J. S.,** The use of enzyme linked immunosorbent assay to detect a nuclear polyhedrosis virus in *Heliothis armigera* larvae, *J. Gen. Virol.,* 40, 465, 1978.

46. **Ignoffo, C. M. and Hink, W. F.,** Propagation of arthropod pathogens in living systems, in *Microbial Control of Insects and Mites,* Burges, H. D. and Hussey, N. W., Eds., Academic Press, New York, 1971, 501.

47. **Podgwaite, J. D.,** NPV production and quality control, *USDA For. Serv. Tech. Bull.,* No. 1584, 461, 1981.

48. **Smirnoff, W. A.,** Preparation and application of viral material, in biological control of the Jack pine sawfly, *For. Chron.,* 40, 187, 1964.

49. **Huber, J.,** The baculoviruses of *Cydia pomonella* and other tortricids, *Proc. 3rd Int. Colloq. Invertebr. Pathol.,* Brighton, U.K., 1982, 119.

50. **Matthiessen, J. N., Christian, R. L., Grace, T. D. C., and Filshie, B. K.,** Large-scale field propagation and the purification of the granulosis virus of the potato moth *Phthorimaea operculella* (Zeller) (Lepidoptera:Gelechiidae), *Bull. Entomol. Res.,* 68, 385, 1978.

51. **Liu, N. C. and Liang, D. R.,** Control of *Pieris rapae* larvae with granulosis virus in PRC, *Proc. 15th Annu. Meet. of Sco. Invertebr. Pathol.,* Brighton, U.K., 1982, 159.

52. **Bucher, G. E. and Turnock, W. J.,** Dosage responses of the larval instars of the bertha armyworm, *Mamestra configurata,* (Lepidoptera:Noctuidae) to a native nuclear polyhedrosis, *Can. Entomol.,* 115, 341, 1983.

53. **Stairs, G. R.,** The development of nuclear-polyhedrosis virus in ligatured larvae of the greater wax moth, *Galleria mellonella, J. Invertebr. Pathol.,* 15, 60, 1970.

54. **Shapiro, M., Bell, R. A., and Owens, C. D.,** *In vivo* mass production of gypsy moth nucleopolyhedrosis virus, *USDA For. Serv. Tech. Bull.,* No. 1584, 633, 1981.

55. **Shapiro, M.,** personal communication, 1983.

56. **Evans, H. F. and Allaway, G. P.,** Dynamics of baculovirus growth and dispersal in *Mamestra brassicae* L. (Lepidoptera:Noctuidae) larval populations introduced into small cabbage plots, *Appl. Environ. Microbiol.,* 45, 493, 1983.

57. **Kalmakoff, J. and Crawford, A. M.,** Enzootic virus control of *Wiseana* spp. in a pasture environment, in *Microbial and Viral Pesticides,* Kurstak, E., Ed., Marcel Dekker, New York, 1982, 435.

58. **Thompson, C. G. and Scott, D. W.,** Production and persistence of the nuclear polyhedrosis virus of the douglas-fir tussock moth, *Orgyia pseudotsugata* (Lepidoptera:Lymantriidae), in the forest ecosystem, *J. Invertebr. Pathol.,* 33, 57, 1979.

59. **Jaques, R. P.,** The inactivation of foliar deposits of viruses of *Trichoplusia ni* (Lepidoptera:Noctuidae) and *Pieris rapae* (Lepidoptera:Pieridae) and tests on protectant additives, *Can. Entomol.,* 104, 1985, 1972.

60. **Ignoffo, C. M., Parker, F. D., Boening, O. P., Pinnell, R. E., and Hostetter, D. L.,** Field stability of the *Heliothis* nucleopolyhedrosis virus on corn silks, *Environ. Entomol.,* 2, 302, 1973.

61. **Jaques, R. P.,** Tests on microbial and chemical insecticides for control of *Trichoplusia ni* and *Pieris rapae* on cabbage, *Can. Entomol.,* 105, 21, 1973.

62. **Entwistle, P. F. and Evans, H. F.,** Viral control, in *Comprehensive Insect Physiology, Biochemistry and Pharmacology,* Vol. 12, Gilbert, L. I. and Kerkut, G. A., Eds., Pergamon Press, Oxford, in press.

63. **Tatchell, G. M.,** The effects of a granulosis virus infection and temperatures on the food consumption of *Pieris rapae* (Lep.:Pieridae), *Entomophaga,* 26, 291, 1981.

64. **Harper, J. D.,** Food consumption by cabbage loopers infected with nuclear polyhedrosis virus, *J. Invertebr. Pathol.,* 21, 191, 1973.

65. **Evans, H. F. and Entwistle, P. F.,** The development of infection during a virus epizootic in spruce sawfly populations in mid-Wales, in *Proc. 1st Int. Colloq. Invertebr. Pathol.,* Kingston, Canada, 1976, 350.

66. **Clark, E. C.,** Survival and transmission of a virus causing polyhedrosis in *Malacosoma fragile, Ecology,* 37, 728, 1956.
67. **Kaya, H. K. and Anderson, J. F.,** Biotic mortality factors in dark tussock moth populations in Connecticut, *Environ. Entomol.,* 5, 1141, 1976.
68. **Evans, H. F.,** unpublished data, 1983.
69. **Entwistle, P. F., Adams, P. H. W., and Evans, H. F.,** Employing defensive regurgitation to estimate levels of nuclear polyhedrosis virus infection in *Gilpinia hercyniae* (Hymenoptera:Diprionidae) larvae, *J. Invertebr. Pathol.,* 41, 262, 1983.
70. **Zelazny, B.,** *Oryctes rhinoceros* populations and behaviour influenced by a baculovirus, *J. Inveretebr. Pathol.,* 29, 210, 1977.
71. **Evans, H. F.,** unpublished data, 1983.
72. **Capinera, J. L., Kirouac, S. P., and Barbosa, P.,** Phagodeterrency of cadaver components of gypsy moth larvae, *Lymantria dispar, J. Invertebr. Pathol.,* 28, 277, 1976.
73. **Doane, C. C.,** Ecology of pathogens of the gypsy moth, in *Perspectives in Forest Entomology,* Anderson, J. F. and Kaya, H. K., Eds., Academic Press, New York, 1976, 285.
74. **Uthai, K.,** personal communication, 1983.
75. **Kalshoven, L. G. E.,** *The Pests of Crops in Indonesia,* revised by Van der Laan, P. A., P. T. Ichtiar Bava-van Hoeve, Jakarta, 1981, 701.
76. **Ehler, L. E.,** Natural enemies of cabbage looper on cotton in the San Joaquin Valley, *Hilgardia,* 45, 73, 1977.
77. **Ignoffo, C. M.,** Specificity of insect viruses, *Bull. Entomol. Soc. Am.,* 14, 265, 1968.
78. **Ignoffo, C. M.,** Evaluation of *in vivo* specificity of insect viruses, in *Baculoviruses for Insect Pest Control: Safety Considerations,* Summers, M., Engler, R., Falcon, L. A., and Vail, P., Eds., American Society of Microbiology, Washington, D.C., 1975, 52.
79. **Ignoffo, C. M. and Couch, T. L.,** The nucleopolyhedrosis virus of *Heliothis* species as a microbial insecticide, in *Microbial Control of Pests and Plant Diseases 1970-1980,* Burges, H. D., Ed., Academic Press, New York, 1981, 329.
80. **Witt, D. J. and Janus, C. A.,** Aspects of the cross transmission to *Galleria mellonella* of a baculovirus from the alfalfa looper, *Autographa californica, J. Invertebr. Pathol.,* 27, 65, 1976.
81. **Andrews, R. E., Jr., Spence, K. D., and Miller, L. K.,** Virulence of cloned variants of *Autographa californica* nuclear polyhedrosis virus, *Appl. Environ. Microbiol.,* 39, 932, 1980.
82. **Nathans, D. and Smith, H. O.,** Restriction endonucleases in the analysis and restructuring of DNA molecules, *Ann. Rev. Biochem.,* 44, 273, 1975.
83. **Tompkins, G. J., Vaughn, J. L., Adams, J. R., and Reichelderfer, C. F.,** Effects of propagating *Autographa californica* nuclear polyhedrosis virus and its *Trichoplusia ni* variant in different hosts, *Environ. Entomol.,* 10, 801, 1981.
84. **Shapiro, M.,** *In vivo* mass production of insect viruses, in *Microbial and Viral Pesticides,* Kurstak, E., Ed., Marcel Dekker, New York, 1982, 463.
85. **Evans, H. F. and Harrap, K. A.,** Persistence of insect viruses, in *Virus Persistence,* Mahy, B. W. J., Minson, A. C., and Darby, G., Eds., Cambridge University Press, Cambridge, U.K., 1982.
86. **Cunningham, J. C.,** Serological and morphological identification of some nuclear-polyhedrosis and granulosis viruses, *J. Invertebr. Pathol.,* 11, 132, 1968.
87. **Reed, E. M.,** Factors affecting the status of a virus as a control agent for the potato moth ((*Phthorimaea operculella* (Zell.) (Lep. Gelechiidae)), *Bull. Entomol. Res.,* 61, 207, 1971.
88. **Stairs, G. R.,** Artificial initiation of virus epizootics in Forest tent caterpillar populations, *Can. Entomol.,* 97, 1059, 1965.
89. **Young, E. C.,** The epizootiology of the pathogens of the coconut palm rhinoceros beetle, *J. Invertebr. Pathol.,* 24, 82, 1965.
90. **Gregory, P. H.,** Interpreting plant disease gradients, *Annu. Rev. Phytopathol.,* 6, 189, 1968.
91. **Entwistle, P. F.,** Passive carriage of baculoviruses in forests, in *Proc. 3rd Int. Colloq. Invertebr. Pathol.,* Brighton, U.K., 1982, 344.
92. **Reed, E. M. and Springett, B. P.,** Large-scale field testing of a granulosis virus for the control of potato moth *(Phthorimaea operculella* (Cell.) (Lep., Gelechiidae)), *Bull. Entomol. Res.,* 61, 223, 1971.
93. **Entwistle, P. F., Evans, H. F., and Adams, P. H. W.,** unpublished data, 1983.
94. **Neilson, M. M. and Elgee, D. E.,** The method and role of vertical transmission of a nuclear polyhedrosis virus in the European spruce sawfly, *Diprion hercyniae, J. Invertebr. Pathol.,* 12, 132, 1968.
95. **Entwistle, P. F., Adams, P. H. W., and Evans, H. F.,** Epizootiology of a nuclear polyhedrosis virus in European spruce sawfly *(Gilpinia hercyniae):* the rate of passage of infective virus through the gut of birds during cage tests, *J. Invertebr. Pathol.,* 31, 307, 1978.
96. **Martignoni, M. E. and Milstead, J. E.,** Trans-ovum transmission of the nuclear polyhedrosis virus of *Colias eurytheme* Boisduval through contamination of the female genitalia, *J. Insect Pathol.,* 4, 113, 1962.

97. **Hamm, J. J. and Young, J. R.,** Mode of transmission of nuclear-polyhedrosis virus to progeny of adult *Heliothis zea, J. Invertebr. Pathol.,* 24, 70, 1974.

98. **Gard, I. E. and Falcon, L. A.,** Autodissemination of entomopathogens: virus, in *Characterization, Production and Utilization of Entomopathogenic Viruses,* Ignoffo, C. M., Martignoni, M. E., and Vaughn, J. L., Eds., American Society of Microbiology, Washington, D.C., 1978, 17.

99. **Elmore, J. C. and Howland, A. F.,** Natural versus artificial dissemination of nuclear-polyhedrosis virus by contaminated adult cabbage loopers, *J. Insect Pathol.,* 6, 430, 1964.

100. **Melamed-Madjar, V. and Raccah, B.,** The transstadial and vertical transmission of granulosis virus from the corn borer *Sesamia nonagrioides, J. Invertebr. Pathol.,* 33, 259, 1979.

101. **Tatchell, G. M.,** The transmission of a granulosis virus following the contamination of *Pieris brassicae* adults, *J. Invertebr. Pathol.,* 37, 210, 1981.

102. **Bedford, G. O.,** Biology, ecology, and control of palm rhinoceros beetles, *Annu. Rev. Entomol.,* 25, 309, 1980.

103. **Geier, P. W. and Oswald, L. T.,** The light brown apple moth, *Epiphyas postvittana* (Walker), 1. Effects associated with contaminations by a nuclear polyhedrosis virus on the demographic performance of a laboratory strain, *Aust. J. Ecol.,* 2, 9, 1977.

104. **Zelazny, B.,** Studies on *Rhabdionvirus.* II. Effect on adults of *Oryctes rhinoceros, J. Invertebr. Pathol.,* 22, 122, 1973.

105. **Tanada, Y. and Omi, E. M.,** Epizootiology of virus diseases in three lepidopterous insect species of alfalfa, *Res. Popul. Ecol.,* 16, 59, 1974.

106. **James, R. A.,** The Effect of Nuclear Polyhedrosis Virus and Other Factors on the Reproductive Biology of the European Spruce Sawfly, *Gilpinia hercyniae* (Htg.) (Hymenoptera:Symphyta), M.S. thesis, Imperial College of London, London, 1974.

107. **Evans, H. F. and Entwistle, P. F.,** Epizootiology of the nuclear polyhedrosis virus of European spruce sawfly with emphasis on persistence of virus outside the host, in *Microbial and Viral Pesticides,* Kurstak, E., Ed., Marcel Dekker, New York, 1982, 449.

108. **Elgee, E.,** Persistence of a virus of the white-marked tussock moth on balsam fir foliage, *Bi-mon. Res. Notes,* 31, 33, 1975.

109. **Evans, H. F.,** The ecology of *Mamestra brassicae* NPV in soil, in *Proc. 3rd Int. Colloq. Invertebr. Pathol.,* Brighton, U.K., 1982, 307.

110. **Jaques, R. P.,** Persistence, accumulation, and denaturation of nuclear polyhedrosis and granulosis viruses, in *Baculoviruses for Insect Pest Control: Safety Considerations,* Summers, M., Engler, R., Falcon, L. A., and Vail, P., Eds., American Society of Microbiology, Washington, D.C., 1975, 90.

111. **Thompson, C. G., Scott, D. W., and Wickman, B. E.,** Long term persistence of the nuclear polyhedrosis virus of the Douglas-fir tussock moth, *Orgyia pseudotsugata,* (Lepidoptera:Lymantriidae) in forest soil, *Environ. Entomol.,* 10, 254, 1981.

112. **Jaques, R. P.,** The persistence of a nuclear-polyhedrosis virus in soil, *J. Insect Pathol.,* 6, 251, 1964.

113. **Payne, C. C.,** Insect viruses as control agents, *Parasitology,* 84, 35, 1982.

114. **Young, S. Y. and Yearian, W. C.,** Soil application of *Pseudoplusia* NPV: persistence and incidence of infection in soybean looper caged on soybean, *Environ. Entomol.,* 8, 860, 1979.

115. **Baugher, D. B. and Yendol, W. G.,** Foliar and soil applications of nuclear polyhedrosis virus to control *Trichoplusia ni* larvae on cabbage, *Proc. 1st Int. Colloq. Invertebr. Pathol.,* Kingston, Canada, 1976, 354.

116. **Entwistle, P. F.,** New perspectives in pest control with pathogenic viruses, *Land,* 1, 84, 1974.

117. **Vago, C., Fosset, J., and Bergoin, M.,** Dissémination des virus de polyédres par les Ephippigères prédateurs d'insectes, *Entomophaga,* 11, 177, 1966.

118. **Beekman, A. G. B.,** The infectivity of polyhedra of nuclear polyhedrosis virus (N.P.V.) after passage through gut of an insect predator, *Experientia,* 36, 858, 1980.

119. **Cooper, D. J.,** The role of predatory Hemiptera in disseminating a nuclear polyhedrosis virus of *Heliothis punctigera, J. Aust. Entomol. Soc.,* 20, 145, 1981.

120. **Biever, K. D., Andrews, P. L., and Andrews, P. A.,** Use of a predator, *Podisus maculiventris,* to distribute virus and initiate epizootics, *J. Econ. Entomol.,* 75, 150, 1982.

121. **Gitay, H. and Polson, A.,** Isolation of a granulosis virus from *Heliothis armigera* and its persistence in avian faeces, *J. Invertebr. Pathol.,* 17, 288, 1971.

122. **Lautenschlager, R. A., Podgwaite, J. D., and Watson, D. E.,** Natural occurrence of the nucleopolyhedrosis virus of the gypsy moth, *Lymantria dispar,* (Lep.:Lymantriidae) in wild birds and mammals, *Entomophaga,* 25, 261, 1980.

123. **Entwistle, P. F., Adams, P. H. W., and Evans, H. F.,** Epizootiology of a nuclear-polyhedrosis virus in European spruce sawfly *(Gilpinia hercyniae):* the status of birds as dispersal agents of the virus during the larval season, *J. Invertebr. Pathol.,* 29, 354, 1977.

124. **Hostetter, D. L. and Biever, K. D.,** The recovery of virulent nuclear polyhedrosis virus of the cabbage looper, *Trichoplusia ni,* from feces of birds, *J. Invertebr. Pathol.,* 15, 173, 1970.

125. **Lautenschlager, R. A. and Podgwaite, D. J.,** Passage of nucleopolyhedrosis virus by avian and mammalian predators of the gypsy moth, *Lymantria dispar, Environ. Entomol.,* 8, 210, 1979.

126. **Wigglesworth, V. B.,** *The Principles of Insect Physiology,* 7th ed., Chapman and Hall, London, 1972, 872.

127. **Young, J. Z.,** *The Life of Mammals. Their Anatomy and Physiology,* Oxford University Press, Oxford, 1975.

128. **Raimo, B., Reardon, R. C., and Podgwaite, J. D.,** Vectoring gypsy moth nuclear polyhedrosis virus by *Apanteles melanoscelus* (Hym.:Braconidae), *Entomophaga,* 22, 207, 1977.

129. **Levin, D. B., Laing, J. E., and Jaques, R. P.,** Transmission of granulosis virus by *Apanteles glomeratus* to its host *Pieris rapae, J. Invertebr. Pathol.,* 34, 317, 1979.

130. **Irabagon, T. A. and Brooks, W. M.,** Interaction of *Campoletis sonorensis* and a nuclear polyhedrosis virus in larvae of *Heliothis virescens, J. Econ. Entomol.,* 67, 229, 1974.

131. **Beegle, C. C. and Oatman, E. R.,** Effect of a nuclear polyhedrosis virus on the relationship between *Trichoplusia ni* (Lepidoptera:Noctuidae) and the parasite, *Hyposoter exiguae* (Hymenoptera:Ichneumonidae), *J. Invertebr. Pathol.,* 25, 59, 1975.

132. **Mohamed, M. A., Coppel, H. C., Hall, D. J., and Podgwaite, J. D.,** Field release of virus-sprayed adult parasitoids of the European pine sawfly (Hymenoptera:Diprionidae) in Wisconsin, *G. Lakes, Entomol.,* 14, 177, 1981.

133. **Reardon, R. C. and Podgwaite, J. D.,** Disease-parasitoid relationships in natural populations of *Lymantria dispar* (Lep.:Lymantriidae) in the Northeastern United States, *Entomophaga,* 21, 333, 1976.

134. **Stairs, G. R.,** Transmission of virus in tent caterpillar populations, *Can. Entomol.,* 98, 1100, 1966.

135. **Vail, P. V.,** Cabbage looper nuclear polyhedrosis virus-parasitoid interactions, *Environ. Entomol.,* 10, 517, 1981.

136. **Kaya, H. K.,** Parasites and predators as vectors of insect diseases, *Proc. 3rd Int. Colloq. Invertebr. Pathol.,* Brighton, U.K., 1982, 39.

137. **Beegle, C. C. and Oatman, E. R.,** Differential susceptibility of parasitized and non-parasitized *Trichoplusia ni* larvae to a nuclear polyhedrosis virus, *J. Invertebr. Pathol.,* 24, 188, 1974.

138. **Hukuhara, T. and Namura, H.,** Distribution of a nuclear-polyhedrosis virus of the fall webworm, *Hyphantria cunea,* in soil, *J. Invertebr. Pathol.,* 19, 308, 1972.

139. **Thomas, E. D., Reichelderfer, C. F., and Heimpel, A. M.,** The effect of soil pH on the presistence of cabbage looper nuclear polyhedrosis virus in soil, *J. Invertebr. Pathol.,* 21, 21, 1973.

140. **Jaques, R. P. and Harcourt, D. G.,** Viruses of *Trichoplusia ni* and *Pieris rapae* in soil in fields of crucifers in southern Ontario, *Can. Entomol.,* 103, 1285, 1971.

141. **Kinsinger, R. A. and McGaughey, W. H.,** Stability of *Bacillus thuringiensis* and a granulosis virus of *Plodia interpunctella* on stored wheat, *J. Econ. Entomol.,* 69, 49, 1976.

142. **Ignoffo, C. M., Hostetter, D. L., Sikorowski, P. P., Sutter, G., and Brooks, W. M.,** Inactivation of representative species of entomopathogenic viruses, a bacterium, fungus, and protozoan by an ultraviolet light source, *Environ. Entomol.,* 6, 411, 1977.

143. **Elleman, C. J. and Entwistle, P. F.,** A study of glands on cotton responsible for the high pH and cation concentration of the leaf surface, *Ann. Appl. Biol.,* 100, 553, 1982.

144. **Young, S. Y., Yearian, W. C., and Kim, K. S.,** Effect of dew from cotton and soybean foliage on activity of *Heliothis* nuclear polyhedrosis virus, *J. Invertebr. Pathol.,* 30, 237, 1977.

145. **McLeod, P. J., Yearian, W. C., and Young, S. Y.,** Inactivation of *Baculovirus heliothis* by ultraviolet irradiation, dew and temperature, *J. Invertebr. Pathol.,* 30, 237, 1977.

146. **David, W. A. L. and Gardiner, B. O. C.,** Persistence of a granulosis virus of *Pieris brassicae* on cabbage leaves, *J. Invertebr. Pathol.,* 8, 180, 1966.

147. **Neilson, M. M. and Morris, R. F.,** The regulation of European spruce sawfly numbers in the Maritime Provinces of Canada from 1939 to 1963, *Can. Entomol.,* 96, 773, 1964.

148. **Entwistle, P. F., Evans, H. F., and Adams, P. H. W.,** unpublished data, 1983.

149. **Fuxa, J. R.,** Prevalence of viral infections in populations of fall armyworm, *Spodoptera frugiperda,* in southeastern Louisiana, *Environ. Entomol.,* 11, 239, 1982.

150. **Steinhaus, E. A. and Martignoni, M. E.,** *An Abridged Glossary of Terms Used in Invertebrate Pathology,* 2nd ed., USDA For. Serv. Pac. Northwest For. and Range Exp. Sta., Portland, 1970, 38.

151. **Ehler, L. E. and van den Bosch, R.,** An analysis of the natural biological control of *Trichoplusia ni* (Lepidoptera:Noctuidae) on cotton in California, *Can. Entomol.,* 106, 1067, 1974.

152. **Briese, D. T.,** The incidence of parasitism and disease in field populations of the potato moth *Phthorimaea operculella* (Zeller) in Australia, *J. Aust. Entomol. Soc.,* 20, 319, 1981.

153. **Hamilton, J. T.,** Seasonal abundance of *Pieris rapae* (L.), *Plutella xylostella* (L.) and their diseases and parasites, *Gen. Appl. Entomol.,* 11, 59, 1979.

154. **Federici, B. A.,** Baculovirus epizootic in a larval population of the clover cutworm, *Scotogramma trifolii,* in Southern California, *Environ. Entomol.,* 7, 423, 1978.

155. **Pimentel, D.,** Natural control of caterpillar populations on cole crops. *J. Econ. Entomol.,* 54, 889, 1961.
156. **Hofmaster, R. N.,** Seasonal abundance of the cabbage looper as related to light trap collections, precipitation, temperature and the incidence of a nuclear polyhedrosis virus, *J. Econ. Entomol.,* 54, 796, 1961.
157. **Franz, J. M.,** Influence of environment and modern trends in crop management on microbial control, in *Microbial Control of Insects and Mites,* Burges, H. D. and Hussey, N. W., Eds., Academic Press, New York, 1971, 407.
158. **Young, S. Y. and Yearian, W. C.,** Persistence of *Heliothis* NPV on foliage of cotton, soybean and tomato, *Environ. Entomol.,* 3, 253, 1974.
159. **Klimetzek, D.,** Die Zeitfolge des Auftretens nadel-fressender Kiefernraupen in der Pfalz seit 1810 und die Ursachen ihres Rückganges in neuerer Zeit, *Z. Angew Entomol.,* 71, 414, 1972.
160. **Klimetzek, D.,** *Insekten-Grosschädlinge an Kiefer in Nordbayern und der Pfalz: Analyse und Vergleich 1810-1970,* Band 2, Freiburger Waldschutz-Abhandlungen, Freiburg im Breisgau, 1979, 173.
161. **Kaya, H. K.,** Insect pathogens in natural and microbial control of forest defoliators, in *Perspectives in Forest Entomology,* Anderson, J. F. and Kaya, H. K., Eds., Academic Press, New York, 1976, 251.
162. **Prebble, M. L. and Graham, K.,** The current outbreak of defoliating insects in coast hemlock forests of British Columbia. II. Factors of natural control, *Br. Columbia Lumberm.,* 29, 37, 1945.
163. **Miller, C. A.,** The black-headed budworm in eastern Canada, *Can. Entomol.,* 98, 592, 1966.
164. **Klomp, H.,** The dynamics of a field population of the pine looper, *Bupalus piniarius* (Lep., Geom.), *Adv. Ecol. Res.,* 3, 297, 1966.
165. **Bucher, G. E.,** Biotic factors of control of the European fir budworm, *Choristoneura murinana* (Hbn.) (N. Comb.), in Europe, *Can. J. Agric. Sci.,* 33, 448, 1953.
166. **Keremidchiev, M. T.,** Dynamic of outbreaks of the gypsy moth *(Lymantria dispar* L) in the People's Republic of Bulgaria, *Proc. 13th Int. Congr. Entomol.,* Nauk, Leningrad, 1972, 51.
167. **Campbell, R. W.,** Population dynamics: historical review, *USDA For. Serv. Tech. Bull.,* No. 1584, 1981.
168. **Katagiri, K.,** Review on microbial control of insect pests in forests in Japan, *Entomophaga,* 14, 203, 1969.
169. **Zethner, O.,** Control experiments on the nun moth *(Lymantria monacha* L.) by nuclear-polyhedrosis virus in Danish coniferous forests, *Z. Angew. Entomol.,* 81, 192, 1976.
170. **Lyons, L. A.,** On the population dynamics of *Neodiprion* sawflies, *U. Minn. Agr. Exp. Sta. Tech. Bull.,* 310, 48, 1977.
171. **Auer, C.,** Erste Ergenbnisse einfacher stochasticker Modelluntersuchungen über die Ursachen der Populations beuegung desgrauen Lärchenwicklers *Zeiraphera diniana,* Gn (= *Z. griseana* Hb) im oberengadin 1949/66, *Z. Angew. Entomol.,* 62, 202, 1968.
172. **Baltensweiler, W.,** *Zeiraphera griseana* Hübner (Lepidoptera:Tortricidae) in the European Alps. A contribution to the problem of cycles, *Can. Entomol.,* 96, 792, 1964.
173. **Anderson, R. M. and May, R. M.,** Infectious diseases and population cycles of forest insects, *Science,* 210, 658, 1980.
174. **Bird, F. T. and Elgee, D. E.,** A virus disease and introduced parasites as factors controlling the European spruce sawfly, *Diprion hercyniae* (Htg) in central New Brunswick, *Can. Entomol.,* 89, 371, 1957.
175. **Neilson, M. M. and Elgee, D. E.,** An unusual increase in spruce sawfly numbers in New Brunswick, *Bim. Prog. Rep. For. Entomol. Pathol. Br. Dept. For. Can.,* 21, 1, 1965.
176. **Neilson, M. M., Martineau, R., and Rose, A. H.,** *Diprion hercyniae* (Hartig.), European spruce sawfly (Hymenoptera:Diprionidae): biological control programmes in Canada 1959-1968, Commonw. Inst. Biol. Control Tech. Bull., No. 4, 136, 1971.
177. **Campbell, R. W.,** The gypsy moth and its natural enemies, *U.S. Dep. Agric. For. Serv. Agric. Infor. Bull.,* 381, 27, 974.
178. **Martignoni, M. E. and Schmid, P.,** Studies on the resistance to virus infections in natural populations of Lepidoptera, *J. Insect Pathol.,* 3, 62, 1961.
179. **Briese, D. T.,** Resistance of insect species to microbial pathogens, in *Pathogenesis of Invertebrate Microbial Diseases,* Davidson, E. A., Ed., Allanheld, Osmun, Totowa, N.J., 1981, 511.
180. **Martignoni, M. E.,** Contributo alla conoscenza di una granulosi di *Eucosma griseana* (Hübner) (Tortricidae, Lepidoptera) quale fattore linitaure il pullulanemts dell'insetts nella Engadina alta, *Mitt. Schweiz. Anst. Forstl. Versuchw.,* 32, 371, 1957.
181. **Pinnock, D. E. and Brand, R. J.,** A quantitative approach to the ecology of the use of pathogens for insect control, in *Microbial Control of Pests and Plant Diseases 1970-1980,* Burges, H. D., Ed., Academic Press, New York, 1981, 655.
182. **Brand, R. J. and Pinnock, D. E.,** Application of biostatistical modelling to forecasting the results of microbial control trials, in *Microbial Control of Pests and Plant Diseases 1970-1980,* Burges, H. D., Ed., Academic Press, New York, 1981, 667.
183. **Pinnock, D. E., Brand, R. J., Milstead, J. E., Kirby, M. E., and Coe, N. F.,** Development of a model for prediction of target insect mortality following field appliction of a *Bacillus thuringiensis* formulation, *J. Invertebr. Pathol.,* 31, 31, 1978.

184. **Entwistle, P. F. and Evans, H. F.,** unpublished data, 1983.
185. **Valentine, H. T. and Podgwaite, J. D.,** Modeling the role of NPV in gypsy moth population dynamics, in *Proc. 3rd Int. Colloq. Invertebr. Pathol.,* Brighton, U.K., 1982, 353.
186. **Ignoffo, C. M.,** Strategies to increase the use of entomopathogens, *J. Invertebr. Pathol.,* 31, 1, 1978.
187. **Zelazny, B.,** Transmission of a baculovirus in populations of *Oryctes rhinoceros, J. Invertebr. Pathol.,* 27, 221, 1976.
188. **Waggoner, P. E.,** Weather, space, time, and chance of infection, *Phytopathology,* 52, 1100, 1962.

Chapter 5

INTERACTIONS BETWEEN BACULOVIRUSES AND OTHER ENTOMOPATHOGENS, CHEMICAL PESTICIDES, AND PARASITOIDS

James D. Harper

TABLE OF CONTENTS

I. INTRODUCTION

This chapter provides an overview of knowledge on the subject of naturally occurring and induced interactions between baculoviruses and other entomopathogens, chemical insecticides, and parasitoids. As the science of insect pathology developed in the present century, a number of examples of infection of the same individual by two or more pathogens began to be reported. Several reviews of the subject have been published.[1-5] Many of the early examples of these multiple infections involved baculoviruses. A natural extension of the recognition of the presence of multiple infections was the development of investigations into the nature of their joint actions in the host. Studies on interactions involving baculoviruses and other pathogens in the same host or host population have been reviewed by Vago,[2] Aizawa,[1] Tanada,[5-8] Smith,[3,9] and Krieg.[4] Studies on interactions of entomopathogens have also been extended to include pathogen-chemical interactions. Much of the early work in this area concerned noninsecticidal chemicals as virus-inducing agents (see Aruga[10]). Later work concentrated principally on insecticide-pathogen interactions. In a 1971 review of this subject, Benz[11] indicated that reports of virus-insecticide combinations were particularly scarce. Since that review, many studies of baculovirus-chemical insecticide interactions have been published and will be reviewed in this chapter. Interactions also occur between parasitoids and entomopathogens when present in the same host, and studies of these interactions have been reviewed by Tanada.[5,6] Again, this subject has received much more attention in the last decade than ever before.

II. THE OCCURRENCE AND NATURE OF BACULOVIRUS INTERACTIONS WITH OTHER PATHOGENS

The subject of baculovirus-pathogen interactions is complex. Subjects under this broad topic which have been examined are extremely varied. They range from recognition of naturally occurring dual infections in individual insects in field populations to a precise elucidation of the biochemical mechanisms involved within the cells of individuals suffering from such infections. In between these extremes lie analyses of naturally and intentionally induced dual infections at cell, tissue, organism, and population levels of organization. This section will examine those studies which provide insights into the natural occurrence of multiple infections and the nature of multiple infections in the individual host. Studies of population mortality resulting from deliberately induced multiple infections will be discussed in the following section.

A. Natural Occurrence

Naturally occurring multiple infections involving baculoviruses and other pathogens (including other baculoviruses) have long been known (see Steinhaus[12]). Members from nearly every entomopathogen group have been reported in naturally occurring, multiple infections with baculoviruses.

Paillot[13] reported *Euxoa segetum* to be simultaneously infected with both granulosis virus (GV) and nuclear polyhedrosis virus (NPV) or with GV and an undetermined mycosis. Other naturally occurring multiple infections of GV and NPV have been reported in *Pieris rapae*,[14] *Pseudaletia unipuncta*,[15,16] *Nephelodes emmedonia*,[17] *Spodoptera frugiperda*,[18] *Trichoplusia ni*,[19] *S. exigua* and *Autographa californica*,[20] and several additional noctuid species.[21]

Whereas two NPVs are known to be associated with specific natural host populations (e.g., Heimpel and Adams,[22] Hughes and Addison[23]) and probably occur in the same individuals, no definite records of multiple infection are available. Laboratory-induced multiple infections with two NPVs are known, however, and will be discussed later.

Smith and Xeros[24] conducted extensive tests with an NPV-cytoplasmic polyhedrosis virus

(CPV) combination found in *Pyrameis cardui* larvae which were "purchased from a dealer" and found to be multiply infected. Naturally occurring NPV and CPV multiple infections have been reported by Smith et al.[25] in *Sphinx ligustri,* by Biliotti et al.[26] in the pine processionary caterpillar, by Tanada and Omi[20] in *Spodoptera exigua,* and by Vago and Vasiljevic[27] in *Hyphantria cunea.* Additionally, Vago[28] reported individual *Pieris brassicae* to be multiply infected with GV and CPV. Bird and Whalen[29] found no CPV + NPV infections in over 1000 field-collected *Choristoneura fumiferana* larvae examined, but obtained multiple infections with inocula from these larvae.

Natural multiple infection of insects with baculoviruses and microsporidia were reported in *Pseudaletia unipuncta,*[16] *Agrotis segetum, Triphaena pronuba, Heliothis zea,* and *H. armigera,*[21] and *Lymantria dispar.*[30] Bergold[31] found GV, CPV, and microsporidia in *C. fumiferana* in a laboratory culture, but did not state definitely that multiple infection occurred.

More complex multiple infections are possible. Hess et al.[32] reported an example of multiple infection in which four distinct viruses occurred in the same cell. These viruses included a NPV, a CPV, and two nonoccluded icosahedral viruses. A fifth nonoccluded virus was present in the same individuals. The authors indicated that the inocula used in their tests were from a "natural source", but the complete history of each inoculum was not given.

It is clear from the above examples that multiple infections of baculoviruses and other microorganisms can occur frequently. The possibilities are perhaps limited only by the absence of the correct conditions of host and pathogen occurrence in time and space, and the ability of trained observers to recognize and record such infections.

Few of the above reports provide more than records of the occurrence of multiple infections. Quantitative studies in natural populations are very limited. Tanada and Chang[16] quantified the incidence of NPV and GV, and of NPV and microsporidiosis in *P. unipuncta* populations in turf in Hawaii. They suggested that studies of these three pathogens would be of interest because of the effects of their interactions on the epizootic wave. Tanada and Omi[20] also quantified multiple infection of *S. exigua* and *Autographa californica* with the NPV and CPV noted above. Despite the presence and often high incidence of two or three different viruses in each species, incidence of multiple infection never exceeded 3%.

B. Studies on the Nature of Baculovirus-Pathogen Interactions in the Host

Although the published information on naturally occurring multiple infections and their effects on host populations are limited, numerous laboratory studies have been conducted on the nature of multiple infections involving baculoviruses. Such studies are necessary for developing a clear understanding of the potential for enhancing the activity of baculoviruses applied as microbial insecticides, either alone against pest populations which may already contain entomopathogens, or in combination with other entomopathogens.

1. Latency

The question of occurrence and nature of latency of entomopathogenic viruses is still largely unresolved. Several researchers have suggested that latent viruses have been activated in host insects following infection by baculoviruses. An early observation by Bergold[31] indicated that a latent NPV could be activated by feeding larvae of *C. fumiferana* a noninfectious GV from *Cacoecia murinana.* Lipa[21] obtained similar results when he fed NPV from *Agrotis segetum* to *A. c-nigrum.* Although the latter species was not susceptible to the NPV from *A. segetum,* a high level of CPV infection occurred following NPV treatment. Lipa interpreted these results as NPV activation of the CPV. Lipa recorded similar results following treatment of *Scotogramma trifolii* with a GV.[21] Yadava[33] also reported stimulation of a latent CPV of *Lymantria monacha* by feeding NPV in combination with several inorganic chemicals. Amargier et al.[34] found densonucleosis virus occurring in *Galleria mellonella*

following infection with NPV. They suggested that the former could have been present in a latent state and may have been activated by the presence of the latter virus.

Such interactions are subject to controversy, particularly with respect to the question of inoculum purity. As mentioned earlier, Hess et al.[32] found a presumably pure preparation of NPV to contain five different viruses.

2. Multiple Infection Studies
a. Competition for Replication Sites

In general, the results of mixed infection studies with baculoviruses indicate that insects can be simultaneously infected, but that specific sites of replication tend to be partitioned between or among the pathogens involved. This partitioning can occur between tissues as shown for two NPVs.[35] It can also occur between cells within the same tissue as shown for two NPVs,[22,36] for a GV and an NPV,[37-39] for a densonucleosis virus and an NPV,[40] and for an iridescent virus and an NPV.[41] The partitioning can also occur between the nucleus and cytoplasm within the same cell with a GV and an NPV,[42] with an NPV and *Nosema* spp.,[43] and with iridescent viruses and NPVs.[40,44-46] In only a few cases has a baculovirus been reported to occur simultaneously in the nucleus of a cell with a second virus.[32,34]

The reasons for the partitioning of replication sites by viruses in insects are not well understood, although they are important to our understanding of multiple infection and its practical implications. From the examples just cited, it is clear that infection of an insect tissue, cell, or subcellular region by one pathogen may result in failure of a second pathogen to be able to utilize the same site. Several factors, either alone or in combination, may be involved and will be discussed.

b. Sequence of Infection

One of the most obvious factors associated with this partitioning is that of priority of infection. Ritter and Tanada[36] proposed that one agent may enter a given cell faster than another and transform the cell in such a way as to make it refractive to infection by the second. Lowe and Paschke[37] proposed that an "interferon like" substance may be produced to prevent NPV infection in susceptible tissues following establishment of a GV infection. Kurstak and Garzon[40] suggested a similar possibility for densovirus inhibition of NPV. Kelly[41] suggested that the presence of one virus in the replicating stage was capable of preventing infection of cultured cells by a second virus. Even partially heat inactivated virus (Type 22 iridescent virus or *Trichoplusia ni* NPV) was capable of causing this inhibition, although to a much reduced degree. This may explain the results of a number of studies which show that sequence of infection is often important in determining which pathogen will successfully replicate in host tissues when the host is multiply infected.

Inoculation of *Choristoneura fumiferana* with GV 1 or 2 days before inoculation with NPV resulted in replication of both.[42] The degree of infection by each was related to the interval between inoculations. When fed simultaneously in "equal" dosages, only NPV infection was found. Larvae of *T. ni* infected with GV 5 to 7 days before being given NPV produced only GV infection, while simultaneous inoculation resulted in about equal infection levels.[37] NPV incidence was low in *C. fumiferana* and *Malacosoma disstria* when administered simultaneously with CPV but increased if administered prior to CPV.[47] Injac[48] found that the incidence of GV or NPV in *Hyphantria cunea* was higher for whichever was administered first. Densovirus replication in cell culture was mostly eliminated if infection was attempted 24 to 26 hr following NPV infection.[40] Kimura and McIntosh[46] found a similar relationship between *Chilo* iridescent virus (CIV) and NPV. A 24-hr incubation period for CIV prevented NPV development, but CIV did not prevent NPV development if NPV had at least a 2-hr advantage. Ritter and Tanada's[36] HNPV strain of NPV occurred in mixed infections with their TNPV strain in *P. unipuncta* if fed prior to the TNPV strain. If fed after the TNPV strain, it did not occur at all. Inhibition of one strain of GV and of NPV by a second GV following pretreatment by the latter was reported by Boucias and Nordin.[49] In mixed infections of multiply (MNPV) and singly (SNPV) embedded NPVs of *Orgyia pseudotsugata,* MNPV tended to be the virus produced. However, SNPV was produced if fed 24 hr prior to MNPV.[50]

c. Temperature

Several reports suggest that the results of multiple infections may be directly related to the temperatures at which hosts or host tissues are incubated and that each pathogen in a combination may have a unique optimal developmental temperature. Kelly[41] found that both *Simulium* iridescent virus (SIV) and NPV were able to reproduce in cultures of *Spodoptera frugiperda* cells. When cell cultures were multiply infected, SIV replicated, even if administered 8 hr after NPV, if temperatures were optimal for the iridescent virus (21°C). If optimal for NPV (28°C), iridescent virus replication was prevented, but NPV infection was also reduced, indicating a fairly complicated interaction. Shvetsova and Tsai[38] also found that below 25°C, GV appeared to have an advantage over NPV in *Agrotis segetum* larvae. NPV occurred without GV only when larvae were incubated at 25°C, and then only in 33% of the test insects. Thus, it appears that temperature can significantly affect the expression of each pathogen in a dual infection. This could have significant impact on the dynamics of natural epizootics when two viruses are present in a host population. Expression of either or both in a given season could be directly related to seasonal temperature regimes.

d. Dosage

The competitiveness of two pathogens in a host or host tissue is also related to relative dosages of each. In general, as the dosage of one agent is increased relative to the second, the relative number of target cells infected by the first will increase. Tanada et al.[35] found that one strain of NPV from *P. unipuncta* was suppressed by a second strain unless dosage of the first was tenfold greater than that of the second. Similarly, Bird[42] found that *C. fumiferana* GV had to be administered in high dosages if it was to be expressed in the presence of NPV. Bird[47] also found that NPV dosages had to be high (in polyhedra counts) if NPV was to be expressed relative to CPV following multiple infection of *C. fumiferana* and *M. disstria.* Following tests combining densonucleosis virus or *Tipula* iridescent virus (TIV) with NPV in cell cultures, Kurstak and Garzon[40] concluded that both viruses could replicate if titers of each were adjusted properly. A nonhomologous GV was able to prevent infection of *H. cunea* by its homologous GV if given at a sufficiently high dosage.[49] Hughes[50] found that an SNPV of *O. pseudotsugata* produced more polyhedra than an MNPV in the same host if given in higher dosages or in equal, but low dosages. Hughes pointed out, however, that polyhedra of MNPVs and SNPVs of *O. pseudotsugata* actually contain very different numbers of virions. Thus, a dosage advantage, while not obvious from dosages expressed in polyhedral numbers, might have caused the higher incidence of MNPV infection noted in Hughes' assays.

e. Mechanical Disruption

Baculoviruses normally invade via the midgut epithelium. Any factor which inhibits or disrupts the normal functioning of this tissue could influence susceptibility to baculoviruses. Lipa[51] suggested that damage to the gut by *Plistophora noctuidae* might have facilitated entry of NPV into the hemocoel of *A. segetum.* In multiple infections involving baculovirus and *Bacillus thuringiensis,* Chancey et al.[52] noted reduced mortality in *T. ni* and suggested that disruption of normal gut function by *B. thuringiensis* may have (1) stopped host feeding and prevented ingestion of a sufficient NPV dosage to cause death, or (2) interfered with gut function, thus preventing virus uptake. Matter and Zohdy[53] reported an increase in

Heliothis armigera mortality following treatment with *B. thuringiensis* and NPV. They suggested that this could be due to the retarded larval development caused by *B. thuringiensis*, which allowed the NPV to develop.

f. Host Age

The age of the host can apparently affect the expression of baculoviruses in multiple infections. Bird,[47] for example, found that *C. fumiferana* and *M. disstria* larvae were less susceptible to a given CPV dosage in later instars than in earlier instars. His combination trials indicated that CPV infection blocked expression of NPV infection in the second, third, and fourth larval instars at selected dosages for both species but not in the first instar of *M. disstria*. Tanada[15] found that even though *Pseudaletia unipuncta* larvae became less susceptible to infection by GV and NPV viruses individually as they developed through their instars, the older larvae were still highly susceptible to a mixture of the viruses. This susceptibility was lower, however, than in the earlier instars.

g. Diet

Lipa[21] reported that higher levels of GV and *Plistophora noctuidae* infection occurred in *H. armigera* larvae fed bean leaves than in larvae fed pea leaves. These results, however, were based on very small samples. Tanada[15] found no differences in susceptibility of *Pseudaletia unipuncta* to GV-NPV mixtures when fed on different monocots.

h. Host

Relative susceptibility of a host to homologous or nonhomologous pathogens in mixtures can influence the expression of each in that host. This was demonstrated by Smith and Xeros[24] when they attempted cross infection of 12 lepidopteran species with a mixture of NPV and CPV isolated from *Vanessa (Pyrameis) cardui*. A total of 11 of the 12 species became infected with polyhedrosis infections. However, not all developed both diseases or responded similarly to both. CPV infections predominated in some species, NPV in others. Multiple infections occurred in most, although the level of polyhedral production of one type was often very low relative to the other. Tanada[54] did not obtain either single or dual infection of four species of Lepidoptera with a mixture of GV and NPV from *P. unipuncta*. Lipa[21] was able to induce double infections of GV plus *Plistophora noctuidae* and GV plus NPV, and triple infections with NPV plus CPV plus *P. noctuidae* in several noctuid species which were not the hosts of origin for one or more of the pathogens involved.

3. Effects of Multiple Infection on Baculovirus Replication Products

The nature of viruses produced following superinfections may be affected by interactions occurring during the course of multiple infections. Injac[48] described abnormalities in both GV and NPV following replication in the same insect. These included abnormal shapes and sizes of GV occlusion bodies and the presence of empty occlusions. While such abnormalities had been noted previously by Bird[42] and Arnott and Smith[55] in infections involving GV only, Injac reported a much higher incidence when multiple infections were involved. They also noted fewer polyhedra in cells infected by NPV, and that the number of virions per envelope within polyhedra was reduced. A similar production of abnormal GV occlusions was reported by Bird[56] following multiple infection of *C. fumiferana* with two "strains" of GV. Iridescent viruses, in combination with NPVs, have resulted in similar problems. Kurstak and Garzon[40] found lowered NPV virion numbers and reduced incorporation of virions into polyhedra in TIV-NPV infected cells. In some cells, there was an accumulation of incomplete virus components and disproportionate deoxyribonucleic acid (DNA) synthesis, resulting in production of nuclear fibrilar material but no virions. The degree of nuclear abnormalities was related to the extent of TIV development in the cytoplasm. At the same time, the TIV virus in the cytoplasm showed hollow particles. Kimura and McIntosh[46] reported that cell cultures infected first with CIV and later with *Autographa californica* NPV produced NPV virions but did not produce occlusion bodies.

At the molecular level, Croizier et al.[57] found that multiple infection of *G. mellonella* with a *G. mellonella* NPV and *A. californica* NPV, followed by four passages of progeny through *G. mellonella,* resulted in loss of parental types and development of a new genome, as revealed through comparative DNA analyses. They suggested that recombination had occurred. This supported previous work by Brown et al.[58] and Lee and Miller[59] who reached similar conclusions using mixtures of various *A. californica* NPV mutants, obtaining progeny with shared or modified characteristics. The nature of the products described above, particularly with respect to abnormalities, could seriously affect the ability of baculoviruses to replicate in natural host populations, particularly if significantly high numbers of defective products were produced. On the other hand, recombination studies might be expected to produce at least some products which could contain improved characteristics of potency, environmental stability, etc.

III. MORTALITY STUDIES INVOLVING BACULOVIRUSES AND OTHER MORTALITY AGENTS

The previous section dealt mainly with the nature of interactions between baculoviruses and other entomopathogenic organisms and how they compete with each other for replication sites within the host under a variety of conditions. This section will examine host mortality in populations under stress by two separate mortality factors. The discussion will deal with interactions between baculoviruses and other agents, the latter including chemical as well as microbial agents. This aspect of interactions is of both fundamental and applied interest. Elucidation of mortality patterns following multiple insults in a host population is very difficult to evaluate under field conditions. Carefully controlled laboratory studies are needed to monitor individual and joint action in such studies. Both types of studies are dependent on effective statistical analyses. Because experimental design and analysis are crucial to such work, a discussion of methodologies is appropriate at this point.

A. Statistical Methodologies

Before discussing specific experimental design and analysis procedures, it is necessary to define several terms as a basis for further discussion and for examining past research results in a somewhat uniform fashion. The terminology utilized in interaction studies is perhaps more extensive and confusing than the physiological, biological, and mathematical phenomena that it describes. If we simplify responses of populations to challenge by two mortality causing agents to (1) *additivity* of effects, (2) *synergism* (those which are greater than additive), and (3) *antagonism* (those which produce less than additive effects), we have a basis for understanding other terminology. It is these three types of responses that are of importance from an applied point of view. From a mode-of-action standpoint, many other types of responses are possible and have been discussed by other authors.[11,60,61] The simplest situation, occurring when two mortality agents are operating in the same host population, involves additivity of effects of separate agents. If each agent acts independently of the other and all hosts are equally susceptible to each agent, then mortality following dual challenge should follow the simple probability formula, $p = p_1 + p_2 - p_1 p_2$, where p is the proportion of the total population responding to dual treatment, p_1 and p_2 are the proportions responding to mortality agents 1 and 2, respectively, and $P_1 P_2$ is the proportion susceptible to both agents, assuming each acts independently.[11,62,63] If p is considered the expected mortality based on separate, experimentally defined values for p_1 and p_2, then any significant deviation in mortality from p when p_1 and p_2, are given simultaneously would indicate synergism (p [actual] > p [expected]) or antagonism (p [actual] < p [expected]).

The most serious problem encountered in the study of interactions between pathogens and other mortality inducing agents, whether these are other pathogens or chemicals, has been the development of analyses which allow the experimenter to state, with some degree of confidence, the exact nature of the interaction. Many of the early studies on combinations provided indications of interactions, but lacked precise control of dosages, making interpretation of the interactions somewhat speculative. Paschke et al.[64] stressed the importance of critical statistical evaluation of bioassay data in interpreting results of pathogen-host interactions and of basing combination tests on known dosage-mortality responses for the individual components of the combination.

As pointed out by Benz[11] and Harper,[65] many tests of pathogen-chemical or pathogen-pathogen combinations have not been designed properly to allow a clear understanding of any interactions which may exist. Both authors discussed the importance of appropriate controls for determining the existence of interactions between two mortality agents.

In order to determine whether interactions occur, as indicated by the formula given above, it is necessary to know the activity of each component separately, but at the same dosage used when they are combined. Harper and Thompson[62] utilized this design for combined dosage studies with DDT and NPV in *Peridroma saucia* larvae. They then used a chi-square test to determine whether the response to combined challenges differed from the response expected if the two agents acted in an additive fashion.

Such tests require that sublethal dosages (herein defined as dosages which result in less than 100% mortality in a bioassay population) be administered if percent mortality is to be the criterion measured. Chi-square comparisons of means with this procedure involve only one degree of freedom. Significant differences between observed and expected values may be hard to detect when combinations involve high or low LD values for either or both mortality agents.

If responses other than percent mortality are of interest, e.g., time to death, food consumption, locomotion, or other behavioral characteristics, it may not be necessary to work with sublethal dosages. However, the experimental design described must be maintained.

When time sequence experiments are utilized, the control dosages must be determined for each specific age group challenged. If, for example, a cohort of larvae is to be challenged at age X with a pathogen and at age X + 2 days with a chemical, the appropriate controls would include the single effects of the pathogen in insects of age X and the single effects of the chemical in insects of age X + 2 days.

Statistical procedures for analyzing responses to a range of combinations have been reported.[60,61] All of these have limitations which restrict their applicability to the types of interactions discussed here. Tammes[66] used a graphical method for comparing combined effects of pesticide mixtures. This method, utilizing bolograms, assumes that additivity of separate effects occurs if the amounts of each which separately produce a given mortality level, e.g., LD_{50}, can be mixed in proportions of their LD_{50} dosages and together provide an LD_{50} response. If less than the expected level of one is required to kill 50% of the test population when mixed with the reduced level of the second, synergism is indicated, and the reverse for antagonism. If several LD_{50} values are determined for agent A at varying levels of agent B, an isobole can be determined which will indicate a tendency toward deviation from additivity. The method requires an arbitrary decision as to how much deviation is required before synergism or antagonism occurs. Bakuniak[67] used this method and considered that synergism occurred when the isobole crossed the line connecting $^1/_2$ LD_{50} values. The technique appears to provide an excellent indicator of the type of interaction involved, but requires multiple bioassays, considerable experimental material, and, as stated, requires making a subjective decision as to when a response curve represents a synergistic or an antagonistic response.

Triggiani[68] produced dosage-mortality response curves for both NPV and *B. thuringiensis,* used alone as well as in combination, against *L. dispar*. From these curves, LD_{50} values and their confidence intervals were compared as a means of determining the nature of any interaction involved. Those LD_{50} values which were significantly different would represent nonadditive interactions.

Development of statistical methodologies for analyzing the responses of laboratory populations to single and combined challenges with two mortality factors is currently the most critical need in the field of interaction studies. With current computational abilities, it should be possible to generate three-dimensional response surfaces for interactions between two mortality factors operating simultaneously in the same host population. When these are available, it will be possible to predict the exact nature of any combination of rates based on limited numbers of experimental treatments. It will also be possible to compare responses from combinations of different agents and to begin to classify them. Until such analyses are available, we will continue to have to test many individual dosages and their combinations to obtain a less than complete overall response pattern.

B. Studies of Baculovirus-Pathogen Interactions
1. Controlled Laboratory Studies

Baculoviruses have been used as components of numerous studies of multiple infections involving several other pathogens, including other baculoviruses. Lowe and Paschke,[69] using the concept that the mortality response to two times the LD_{50} dosage of either pathogen should equal the response to the combination of L_{50}s of both, concluded that *T. ni* NPV and GV combinations were additive at LD_{50} dosages assayed in *T. ni*. Similar findings for GV-NPV combination studies with the tortricid, *Adoxophyes orana*, were reported by Fluckiger.[70] Whitlock,[71] however, using a GV and NPV in trials with *H. armigera*, concluded that antagonism (interference) occurred between the two organisms.

Tanada[15] demonstrated an obvious increase in mortality of *Pseudaletia unipuncta* larvae to combined dosages of GV and NPV over the combined individual effects. In the mixed inoculations, dosages of the individual viruses were approximately 50% of the single virus dosages. The nature of this interaction was further elucidated by experiments which showed that the GV was acting as the synergist, and that this activity was expressed even after heat inactivation of the GV.[72] The synergistic activity was apparently strain specific and dosage related. A second *P. unipuncta* GV isolate did not elicit the same response, nor did the synergistic strain when administered at sublethal dosages.[73] This synergistic factor was later found to be associated with the occlusion body rather than the virion.[74] Neither Lowe and Paschke,[69] nor Whitlock[71] could demonstrate this synergistic response in GV-NPV combinations using *T. ni* and *H. armigera*, respectively. In both studies normal and heat-inactivated viruses were tested.

Bird's[47] indirect method of reporting dosage responses via time in days to 50% mortality by instar and dosage led Bird to conclude that interference (antagonism), no interference (additivity), or synergism were all possible outcomes of CPV plus NPV infection in *C. fumiferana* and *M. disstria*; the observed outcomes were dependent on both dosages used and age of host. No dosage-mortality response curves were presented to allow testing of these conclusions, however.

Results have varied considerably in assays where baculoviruses and bacteria were combined. Chancey et al.[52] presented data on specific dosages of *B. thuringiensis* and *T. ni* NPV that suggest an apparent antagonism between the two in *T. ni* larvae, but only in selected dosage combinations. Statistical comparisons were not made. McVay et al.[75] utilized a chi-square test to determine differences in observed and expected responses of *T. ni* larvae to *B. thuringiensis* and NPV combinations and found only additive effects at the dosages tested. Lipa et al.[76] utilized carefully controlled dosage experiments to determine interactions between *B. thuringiensis* and NPV in *S. exigua*. By comparing expected and observed LD_{50}

values from combination tests, they concluded that a synergistic interaction occurred in the mixed infection. They confirmed this through development of "interaction coefficients" and use of the isobole method. While Schmid[77] found GV-*B. thuringiensis* combinations to be somewhat antagonistic in laboratory studies when administered simultaneously to larval *Zeiraphera diniana,* the same study demonstrated the two agents to be synergistic if the virus was already present in the host at sublethal levels prior to administration of the bacterium.

Combinations of *Heliothis* NPV or *Autographa californica* NPV with *B. thuringiensis* in assays conducted by Luttrell et al.[78] against both *H. zea* and *H. virescens* did not show any beneficial interaction. Although not statistically analyzed, their data suggested that each virus-bacterium combination used was somewhat antagonistic in each host species. Trials by Stelzer[79] suggested that some advantage might be obtained by utilizing combinations of *B. thuringiensis* and NPV for control of *Malacosoma fragile,* but the experimental design used did not provide suitable controls to test this hypothesis statistically. Data presented by Triggiani,[68] however, do indicate a synergistic response in *L. dispar* when fed a combination of NPV and *B. thuringiensis.* In studies with other bacterial species, To et al.[80] found no synergistic response when *T. ni* was simultaneously infected with a pathogenic isolate of *Bacillus cereus* and NPV. Similarly, Schmid[77] found *B. sphaericus* and GV to be additive against *Z. diniana,* with some evidence for a weak synergistic action at one combination of dosages tested. Vago,[81] while studying NPV infection in *Thaumetopoea pityocampa* larvae, noted that septicemia caused by *Bacterium paracoli* progressed more rapidly in larvae which were already infected with virus.

In such cases, one pathogen might obscure the presence of the second due to its dominant numbers or more obvious symptoms. Such cases probably occur frequently when microbial insecticides are applied to control pest populations, undoubtedly resulting in incorrect or at least incomplete diagnoses of the true causes of death in individual cadavers sampled.

Bacteria normally develop in larvae which have died from baculovirus infection. Because it has been suggested that highly purified baculovirus occlusion bodies are less infectious than unpurified preparations, several investigators have tested the contaminating bacteria for pathogenicity. Tanada[14] isolated such bacteria from GV infected *Pieris rapae* and found them to be nonpathogenic. He did not test for interaction with the GV. Magnoler[82] found purified *L. dispar* NPV occlusion bodies less lethal than those which were nonpurified. He suggested the activity of saprophytic bacteria or their toxic metabolites as one possible explanation for increased pathogenicity of the nonpurified NPV. Lipa[21] fed *Brevibacterium maris* and NPV to *Agrotis segetum.* While the bacterium caused little mortality alone, mortality of the mixture was increased over three times that of virus alone in second-instar larvae, indicting an interaction. Considering that commercial preparations of baculoviruses normally contain levels of saprophytic bacteria, this subject certainly deserves more attention.

A well designed and thoroughly analyzed study by Fuxa[83] indicated that *Heliothis* NPV and *Vairimorpha necatrix,* when fed to *H. zea* larvae, were antagonistic at most dosage combinations tested. Fuxa utilized both isoboles and Z-tests to analyze results.

Although dual infections may not necessarily result in enhanced levels of mortality, the rate of mortality of dual infections as compared to that of single infections may be altered in some cases. This could certainly have practical implications. Shvetsova and Tsai[38] determined that dual infections in *A. segetum* by GV and NPV resulted in mortality rates similar to those of the faster acting NPV. No time advantage was gained with the combination. Lowe and Paschke[69] reported similar results with *Trichoplusia ni* NPV and GV. Combinations showed mean mortality times intermediate between the faster acting NPV and the slower GV. CPV-NPV combinations killed *C. fumiferana* larvae at about the same rate as the more dominant CPV,[47] as did CPV plus CIV combinations in *Bombyx mori.*[45] Combinations of *Nosema* sp. and NPV in *Hyphantria cunea* resulted in increases in LT_{50} values[84] as did two

GVs in the same species.[49] Young et al.[85] reported similar results with *Bacillus thuringiensis*-NPV combinations in *T. ni.*

2. Field Trials

Field trials involving combinations of baculoviruses and other pathogens have been relatively few in number. The majority have involved combinations of NPVs or GVs with *B. thuringiensis*. As early as 1959, McEwen and Hervey[86] applied these combinations for control of *T. ni* and *P. rapae* on cabbage. Their tests were established to obtain control through the separate actions of each of the three pathogens on the basis that they would partition the host complex and yield an additive mortality response. This, in fact, was shown. Each virus reduced its respective host population. The commercial *B. thuringiensis* strain utilized in 1959 was not extremely effective against *T. ni.* In combination with the *P. rapae* GV, *B. thuringiensis* gave poor *T. ni* control. In combination with *T. ni* NPV, both species were controlled because *P. rapae* was more susceptible to *B. thuringiensis* than *T. ni.* Semel[87] made single applications of *T. ni* NPV, *B. thuringiensis,* or these two pathogens in combination to determine the possibility of extended control. Again, the early formulations of *B. thuringiensis* had only a small impact on *T. ni,* and the combination produced effects no different from those of the virus alone. The virus apparently cycled in the treated plots, further reducing the *T. ni* population. Results where the virus and bacterium were combined were identical, indicating no obvious interaction.

Baculoviruses infecting *T. ni* and *P. rapae* were applied in various combinations with *B. thuringiensis* for control of this pest complex on cabbage by Jaques[88,89] and Jaques and Laing.[90] While control of the pest complex was generally good, the tests were not established in such a way as to determine interactions. Each virus suppressed its specific host, and *B. thuringiensis* affected both as well as other pests which may have been present. The overall result was adequate total population reduction and damage prevention.[88]

Stelzer[91] applied *B. thuringiensis* plus *Malacosoma fragile incurva* NPV to mixed cottonwood-willow forest infested with *M. fragile incurva.* No single component treatments were made so interactions could not be determined. The bacterium could be detected in combination with NPV at 9 days posttreatment, but only virus was found in later samples, and no measurement of the role of each component was possible. Stelzer et al.[92] treated mixed Douglas fir-grand fir forest plots with either a virus, *B. thuringiensis,* or a mixture of the virus and *B. thuringiensis* that contained an amount of each that was sufficient for *Orgyia pseudotsugata* control. Population reduction occurred more rapidly with the combination treatment, and continued cycling of the virus was not impeded. The ultimate defoliation level was equal for plots treated with either component separately, or the combination. Both pathogens were present in some cadavers in the plots treated with the combination, but at day 7 posttreatment, the bacterium was the predominant mortality factor. The data suggested that initially *B. thuringiensis* was responsible for most larval death followed by a prolonged phase, days 12 to 35 posttreatment, when the virus was responsible for most larval mortality.

Trials by Schmid[77] using *B. thuringiensis* and GV for control of *Z. diniana* showed an antagonistic response between the two. Control was much higher in plots treated with the virus alone. When the bacterium was added to the same virus dosage, total mortality was reduced significantly.

Oatman et al.[93] controlled *Heliothis zea* on sweet corn effectively with multiple applications of NPV and *B. thuringiensis* applied to silks. However, no rate information was presented for the combinations used and it is impossible to critically evaluate the results reported. Combining *B. thuringiensis* with *H. zea* NPV in baits did not improve control of *H. zea* or *T. ni* on tomatoes over use of the baited *B. thuringiensis* alone.[94] In one field trial on cotton, Bell and Romine[95] obtained significantly higher yields when *B. thuringiensis* was added to an adjuvant mixture containing *Autographa californica* NPV in comparison to the virus-adjuvant used alone at the same rate. In a second trial, no improvement was noted, although reduced rates of virus were used in the combination treatments, so valid comparisons of components were not possible.

Combinations of *T. ni* NPV or *A. californica* NPV with *P. rapae* GV for control of the *T. ni* and *P. rapae* pest complex on cabbage have been field-tested by Jaques[88,89,96] and Jaques and Laing.[90] The combinations were used to provide control of each pest by one of the viruses and were not established to test for interactions. In general, results of these reports support the concept of control of a pest complex through mixtures of mortality agents and exemplify the interspecific economic synergism concept of Benz.[11]

It appears that no clear generalizations can be made regarding the desirability of using baculoviruses in combination with other pathogens for insect control. The most frequently observed response, both in the laboratory and field, seems to be that of additivity of effects. Several deviations from additivity have been noted, suggesting that each host-baculovirus-pathogen combination may behave quite differently, at least under the various conditions utilized in the trials reported to date. The cases in which nonadditivity has been reported have not involved major shifts from expected or additive responses. However, because several instances of antagonism have been reported, especially in the case of NPV or GV in combination with *B. thuringiensis*, caution should be exercised in proceeding with such combinations for host population regulation without adequate field testing.

C. Studies of Baculovirus-Chemical Interactions
1. Controlled Laboratory Studies

The use of chemical insecticides in combination with baculoviruses has received considerable attention since publication of Benz's[11] review on the subject. As Benz recognized, much of the work, especially field trials, to that date was empirical in nature and lacked controls necessary to gain complete understanding of any interactions that might be involved.

Early research emphasis centered on the question of compatibility. (See reviews by Benz[11] and Ignoffo and Montoya.[97]) With only a few exceptions, combinations of viruses and pesticides have been found to be compatible.

Theoretically chemical-baculovirus combinations should provide several distinct advantages for insect pest management programs. As noted by several authors,[4,11,65] these include the potential for reducing amounts of each agent used. Such reductions would mean potentially lower costs, lower environmental impacts, less damage to (and thus conservation of) beneficial organisms, and reduced selection pressure leading to the development of resistance to each agent. When pests in addition to the pathogen host are present, the chemical can provide control.

As studies in the area of combined mortality agents have progressed, particularly in the last 15 years, more investigators have recognized the importance of applying statistical tests to their data. In so doing, a variety of responses which have been obtained with combinations of chemicals and baculoviruses, have not been identified as providing any obvious patterns. Harper and Thompson[62] found that the overall response of *Peridroma saucia* larvae to DDT-NPV combinations was additive, regardless of dosages used, whether applied simultaneously, or whether one agent was administered 24 hr before the other. In several combinations, a higher than expected mortality rate was caused by either DDT or NPV, but no patterns or trends were evident. Schnyder[98] conducted somewhat similar tests with *Sterrha seriata* and concluded that sublethal dosages of DDT tended to reduce incidence of NPV, except in one test where an apparent synergistic response occurred. Statistical analyses of these data were not included. The results of Girardeau and Mitchell[99] are particularly interesting. By pre-stressing *T. ni* larvae with NPV using levels causing less than 10% mortality, they found that susceptibility to endrin, endosulfan, and trichlorfon was significantly increased if these chemicals were applied 24 or 48 hr after the virus. This increase was not seen when the two agents were applied simultaneously. Their data were analyzed by comparing LD_{50} values for each chemical used alone, simultaneously, or at 24 or 48 hr postvirus treatment.

Hunter et al.[100] tested the compatibility of malathion and the GV of *Plodia interpunctella*. In laboratory trials against this species, the combination appeared to provide additive control, especially as efficacy of the virus declined in storage. Loss of virus activity was made up for by presence of the chemical which alone killed <50% of the test population. Use of the combination against two other commonly associated stored product pests demonstrated the value of a mixture for control of pest complexes in which some species are not susceptible to the pathogen used. Both *Tribolium castaneum* and *Oryzaephilus mercator* were well controlled by the malathion, whether in mixture or not. Thus the mixture provided better control of this three-species complex than either agent alone.

Komolpith and Ramakrishnan[101] designed a good test for interactions between each of four chemical insecticides and *B. thuringiensis* and the NPV of *Spodoptera litura* in that host. By applying the criteria for interactions as given by Benz,[11] but failing to statistically compare observed and expected values, they drew several conclusions as to the effects of combinations which cannot be validated. Following application of a chi-square test to their data, most of the responses appear to be additive. As concluded by the authors, the pyrethrin-NPV combination did show synergistic activity. When Harpaz and Raccah[102] combined *S. littoralis* NPV with Phosfon, a plant growth retardant, they obtained a synergistic response in terms of larval mortality. This suggests that attention should be given, not only to microbial insecticide and chemical insecticide combinations, but to possible interactions between baculoviruses and any chemical agent which may be applied in the same system while the virus is active in that system.

Other authors have reported properly designed experiments with appropriate analyses. Luttrell et al.[103] demonstrated simple additivity between Elcarb (*Heliothis* NPV) and permethrin, EPN-methyl parathion, and methomyl when the chemicals were administered simultaneously or at 24 or 48 hr after the virus. Savanurmath and Mathad[104] bioassayed fenitrothion and NPV against the armyworm *Mythimna (Pseudaletia) separata* and found several cases of synergism, particularly when insecticidal treatment was made 7 days after treatment with the virus. An identical experiment was conducted by the same authors[105] using endosulfan and NPV against the same host insect with very similar results. No actual data were presented, however, and it was impossible to check the results. The authors again reported that the most common situation providing synergism was in assays involving insecticide applied 7 days after the virus. It appears that insects under virus stress were more susceptible to the insecticides than healthy insects.

Mohamed et al.[106] assayed 16 chemical pesticides in combination with *Heliothis* NPV against both first- and third-instar *H. virescens* on meridic diet. All tests involved simultaneous challenges. Most combinations resulted in additive responses. However, several resulted in either antagonistic or synergistic responses when evaluated at 7 days posttreatment or at pupation, and responses to a combination sometimes differed between the evaluation dates and between the instar cohorts utilized. In a second test, Mohamed et al.[107] utilized eight pesticides in combination with *Heliothis* NPV on field-gathered cotton leaves which were then fed to larvae in the laboratory. Again, most combinations resulted in additive responses, but several, particularly using first-instar larvae, resulted in antagonistic or synergistic responses. Only the fungicide fentin hydroxide affected the third-stage larvae (antagonism). This was somewhat surprising, since the same compounds resulted in a synergistic response when assayed on larval medium in the previous study. There was almost no agreement among those combinations showing synergistic or antagonistic responses between these two studies.

In addition to pesticides, several inorganic chemicals and other agents have been shown to interact synergistically when combined with baculoviruses. Among these are copper sulfate,[108,109] boric acid,[110,111] sodium silicate, plant ashes,[110] zinc sulfate, iron sulfate, magnesium nitrate,[109] and sorbic acid.[111] Many of these compounds, including boric and sorbic acids, sodium silicate, plant ashes, and copper sulfate tended to cause earlier mortality in combination with virus than was caused by virus infection alone.

2. Field Trials

Early field trials with mixtures of baculoviruses and chemicals were established principally to achieve control of pest complexes, i.e., one pest with a baculovirus, the others with a chemical. This concept was suggested by Genung[112] as a method of overcoming the specificity of *Trichoplusia ni* NPV used on cabbage infested simultaneously with other pest species. Genung's trials combining toxaphene and NPV at half the rates used alone provided control equal to that of either material used alone. Other chemicals were combined with *T. ni* NPV by McEwen and Hervey,[86] Hofmaster and Ditman,[113] Getzin,[114] Wolfenbarger,[115] and Jaques.[88,89,96,116] These tests demonstrated that combinations of baculoviruses with a variety of chemicals could provide acceptable control of the complex of defoliators attacking cabbage or other crucifers, but few tests included designs which allowed analysis of specific combined effects. The single exception was the work of Getzin[114] who used all appropriate controls for his combination treatments. By combining NPV with toxaphene or parathion, he reduced damage and larval counts well below the level obtained by use of either agent alone. The data even suggested a possible synergism between NPV and parathion.

On lettuce, Vail et al.[117] obtained good protection of heads from *T. ni* damage using acephate and *A. californica* NPV in combination. The combination was superior to the individual components in terms of plant protection (as compared to the virus used alone) or larval reduction (as compared to acephate used alone). The combination also performed as well as the chemical standard used in the test. Combinations of NPV and a low rate of DDT against *S. litura* on cauliflower in India appeared to protect plants better than either agent alone.[118] Endosulfan did not appear to enhance activity in this complex.

Combinations of *Heliothis* NPV and chemical insecticides have been field-tested for *H. zea* and *H. virescens* control on cotton.[103,119-122] Again most of these tests[119-121] provide empirical evidence that combinations can be effective against *Heliothis* sp., but without having included specific control treatments, the economic or biological justification for their use is impossible to determine from the data presented. Field tests conducted by Luttrell et al.[103] were well designed to determine the effects of combining agents. They demonstrated no benefit from combining methomyl, permethrin, or EPN-methyl parathion with Elcar®.

The tests for combinations of chlordimeform and *Baculovirus heliothis* (Elcar® and other formulations) reported by Yearian et al.[122] illustrate some of the difficulties in evaluating field combinations for efficacy. In their tests, some combinations provided more (or less) cotton yield than the chemical alone (appropriate controls were used for most combinations), but the results varied widely and the data did not allow clear distinctions between effects caused by the different treatments, despite rather substantial numerical differences in mean yields. Such variation is difficult to control in many field experiments and results in an inability to evaluate combinations of pathogens and chemical insecticides at the level of precision possible in the laboratory.

When endrin/bidrin, trichlorfon, and NPV were applied for control of *S. littoralis* on cotton in the United Arab Republic, a combination of endrin/bidrin plus NPV provided population reductions similar to that provided by the chemical used alone at a higher dosage, and much better than that provided by the NPV used alone at the combination rate.[123] A dosage of endrin/bidrin equivalent to that used in the combination was not tested. Dipterex reduced populations more when used alone than when used at an equal rate in combination with NPV. Statistical comparisons of treatments were not reported, however.

Hamm and Young[124] conducted a complicated but well-controlled experiment which demonstrated the effectiveness of *Heliothis* NPV-DDT combinations applied at different developmental stages of corn for *H. zea* suppression. They also utilized *S. frugiperda* NPV in a second combination test for control of a late-season pest complex. Their results demonstrate how combinations can be used in a management system which takes into account the interrelationships of virus dynamics, host-plant phenology, and host-insect biology. The most effective control of *H. zea* in early season corn was achieved with applications of virus plus DDT to corn tassels. Virus-infected larvae presumably moved to the silks and contaminated them. Continued use of the combination on silks (five applications) provided better protection than any other combination of tassel and silk treatments utilizing single agent applications at either plant development stage.

An interesting experiment for control of *Choristoneura fumiferana* with fenitrothion and its NPV was reported by Morris et al.[125] and Morris.[126] A nonreplicated field plot design provided for efficient application of chemical, chemical plus virus, and virus alone to respective blocks of white spruce and balsam fir in Ontario, Canada. From the data obtained, the authors concluded that the combination treatment appeared to provide the best long-term control because of natural recycling of the virus in the host population. This treatment was also associated with a high incidence of microsporidia infection in the second year postspray and appeared to have no deleterious effect on rates of parasitism in subsequent years.

IV. BACULOVIRUS-PARASITOID INTERACTIONS

Most, if not all, insect species which are known hosts of baculoviruses are also parasitized by one or more species of parasitoids. Interactions between these agents at the population level certainly occur, but very little quantitative documentation exists. Data on interactions at the individual host level has been of increasing interest in recent years. Both levels of interactions have been reviewed by several authors.[6,12,127] Early authors generally considered entomopathogens to be detrimental to the activity, survival, and the capacity of parasitoids to regulate host population levels.[128,129] These authors argued that the mass mortality caused by entomopathogens severely limited the ability of parasitoids to maintain stable populations. Kelsey,[130] however, reported on a natural, temporal partitioning of *Pieris rapae* populations by a GV and several parasitoids which provided adequate regulation in home gardens. The partitioning of host populations by these entomopathogens and parasitoids, their competition for hosts, the impact of this competition on survivorship of each, and the fitness of progeny deserve much consideration.

Laboratory studies have provided many insights into several of these potential interactions. It is clear from many studies on simultaneous parasitism involving parasitoids and baculoviruses that no direct viral infection of the parasitoids by the viruses occurs.[127,130-134] In each of these studies as well as others,[135-137] indirect mortality of parasitoids was demonstrated. In each parasitoid-virus-host system studied, the parasitoids died within the infected host if the host succumbed to infection before parasitoid development was complete. Successful parasitoid emergence was possible from virus-infected hosts, but normally only when parasitism of the host occurred before virus infection. The necessary time interval between parasitism and infection varied according to the species involved. Levin et al.[136] further showed a negative correlation between probability of parasitoid emergence and GV dosage in the host.

Kaya and Brayton[138] also demonstrated that the entomoparasitic nematode, *Neoaplectana carpocapsae*, was not adversely affected by the presence of GV in its host. The nematode reproduced normally in size and numbers in infected *Pseudaletia unipuncta*. Nematode tissues were not infected by the virus, although, as with endoparasitoids, GV was found in the gut lumen of the parasite.

In a series of reports, Kaya[139] and Kaya and Tanada[140-143] described a toxin produced in GV-infected Lepidoptera which killed parasitoids developing in virus-infected larvae. The response was recorded from GVs of several hosts and occurred in several host species. The importance of such a response under natural conditions has not been investigated.

Vail[134] demonstrated clearly that the rate of parasitism under field conditions can be greatly underestimated because of this negative interaction. Vail found up to 100% of NPV killed *T. ni* larvae to contain dead *Voria ruralis* parasitoids in some field collections in California.

Some parasitoids are apparently capable of surviving in virus-killed host cadavers. Biliotti[144] reported emergence of larval tachinids from several hosts up to 7 days after they died from polyhedral infections. Parasitoids emerging from virus-infected or virus-killed hosts are often of reduced size and shorter lived than normal.[132,144] Beegle and Oatman[127] reported that *Hyposoter exiguae* spent less time in NPV-infected *T. ni* hosts than in healthy hosts, but were otherwise unaffected.

Given that parasitoids can be adversely affected by active baculoviruses in their hosts, are mechanisms of avoidance of such hosts possible? No definitive studies are known, but several observations have been reported. Niklas[145] found that diseased, late-instar nun moth larvae were not parasitized by tachinids which normally utilized only late-instar hosts. Elsey and Rabb[137] concluded that the tachinid, *Voria ruralis,* parasitized NPV infected and non-infected *T. ni* equally, while Gosswald[146] stated that *Sarcophaga shutzei* parasitized only virus-infected hosts. It appears that an entire spectrum of behavioral responses exists which may represent a variety of adaptations on the part of the parasitoids. The latter observation[146] might suggest that some benefit is derived by certain parasitoid species from host infection.

Regardless of the impact on the parasitoid, the baculovirus appears to derive one great advantage from their interaction; that of transmission. Transmission occurs in two ways — through oviposition activities of female parasitoids and through contamination of host substrate by parasitoids contaminated during their development in virus-infected hosts. Transmission brought about as a result of oviposition has been demonstrated in numerous controlled studies.[127,130-133,147,148] Transmission occurs when a female stings healthy caterpillars after stinging a virus-infected host, presumably contaminating her ovipositor in the process. A tachinid which oviposits on the external surface of its host did not transmit virus in this fashion.[134] Interestingly, most of these authors found that adult female parasitoids which had emerged from baculovirus-infected hosts do not subsequently transmit infections to new hosts through oviposition. Transmission, when it was demonstrated from either male or female parasitoids after emergence, was mediated through substrate contamination. Parasitoids contaminated host food as they walked, defecated, or voided meconia.[127,133,134,149]

Both of these mechanisms undoubtedly occur in nature, although such transmission is very difficult to document. Early workers (see review by Steinhaus[12]) speculated that a baculovirus of the gypsy moth was possibly introduced as a contaminant on parasitoids introduced for control of that pest. This is generally accepted as the route of intercontinental transmission of the NPV of *Gilpinia (Diprion) hercyniae* into Canada.[150] Bird[150] suggested that parasitoids must be extremely important in between-tree transmission of NPVs of several sawfly species, basing conclusions on studies of epizootics in field populations. Smith et al.[149] also cited indirect evidence of transmission of GV by parasitoids in field populations of *Harrisina brillians.*

Given the interactions described above, the question arises as to whether artificial baculovirus applications might be disruptive to parasitoid populations. In studies by Kelsey,[130] Vail et al.,[151] and Hamm and Hare,[152] increases in virus incidence following virus applications resulted in decreases in percent parasitism in three distinctly different agroecosystems. It would appear that one mortality factor is simply being displaced by another in these cases. Whether imbalances that occur due to this shift are ecologically disruptive, i.e., are similar in nature to those caused by insecticides, is a question that merits further investigation.

The rapidly increasing volume of literature indicates quite clearly that important interactions do occur between parasitoids and baculoviruses. There is little doubt that we greatly underestimate the importance of parasitoids in the generation and maintenance of epizootics. Quantification of the vectoring capacity under actual field conditions may be extremely important in the success or failure of baculoviruses applied for pest control purposes. At the same time, recognition of potential negative interactions on parasitoid and predator populations should be considered. Although such negative interactions are unlikely to be of the magnitudes encountered with the use of broad spectrum chemical insecticides, they should be evaluated. An integrated approach to the use of baculoviruses seems appropriate. Sound quantitative studies are needed to effectively achieve this integration with predictable results.

V. SUMMARY

The subject of interactions between baculoviruses and other mortality agents has many facets. This chapter has attempted to summarize much of the known information. While not an exhaustive review, because of the inaccessibility of some literature, it provides insights into the principles involved and future research needs.

Baculoviruses interact with other pathogens. The nature of these interactions has been at least partially elucidated at varying levels of biological organization. The vast number of potentially interacting combinations of baculoviruses and other organisms suggests that this could be a never ending task. Development of improved statistical analyses for describing interactions could alleviate much of this burden. Responses of insects to two mortality agents should follow specific mathematical models. Models for different hosts may differ, but once established, the correct model for a given combination of host and mortality factors should be determinable from a relatively small amount of experimental data. Carefully controlled studies at the tissue level are needed to determine how *Bacillus thuringiensis* interferes with baculoviruses infection, as shown in numerous studies cited earlier. Studies are needed to determine why tissues are partitioned by two organisms simultaneously infecting the same insect and why an insect becomes refractory to infection by one virus after becoming infected by another. Several hypotheses have been proposed, but few tested. Answers to such questions could have very real practical implications. Have some of our applied field studies with entomopathogens shown variable results because of interactions? Populations with a moderate to high level of naturally occurring disease incidence may respond very differently to chemical or microbial control agents than do "healthy" populations. Most data available on the subject are empirical in nature, and thus controlled studies are needed.

In general we can say that the interactions between parasitoids and baculoviruses are less severe than those between chemical pesticides and parasitoids. Even so, reduction of a host population by an entomopathogen affects parasitoids indirectly by reducing the number of potential hosts. The immediate and long-term effects of this reduction on the parasitoid species, and more importantly on the pest population's stability, are almost totally left to speculation at this time.

If we are to be able to realize the potential of baculoviruses in integrated pest management programs, we must conduct considerably more research on their interactions with pathogens, chemicals, and parasitoids. Interactions must be thoroughly understood at the most basic level if we are to benefit from their positive potentials and avoid their negative effects.

REFERENCES

1. **Aizawa, K.,** The nature of infections caused by nuclear-polyhedrosis viruses, in *Insect Pathology — An Advanced Treatise,* Vol. I, Steinhaus, E. A., Ed., Academic Press, New York, 1963, chap. 12.
2. **Vago, C.,** Predispositions and interrelations in insect diseases, in *Insect Pathology — An Advanced Treatise,* Vol. I, Steinhaus, E. A., Ed., Academic Press, New York, 1963, chap. 11.
3. **Smith, K. M.,** *Insect Virology,* Academic Press, New York, 1967, chap. 10.
4. **Krieg, A.,** Interactions between pathogens, in *Microbial Control of Insects and Mites,* Burges, H. D. and Hussey, N. W., Eds., Academic Press, New York, 1971, chap. 21.
5. **Tanada, Y.,** Ecology of insect viruses, in *Perspectives in Forest Entomology,* Anderson, J. F. and Kaya, H. K., Eds., Academic Press, New York, 1976, chap. 18.
6. **Tanada, Y.,** Epizootiology of insect viruses, in *Biological Control of Insect Pests and Weeds,* DeBach, P., Ed., Reinhold, New York, 1964, chap. 19.
7. **Tanada, Y.,** Factors affecting the susceptibility of insects to viruses, *Entomophaga,* 10, 139, 1965.
8. **Tanada, Y.,** Interactions of insect viruses, with special emphasis on interference, in *The Cytoplasmic-Polyhedrosis Virus of the Silkworm,* Aruga, H. and Tanada, Y., Eds., University of Tokyo Press, Tokyo, 1971, chap. 10.
9. **Smith, K. M.,** *Virus-Insect Relationships,* Longman, New York, 1976, chap. 13.
10. **Aruga, H.,** Induction of virus infections, in *Insect Pathology — An Advanced Treatise,* Vol. I, Steinhaus, E. A., Ed., Academic Press, New York, 1963, chap. 15.
11. **Benz, G.,** Synergism of micro-organisms and chemical insecticides, in *Microbial Control of Insects and Mites,* Burges, H. D. and Hussey, N. W., Eds., Academic Press, New York, 1971, chap. 14.
12. **Steinhaus, E. A.,** The effects of disease on insect populations, *Hilgardia,* 23, 197, 1954.
13. **Paillot, A.,** Nouveau type de maladies à polyèdres ou polyédries observé chez les Chenilles d' *Euxoa (Agrotis) segetum* Schiff, *C. R. Acad. Sci.,* 202, 254, 1936.
14. **Tanada, Y.,** Descriptions and characteristics of a granulosis virus of the imported cabbage-worm, *Proc. Hawaiian Entomol. Soc.,* 15, 235, 1953.
15. **Tanada, Y.,** Some factors affecting the susceptibility of the armyworm to virus infections, *J. Econ. Entomol.,* 49, 52, 1956.
16. **Tanada, Y. and Chang, G. Y.,** An epizootic resulting from a microsporidian and two virus infections in the armyworm, *Pseudaletia unipuncta* (Haworth), *J. Insect Pathol.,* 4, 129, 1962.
17. **Steinhaus, E. A.,** New records of insect-virus diseases, *Hilgardia,* 26, 417, 1957.
18. **Steinhaus, E. A. and Marsh, G. A.,** Report of diagnoses of diseased insects 1951-1961, *Hilgardia,* 33, 349, 1962.
19. **Paschke, J. D. and Hamm, J. J.,** Granulosis-polyhedrosis complexes in loopers, *Proc. N. Cent., Branch-ESA,* 17, 148, 1962.
20. **Tanada, Y. and Omi, E. M.,** Epizootiology of virus diseases in three lepidopterous insect species on alfalfa, *Res. Popul. Ecol.,* 16, 59, 1974.
21. **Lipa, J. J.,** Studies on Interactions on Various Pathogens in One Insect Host (Cutworms), Final Rep., E21-Ent-17, FG-Po-194, Institute of Plant Protection, Poznan, Poland, 1971.
22. **Heimpel, A. M. and Adams, J. R.,** A new nuclear polyhedrosis of the cabbage looper, *Trichoplusia ni,* *J. Invertebr. Pathol.,* 8, 340, 1966.
23. **Hughes, K. M. and Addison, R. B.,** Two nuclear polyhedrosis viruses of the Douglas-fir tussock moth, *J. Invertebr. Pathol.,* 16, 196, 1970.
24. **Smith, K. M. and Xeros, N.,** Studies on the cross-transmission of polyhedral viruses: experiments with a new virus from *Pyrameis cardui,* the painted lady butterfly, *Parasitology,* 43, 178, 1953.
25. **Smith, K. M., Wyckoff, R. W. G., and Xeros, N.,** Polyhedral virus diseases affecting the larvae of the privet hawk moth *(Sphinx ligustri), Parasitology,* 42, 287, 1953.
26. **Biliotti, E., Grison, P., and Vago, C.,** Essai d'utilisation des polyèdres isolés de la processionaire due Pin, comme methode de lutte biologique contre cet insecte, *C. R. Acad. Sci.,* 243, 206, 1956.
27. **Vago, C. and Vasiljevic, L.,** Polyédrie cytoplasmique chez L'écaille fileuse *(Hyphantria cunea* Drury, Lep. Arctiidae), *Entomophaga,* 3, 197, 1958.
28. **Vago, C.,** On the pathogenesis of simultaneous virus infections in insects, *J. Insect Pathol.,* 1, 75, 1959.
29. **Bird, F. T. and Whalen, M. M.,** A nuclear and a cytoplasmic polyhedral virus disease of the spruce budworm, *Can. J. Zool.,* 32, 82, 1954.
30. **Saftoiu, A. and Caloianu-Iordachel, M.,** New ultrastructural data on the development of the protozoan *Nosema lymantria* (Weiser) (Microsporidia) intracellular parasite on *Lymantria dispar* L. (Lepidoptera), *Trav. Mus. Hist. Nat. "Grigore Antipa,"* 19, 83, 1978.
31. **Bergold, G. H.,** The polyhedral disease of the spruce budworm, *Choristoneura fumiferana* (Clem.) (Lepidoptera: Tortricidae), *Can. J. Zool.,* 29, 17, 1951.
32. **Hess, R. T., Summers, M. D., and Falcon, L. A.,** A mixed virus infection in midgut cells of *Autographa californica* and *Trichoplusia ni* larvae, *J. Ultrastruct. Res.,* 65, 253, 1978.

33. **Yadava, R. L.,** Studien über den Einfluss von Wasserglas und Borsäure als chemische Stressoren für die künstlich applizierte kern-und die latente Cytoplasmapolyedrose der Nonne (*Lymantria monacha* L.); mit einer Anmerkung uber die gegenseitige Beeinflussung beider Krankheiten, *Z. Angew. Entomol.,* 65, 175, 1970.

34. **Amargier, A., Meynadier, G., and Vago, C.,** Un complexe de viroses: polyédrie nucléaire et densonucléose chez le Lépidoptère *Galleria mellonella* L., *Mikroskopie,* 23, 245, 1968.

35. **Tanada, Y., Hukuhara, T., and Chang, Y.,** A strain of nuclear-polyedrosis virus causing extensive cellular hypertrophy, *J. Invertebr. Pathol.,* 13, 394, 1969.

36. **Ritter, K. S. and Tanada, Y.,** Interference between two nuclear polyhedrosis viruses of the armyworm, *Pseudaletia unipuncta* [Lep.: Noctuidae], *Entomophaga,* 23, 349, 1978.

37. **Lowe, R. E. and Paschke, J. D.,** Pathology of a double viral infection of *Trichoplusia ni, J. Invertebr. Pathol.,* 12, 438, 1968.

38. **Shvetsova, O. I. and Tsai, H.,** Virus diseases of *Agrotis segetum* Schiff. and *Hadena sordida* BKh. (Lepidoptera, Noctuidae) under conditions of simultaneous infection with granulosis and polyhedral disease, *Entomol. Rev.,* 41, 486, 1962.

39. **Watanabe, H. and Kobayashi, M.,** Histopathology of a granulosis in the larva of the fall webworm, *Hyphantria cunea, J. Invertebr. Pathol.,* 16, 71, 1970.

40. **Kurstak, E. and Garzon, S.,** Multiple infections of invertebrate cells by viruses, *Ann. N.Y. Acad. Sci.,* 266, 232, 1975.

41. **Kelly, D. C.,** Suppression of baculovirus and iridescent virus replication in dually infected cells, *Microbiologica,* 3, 177, 1980.

42. **Bird, F. T.,** Polyhedrosis and granulosis viruses causing single and double infections in the spruce budworm, *Choristoneura fumiferana* Clemens, *J. Insect Pathol.,* 1, 406, 1959.

43. **Atger, P. and Vago, C.,** Pathogenèse simultanée de virose et de nosemose chez les lépidoptères, *Rev. Pathol. Veg. Entomol. Agric. Fr.,* 39, 205, 1960.

44. **Garzon, S. and Kurstak, E.,** Infection double inhabituelle de cellules d'un arthropode par le virus de la polyédrie nucléaire (VPN) et le virus irisant de *Tipula* (TIV), *C. R. Acad. Sci. Paris,* 275, 507, 1972.

45. **Obha, M.,** Studies on the pathogenesis of *Chilo* iridescent virus. IV. Simultaneous infection of CIV and a nuclear polyhedrosis virus (in Japanese), *Sci. Bull. Fac. Agric. Kyushu Univ.,* 30, 83, 1975.

46. **Kimura, M. and McIntosh, A. H.,** Dual infection of the *Trichoplusia ni* cell line with the *Chilo* iridescent virus (CIV) and *Autographa californica* nuclear polyhedrosis virus, in *Intervertebrate Tissue Culture,* Kurstak, E. and Maramorosch, K., Eds., Academic Press, New York, 1976, chap. 39.

47. **Bird, F. T.,** Infection and mortality of spruce budworm, *Choristoneura fumiferana,* and forest tent caterpillar, *Malacosoma disstria,* caused by nuclear and cytoplasmic polyhedrosis viruses, *Can. Entomol.,* 101, 1269, 1969.

48. **Injac, M.,** Recherches sur les maladies virales du Lépidoptère *Hyphantria cunea* Drury, Docteur-Ingénieur Thesis, Univ. des Sciences Techniques du Languedoc, Montpellier, France, 1973.

49. **Boucias, D. G. and Nordin, G. L.,** Susceptibility of *Hyphantria cunea* infected with the *Diacrisia* granulosis virus to its homologous baculoviruses, *J. Invertebr. Pathol.,* 32, 341, 1978.

50. **Hughes, K. M.,** Some interactions of two baculoviruses of the Douglas-fir tussock moth (Lepidoptera: Lymantriidae), *Can. Entomol.,* 111, 521, 1979.

51. **Lipa, J. J.,** Interaction of microsporidan *Plistophora noctuidae* and nuclear polyhedrosis virus *Borrelinavirus agrotidis* in *Agrotis segetum* (Lepidoptera, Noctuidae), *Proc IV Int. Colloq. on Insect Pathol.,* College Park, Md., 1970, 152.

52. **Chancey, G., Jr., Yearian, W. C., and Young, S. Y.,** Pathogen mixtures to control insect pests, *Ark. Farm Res.,* 22(3), 9, 1973.

53. **Matter, M. M. and Zohdy, N. Z. M.,** Biotic efficiency of *Bacillus thuringiensis* Berl. and a nuclear-polyhedrosis virus on larvae of the American bollworm, *Heliothis armigera* Hbn. (Lepid., Noctuidae), *Z. Angew. Entomol.,* 92, 336, 1981.

54. **Tanada, Y.,** Descriptions and characteristics of a nuclear polyhedrosis virus and a granulosis virus of the armyworm, *Pseudaletia unipuncta* (Haworth) (Lepidoptera, Noctuidae), *J. Insect Pathol.,* 1, 197, 1959.

55. **Arnott, H. J. and Smith, K. M.,** Ultrastructure and formation of abnormal capsules in a granulosis virus of the moth *Plodia interpunctella* (Hbn.), *J. Ultrastruct. Res.,* 22, 136, 1968.

56. **Bird, F. T.,** Effects of mixed infections of two strains of granulosis virus of the spruce budworm, *Choristoneura fumiferana* (Lepidoptera: Tortricidae), on the formation of viral inclusion bodies, *Can. Entomol.,* 108, 865, 1976.

57. **Croizier, G., Godse, D., and Vlak, J.,** Sélection de types viraux dans les infections doubles à *Baculovirus* chez les larves de Lépidoptère, *C. R. Acad. Sci. Paris Ser. D.,* 290, 579, 1980.

58. **Brown, M., Crawford, A. M., and Faulkner, P.,** Genetic analysis of a baculovirus, *Autographa californica* nuclear polyhedrosis virus I. Isolation of temperature-sensitive mutants and assortment into complementation groups, *J. Virology,* 31, 190, 1979.

59. **Lee, H. H. and Miller, L. K.,** Isolation, complementation, and initial characterization of temperature-sensitive mutants of the baculovirus *Autographa californica* nuclear polyhedrosis virus, *J. Virology*, 31, 240, 1979.

60. **Finney, D. J.,** *Probit Analysis*, 3rd ed., Cambridge University Press, Cambridge, U.K., 1971, chap. 11.

61. **Hewlett, P. S.,** Joint action in insecticides, in *Advances in Pest Control Research*, Vol. 2, Metcalf, R. L., Ed., Interscience, New York, 1960, 27.

62. **Harper, J. D. and Thompson, C. G.,** Mortality in *Peridroma saucia* following single and combined challenge with DDT and nuclear polyhedrosis virus, *Proc. IV Int. Colloq. on Insect Pathol.*, College Park, Md., 1970, 307.

63. **Bliss, C. I.,** The toxicity of poisons applied jointly, *Ann. Appl. Biol.*, 26, 585, 1939.

64. **Paschke, J. D., Lowe, R. E., and Giese, R. L.,** Bioassay of the nucleopolyhedrosis and granulosis viruses of *Trichoplusia ni*, *J. Invertebr. Pathol.*, 10, 327, 1968.

65. **Harper, J. D.,** Synergistic interactions of control agents, *Proc. of Workshop on Insect Pest Management with Microbial Agents: Recent Advancements, Deficiencies, and Innovations*, Boyce Thompson Research Institute, Ithaca, N.Y., 1980, 51.

66. **Tammes, P. M. L.,** Isoboles, a graphic representation of synergism in pesticides, *Neth. J. Plant Pathol.*, 70, 73, 1964.

67. **Bakuniak, E.,** The synergism evaluating method of two-component insecticidal mixtures, *Pol. Pismo Entomol.*, 43, 395, 1973.

68. **Triggiani, O.,** Prove di suscettibilita delle larve della *Lymantria dispar* L. (Lep. Lymantriidae) a varie concentrazioni di *Bacillus thuringiensis* Berl. var. *kurstaki* e *Baculovirus* (sottogruppo A) tra di loro combinate, *Entomologica*, 16, 5, 1980.

69. **Lowe, R. E. and Paschke, J. D.,** Simultaneous infection with the nucleopolyhedrosis and granulosis viruses of *Trichoplusia ni*, *J. Invertebr. Pathol.*, 12, 86, 1968.

70. **Fluckiger, C. R.,** Untersuchungen uber drei Baculovirus — Isolate des Schalenwicklers *Adoxophyes orana* F.V.R. (Lep., Tortricidae), dessen Phanologie und erste Feldversuche, als Grundlagen zur mikrobiologischen Bekampfung dieses Obstschadlings, *Mitt. Schw. Entomol. Ges.*, 55, 241, 1982.

71. **Whitlock, V. H.,** Simultaneous treatments of *Heliothis armigera* with a nuclear polyhedrosis and a granulosis virus, *J. Invertebr. Pathol.*, 29, 297, 1977.

72. **Tanada, Y.,** Synergism between two viruses of the armyworm, *Pseudaletia unipuncta* (Haworth) Lepidoptera, Noctuidae), *J. Insect Pathol.*, 1, 215, 1959.

73. **Tanada, Y. and Hukuhara, T.,** A nonsynergistic strain of a granulosis virus of the armyworm, *Pseudaletia unipuncta*, *J. Invertebr. Pathol.*, 12, 263, 1968.

74. **Tanada, Y. and Hukuhara, T.,** Enhanced infection of a nuclear-polyhedrosis virus in larvae of the armyworm, *Pseudaletia unipuncta*, by a factor in the capsule of a granulosis virus, *J. Invertebr. Pathol.*, 17, 116, 1971.

75. **McVay, J. R., Gudauskas, R. T., and Harper, J. D.,** Effects of *Bacillus thuringiensis* nuclear-polyhedrosis virus mixtures on *Trichoplusia ni* larvae, *J. Invertebr. Pathol.*, 29, 367, 1977.

76. **Lipa, J. J., Slizynski, K., Ziemnicka, J., and Bartkowski, J.,** Interactions of *Bacillus thuringiensis* and nuclear polyhedrosis virus in *Spodoptera exigua*, in *Enviromental Quality and Safety*, Suppl. Vol. 3, Coulston, F. and Korte, F., Eds., Thieme Verlag, Stuttgart, 1975, 668.

77. **Schmid, A.,** Interferenz Zwischen dem spezifischen Granulosisvirus und zwei Bakterienparaparaten bei Raupen des Grauen larchenwicklers, *Zeiraphera diniana* (Gn.) *Mitt. Sch. Entomol. Ges.*, 48, 173, 1975.

78. **Luttrell, R. G., Young, S. Y., Yearian, W. C., and Horton, D. L.,** Evaluation of *Bacillus thuringiensis* — spray adjuvant — viral insecticide combinations against *Heliothis* spp. (Lepidoptera: Noctuidae), *Environ. Entomol.*, 11, 783, 1982.

79. **Stelzer, M. J.,** Susceptibility of the great basin tent caterpillar, *Malacosoma fragile* (Stretch), to a nuclear-polyhedrosis virus and *Bacillus thuringiensis* Berliner, *J. Invertebr. Pathol.*, 7, 122, 1965.

80. **To, W. N., Gudauskas, R. T., and Harper, J. D.,** Pathogenicity of *Bacillus cereus* isolated from *Trichoplusia ni* larvae, *J. Invertebr. Pathol.*, 26, 135, 1975.

81. **Vago, C.,** Actions virusales indirectes, *Entomophaga*, 1, 82, 1956.

82. **Magnoler, A.,** The differing effectiveness of purified and nonpurified suspensions of the nuclear-polyhedrosis virus of *Porthetria dispar*, *J. Invertebr. Pathol.*, 11, 326, 1968.

83. **Fuxa, J. A.,** Interactions of the microsporidium *Vairimorpha necatrix* with a bacterium, virus, and fungus in *Heliothis zea*, *J. Invertebr. Pathol.*, 33, 316, 1979.

84. **Nordin, G. L. and Maddox, J. V.,** Effects of simultaneous virus and microsporidian infections on larvae of *Hyphantria cunea*, *J. Invertebr. Pathol.*, 20, 66, 1972.

85. **Young, S. Y., McCaul, L. A., and Yearian, W. C.,** Effect of *Bacillus thuringiensis-Trichoplusia* nuclear polyhedrosis virus mixtures on the cabbage looper, *Trichoplusia ni*, *J. Ga. Entomol. Soc.*, 15, 1, 1908.

86. **McEwen, F. L. and Hervey, G. E. R.,** Microbial control of two cabbage insects, *J. Insect Pathol.*, 1, 86, 1959.

87. **Semel, M.,** The efficiency of a polyhedrosis virus and *Bacillus thuringiensis* for control of the cabbage looper on cauliflower, *J. Econ. Entomol.,* 54, 698, 1961.
88. **Jaques, R. P.,** Tests on microbial and chemical insecticides for control of *Trichoplusia ni* (Lepidoptera: Pieridae) and *Pieris rapae* (Lepidoptera: Peridae) on cabbage, *Can. Entomol.,* 105, 21, 1973.
89. **Jaques, R. P.,** Field efficacy of viruses infectious to the cabbage looper and imported cabbageworm on late cabbage, *J. Econ. Entomol.,* 70, 111, 1977.
90. **Jaques, R. P. and Laing, D. R.,** Efficacy of mixtures of *Bacillus thuringiensis,* viruses, and chlordimeform against insects on cabbage, *Can. Entomol.,* 110, 443, 1978.
91. **Stelzer, M. J.,** Control of a tent caterpillar, *Malacosoma fragile incurva,* with an aerial application of a nuclear-polyhedrosis virus and *Bacillus thuringiensis, J. Econ. Entomol.,* 60, 38, 1967.
92. **Stelzer, M. J., Neisess, J., and Thompson, C. G.,** Aerial applications of a nucleopolyhedrosis virus and *Bacillus thuringiensis* against the Douglas fir tussock moth, *J. Econ. Entomol.,* 68, 269, 1975.
93. **Oatman, E. R., Hall, I. M., Arakawa, K. Y., Platner, G. R., Bascom, L. A., and Beegle, C. C.,** Control of the corn earworm on sweet corn in southern California with a nuclear polyhedrosis virus and *Bacillus thuringiensis, J. Econ. Entomol.,* 63, 415, 1970.
94. **Creighton, C. S., McFadden, T. L., and Cuthbert, R. B.,** Control of caterpillars on tomatoes with chemical and pathogens., *J. Econ. Entomol.,* 64, 737, 1971.
95. **Bell, M. R. and Romine, C. L.,** Tobacco budworm field evaluation of microbial control in cotton using *Bacillus thuringiensis* and a nuclear polyhedrosis virus with a feeding adjuvant, *J. Econ. Entomol.,* 73, 427, 1980.
96. **Jaques, R. P.,** Control of the cabbage looper and the imported cabbage-worm by viruses and bacteria, *J. Econ. Entomol.,* 65, 757, 1972.
97. **Ignoffo, C. M. and Montoya, E. L.,** The effects of chemical insecticides and insecticidal adjuvants on a *Heliothis* nuclear-polyhedrosis virus, *J. Invertebr. Pathol.,* 8, 409, 1966.
98. **Schnyder, U.,** Untersuchung einer Kernpolyedros von *Sterrha seriata* Schrk. *(= Ptychopoda seriata* Schrk., *(= Acidalia virgularia* Hb.) (Geometridae, Lepidoptera) und deren beeinflussbarkeit durch Hunger, DDT, DNOC und Farnesyl-methyl-ather, Diss. No. 3802, Eidgenossischen Technischen Hochschule, Zurich, 1967.
99. **Girardeau, J. H., Jr. and Mitchell, E. R.,** The influence of a sub-acute infection of polyhedrosis virus in the cabbage looper on susceptibility to chemical insecticides, *J. Econ. Entomol.,* 61, 312, 1968.
100. **Hunter, D. K., Collier, S. J., and Hoffman, D. F.,** Compatibility of malathion and the granulosis virus of the Indian meal moth, *J. Invertebr. Pathol.,* 25, 389, 1975.
101. **Komolpith, U. and Ramakrishnan, N.,** Joint action of a baculovirus of *Spodoptera litura* (Fabricius) and insecticides, *J. Entomol. Res.,* 2, 15, 1978.
102. **Harpaz, I. and Raccah, B.,** Nucleopolyhedrosis virus (NPV) of the Egyptian cottonworm, *Spodoptera littoralis* (Lepidoptera, Noctuidae): temperature and pH relations, host range and synergism, *J. Invertebr. Pathol.,* 32, 368, 1978.
103. **Luttrell, R. G., Yearian, W. C., and Young, S. Y.,** Laboratory and field studies on the efficacy of selected chemical insecticide — Elcar *(Baculovirus heliothis)* combinations against *Neliothis* spp., *J. Econ. Entomol.,* 72, 57, 1979.
104. **Savanurmath, C. J. and Mathad, S. B.,** Efficacy of fenitrothion and nuclear polyhedrosis virus combinations against the armyworm *(Mythimna (Pseudaletia) separata* (Wlk.) (Lepidoptera: Noctuidae), *Z. Angew. Entomol.,* 91, 464, 1981.
105. **Savanurmath, C. J. and Mathad, S. B.,** Competence of endosulfan integration with nuclear polyhedrosis virus in the management of armyworm *Mythimna (Pseudaletia) separata* (Wlk.) (Lep., Noctuidae), *Z. Angew. Entomol.,* 93, 413. 1982.
106. **Mohamed, A. I., Young, S. Y., and Yearian, W. C.,** Effects of microbial agent-chemical pesticide mixtures on *Heliothis virescens* (F.) (Lepidoptera: Noctuidae), *Environ. Entomol.,* 12, 478, 1983.
107. **Mohamed, A. I., Young, S. Y., and Yearian, W. C.,** Susceptibility of *Heliothis virescens* (F.) (Lepidoptera: Noctuidae) larvae to microbial agent-chemical pesticide mixtures on cotton foliage, *Environ. Entomol.,* 12, 1403, 1983.
108. **Wellenstein, G. and Lühl, R.,** Bekämpfung schädlicher Raupen mir insektenpathogenen Polyederviren und chemischen Stressoren, *Naturwissenschaften,* 59, 517, 1972.
109. **Lühl, V. R.,** Versuche mit insektenpathogenen Polyederviren und chemischen Stressoren zur Bekämpfung forstschädlicher Raupen, *Z. Angew. Entomol.,* 76, 49, 1974.
110. **Yadava, R. L.,** On the chemical stressors of nuclear-polyhedrosis virus of gypsy moth, *Lymantria dispar* L., *Z. Angew. Entomol.,* 69, 303, 1971.
111. **Shapiro, M. and Bell, R. A.,** Enhanced effectiveness of *Lymantria dispar* (Lepidoptera: Lymantriidae) nucleopolyhedrosis virus formulated with boric acid, *Ann. Entomol. Soc. Am.,* 75, 346, 1982.
112. **Genung, W. G.,** Comparison of insecticides, insect pathogens and insecticide-pathogen combinations for control of cabbage looper *Trichoplusia ni* (Hbn.), *Fla. Entomol.,* 43, 65, 1960.

113. **Hofmaster, R. N. and Ditman, L. P.,** Utilization of a nuclear polyhedrosis virus to control the cabbage looper on cole crops in Virginia, *J. Econ. Entomol.,* 54, 921, 1961.

114. **Getzin, L. W.,** The effectiveness of the polyhedrosis virus for control of the cabbage looper, *Trichoplusia ni, J. Econ. Entomol.,* 55, 442, 1962.

115. **Wolfenbarger, D. A.,** Polyhedrosis-virus-surfactant and insecticide combinations, and *Bacillus thuringiensis*- surfactant combinations, for cabbage-looper control, *J. Invertebr. Pathol.,* 7, 33, 1965.

116. **Jaques, R. P.,** Control of cabbage insects by viruses, *Proc. Entomol. Soc. Ont.,* 101, 28, 1971.

117. **Vail, P. V., Seay, R. E., and DeBold, J.,** Microbial and chemical control of the cabbage looper on fall lettuce, *J. Econ. Entomol.,* 73, 72, 1980.

118. **Chaudhari, S. and Ramakrishnan, N.,** Field efficacy of baculovirus and its combination with sub-lethal dose of DDT and endosulfan on cauliflower against tobacco caterpillar *Spodoptera litura* (Fabricius), *Indian J. Entomol,* 42, 592, 1980.

119. **Ignoffo, C. M., Chapman, A. J., and Martin, D. F.,** The nuclear-polyhedrosis virus of *Heliothis zea* (Boddie) and *Heliothis virescens* (Fabricius). III. Effectiveness of the virus against field populations of *Heliothis* on cotton, corn, and grain sorghum, *J. Invertebr. Pathol.,* 7, 227, 1965.

120. **Chapman, A. J. and Ignoffo, C. M.,** Influence of rate and spray volume of a nucleopolyhedrosis virus on control of *Heliothis* in cotton, *J. Invertebr. Pathol.,* 20, 183, 1972.

121. **Pieters, E. P., Young, S. Y., Yearian, W. C., Sterling, W. L., Clower, D. F., Melville, D. R., and Gilliland, F. R., Jr.,** Efficacy of microbial pesticide and chlordimeform mixtures for control of *Heliothis* spp. on cotton, *Southwestern Entomol.,* 3, 237, 1978.

122. **Yearian, W. C., Luttrell, R. G., Stacey, A. L., and Young, S. Y.,** Efficacy of *Bacillus thuringiensis* and *Baculovirus heliothis*-chlordimeform spray mixtures against *Heliothis* spp. on cotton, *J. Ga. Entomol. Soc.,* 15, 260, 1980.

123. **Hafez, M., Kamel, A. A. M., Mostafa, T. H., and Omar, E. E.,** Field test of combinations of polyhedrosis virus suspensions and certain chemical insecticides for control of the cotton leafworm, *Spodoptera littoralis* (Boisd.), *Bull. Entomol. Soc. Egypt Econ. Ser.,* 4, 65, 1970.

124. **Hamm, J. J. and Young, J. R.,** Value of virus presilk treatment for corn earworm and fall armyworm control in sweet corn, *J. Econ. Entomol.,* 64, 144, 1971.

125. **Morris, O. N., Armstrong, J. A., Howse, G. M., and Cunningham, J. C.,** A 2-year study of virus-chemical insecticide combination in the integrated control of the spruce budworm, *Choristoneura fumiferana* (Lepidoptera: Tortricidae), *Can. Entomol.,* 106, 813, 1974.

126. **Morris, O. N.,** Long term effects of aerial applications of virus-fenitrothion combinations against the spruce budworm, *Choristoneura fumiferana* (Lepidoptera: Tortricidae), *Can. Entomol.,* 109, 9, 1977.

127. **Beegle, C. C. and Oatman, E. R.,** Effect of a nuclear polyhedrosis virus on the relationship between *Trichoplusia ni* (Lepidoptera: Noctuidae) and the parasite, *Hyposoter exiguae* (Hymenoptera: Ichnuemonidae), *J. Invertebr. Pathol.,* 25, 59, 1975.

128. **Howard, L. O. and Fiske, W. F.,** The importation into the United States of the parasites of the gipsy moth and the brown-tail moth, *U.S. Dept. Agr. Bur. Entomol. Bull.,* 91, 1911.

129. **Ullyett, G. C. and Schonken, D. B.,** A fungus disease of *Plutella maculipennis,* Curt., in South Africa, with notes on the use of entomogenous fungi in insect control, *Union S. Afr. Sci. Bull. Dept. Agr. For.,* 218, 1, 1940.

130. **Kelsey, J. M.,** Interaction of virus and insect parasites of *Pieris rapae* L., *Proc. 11th Int. Congr. Entomol.,* 2, 790, 1960.

131. **Laigo, F. M. and Tamashiro, M.,** Virus and insect parasite interaction in the lawn armyworm, *Spodoptera maruita acronyctoides* (Guenee), *Proc. Hawaiian Entomol. Soc.,* 19, 233, 1966.

132. **Laigo, F. M. and Paschke, J. D.,** *Pteromalus puparum* L. parasites reared from granulosis and microsporidiosis infected *Pieris rapae* L. chrysalids, *Phillip. Agric.,* 52, 430, 1968.

133. **Irabagon, T. A. and Brooks, W. M.,** Interaction of *Campoletis sonorensis* and a nuclear polyhedrosis virus in larvae of *Heliothis virescens, J. Econ. Entomol.,* 67, 229, 1974.

134. **Vail, P. V.,** Cabbage looper nuclear polyhedrosis virus-parasitoid interactions, *Environ. Entomol.,* 10, 517, 1981.

135. **Brubaker, R. W.,** Seasonal occurrence of *Voria ruralis,* a parasite of the cabbage looper, in Arizona, and its behavior and development in laboratory culture, *J. Econ. Entomol.,* 61, 306, 1968.

136. **Levin, D. B., Laing, J. E., and Jaques, R. P.,** Interactions between *Apanteles glomeratus* (L.) (Hymenoptera: Braconidae) and granulosis virus in *Pieris rapae* (L.) (Lepidoptera: Pieridae), *Environ. Entomol.,* 10, 65, 1981.

137. **Elsey, K. D. and Rabb, R. L.,** Biology of *Voria ruralis* (Diptera: Tachinidae), *Ann. Entomol. Soc. Am.,* 63, 216, 1970.

138. **Kaya, H. K. and Brayton, M. A.,** Interaction between *Neoaplectana carpocapsae* and a granulosis virus of the armyworm *Pseudaletia unipuncta, J. Nematol.,* 10, 350, 1978.

139. **Kaya, H. K.,** Toxic factor produced by granulosis virus in armyworm larva: effect on *Apanteles militaris, Science,* 168, 251, 1970.

140. **Kaya, H. K. and Tanada, Y.,** Properties of a viral factor toxic to the parasitoid, *Apanteles militaris, J. Insect. Physiol.,* 17, 2125, 1971.
141. **Kaya, H. K. and Tanada, Y.,** Pathology caused by a viral toxin in the parasitoid *Apanteles militaris, J. Invertebr. Pathol.,* 19, 262, 1972.
142. **Kaya, H. K. and Tanada, Y.,** Response of *Apanteles militaris* to a toxin produced in a granulosis-virus-infected host, *J. Invertebr. Pathol.,* 19, 1, 1972.
143. **Kaya, H. K. and Tanada, Y.,** Hemolymph factor in armyworm larvae infected with a nuclear-polyhedrosis virus toxic to *Apanteles militaris, J. Invertebr. Pathol.,* 21, 211, 1973.
144. **Biliotti, E.,** Survie des larves endophages de Tachinaires à une mort prématurée de leur hôte par maladie, *C. R. Acad. Sci.,* 240, 1021, 1955.
145. **Niklas, O.,** Zum Massenwechsel der Tachine *Parasetigena segregata* Rond. (*Phorocera agilis* R. -D) in der Rominter Heide., *Z. Angew. Entomol.,* 26, 63, 1939.
146. **Gosswald, K.,** Physiologische Untersuchunger uber die Einwirkung okologischer Faktoren, besonders Temperatur und Luftfeuchtigkeit, auf die Entwicklung von *Diprion (Lophyrus) pini* L. zur Feststellung der Ursachen des Massenwechsels, *Z. Angew. Entomol.,* 22, 331, 1934.
147. **Thompson, C. G. and Steinhaus, E. A.,** Further tests using a polyhedrosis virus to control the alfalfa caterpillar, *Hilgardia,* 19, 411, 1950.
148. **Levin, D. B., Laing, J. E., and Jaques, R. P.,** Transmission of granulosis virus by *Apanteles glomeratus* to its host *Pieris rapae, J. Invertebr. Pathol.,* 34, 317, 1979.
149. **Smith, O. J., Hughes, K. M., Dunn, P. H., and Hall, I. H.,** A granulosis virus disease of the western grape leaf skeletonizer and its transmission, *Can. Entomol.,* 88, 507, 1956.
150. **Bird, F. T.,** Transmission of some insect viruses with particular reference to ovarial transmission and its importance in the development of epizootics, *J. Insect Pathol.,* 3, 352, 1961.
151. **Vail, P. V., Soo Hoo, C. F., Seay, R. S., Killinen, R. G., and Wolf, W. W.,** Microbial control of lepidopterous pests of fall lettuce in Arizona and effects of chemical and microbial pesticides on parasitoids, *Environ. Entomol.,* 1, 780, 1972.
152. **Hamm, J. J. and Hare, W. W.,** Application of entomopathogens in irrigation water for control of fall armyworms and corn earworms on corn, *J. Econ. Entomol.,* 75, 1074, 1982.

Chapter 6

FORMULATION AND APPLICATION OF BACULOVIRUSES

S. Y. Young III and W. C. Yearian

TABLE OF CONTENTS

I. INTRODUCTION

Prerequisites to the effective utilization of baculoviruses in insect pest management systems are the availability of active, standardized, and stable preparations and application systems that can deliver these preparations to the target so as to optimize their biological activity. Although the importance of formulation and application in effective utilization of baculoviruses has been generally recognized, the research attention they have received has not been proportionate. The development of baculoviruses for use in pest management systems has been primarily as viral insecticides. The methodologies most often utilized have been quite similar to those for chemical insecticides. This approach has advantages, as existing formulation technology for chemical insecticides is extensive, and application equipment is widely available and routinely used. These technologies, however, were developed for formulation and application of chemical compounds that are relatively stable in storage and in the field, and are fast-acting contact poisons. Viruses, on the other hand, are relatively unstable particulate entities that must be consumed to initiate an infection which must be allowed to proceed for days or weeks before the target pests cease to inflict damage. These unique characteristics of baculoviruses may require methodologies that differ from those for conventional insecticides. Such requirements may differ with each virus, target insect, crop, and method of utilization. The development and adoption of new and different technologies must be tempered, however, by their economical feasibilities. The more closely related the method(s) of choice to those routinely use for chemical insecticides, the more likely the product will be economically acceptable and enjoy widespread use.

II. FORMULATION

Requirements for formulation of baculoviruses are similar to those for pesticides in general. The process must preserve biological activity and should add physical properties to the finished product that promote storage stability, tank mixing, application, and coverage. Formulation of viral insecticides has typically been as wettable powders for application as sprays through conventional equipment. Viruses have also been formulated as flowables, dusts, granules, and solid baits.

The definition of formulation used here will be that of Van Valkenburg;[1] the mixture of the candidate pesticide and materials that affect its chemical, physical, and biological properties. Formulations will be further divided into basic and tank mixture as per Couch and Ignoffo[2] and Yearian and Young.[2,3] A basic formulation is that prepared for storage and distribution to the user. A tank mixture is the basic formulation plus any additional materials, including water, added by the user prior to application.

As with chemical insecticides, much of the research on formulation of viruses has been in the private sector and as such is of a proprietary nature and not generally made available to the public. Formulation research by the public sector has usually been on tank mixtures with the aim of improving efficacy of a particular virus for a selected pest-crop system.

A. Standardization

Standardization of the product is essential for the effective formulation of pesticides. Viruses and other biological pesticides represent a special problem in that biological activity can not be standardized by chemical analyses.

Historically baculoviruses have been quantified by determining the concentration of occluded virus bodies (polyhedra or granules) by light microscopy using a hemacytometer. Although this measures the concentration of occluded virus bodies, it may not be a reliable indicator of activity in the preparation and used alone it has serious limitations. For example, Ignoffo and Shapiro found that *Heliothis zea* nuclear polyhedrosis virus (NPV) harvested from dead larvae was significantly more active per polyhedron than virus harvested from larvae prior to death.[4] Shapiro found that activity of *Lymantria dispar* NPV was similarly affected.[4] Virus that has been stored without freezing for a lengthy period may also lose activity that is not reflected by counts of occluded virus. Reduced activity of viral preparations, either during the formulation process or in subsequent storage, also may not be detected by counts of occluded virus.

The exclusive use of occlusion body counts as an estimate of activity has been discouraged, and bioassay of viral preparations should be periodically performed to assure that activity is retained.[5,6] Martignoni states, ''The sole use of polyhedral inclusion body counts has limited value as a measure of activity of a technical product for field application, if it is not accompanied by an activity titration.''[7] During the early years of commercial development of the *H. zea* NPV, it was discovered that some formulations supplied for testing had little activity remaining when used. This problem was circumvented by increased use of bioassay by the producer prior to distribution to researchers, as well as by bioassay by Dr. H. T. Dulmage, Brownsville, Tex. at the beginning and end of each testing season. More recently, the U.S. Environmental Protection Agency has required that commercial viral preparations bear an activity rating on the label. The bioassay of viral products also has limitations. The procedure is time consuming and often less precise than desired.[8] Dulmage and Burgerjon[9] reported on laboratory bioassays for *H. zea* NPV using *H. virescens* as the test insect. The mean coefficient of variation was 0.20 compared to 0.15 for similar *Bacillus thuringiensis* bioassays. Although inclusion body counts and bioassays have limitations, most baculovirus formulations are standardized utilizing both procedures, as this joint use results in a more reliably standardized product than does either method alone. Assay methods are being refined and new, more precise methods are being developed; they will not be discussed here since they are presented in Chapter 1.

The severity of problems with standardization of viral insecticides was emphasized by Dulmage and Burgerjon, and they called for industrial and international standardization.[9] They point out that many of the problems with the early development of *B. thuringiensis* were due to the lack of standardization and that many potential problems with viral insecticides could be avoided by adequate standardization. They proposed standardization procedures that emphasized the use of bioassay.

B. Methods

Baculovirus formulations for application in field trials have often consisted of a filtrate containing virus prepared from macerated cadavers and mixed with water. Provided this type of formulation is stored under refrigeration or frozen, its efficacy in tank mixtures has typically been equal or superior to more elaborate formulations. This approach to formulation is not practical when large quantities are required or when commercial products must be stabilized for storage and used under a wide variety of conditions.

Several approaches have been taken to concentrate basic formulations and improve their stability in storage and distribution. The particulate nature of viruses restricts development of flowable formulations. Thus, most viruses have been formulated as wettable powders. Lyophilization of the filtrate from macerated cadavers, although expensive, has often been used. For example, a lyophilization step is included in the formulation of *L. dispar* NPV (Gypchek) and *Orgyia pseudotsugata* NPV (TM-Biocontrol-1), both of which have been registered by the U.S. Forest Service.[7,10] With TM-Biocontrol-1, cadavers of mature NPV killed *O. pseudotsugata* larvae are blended, and sieved, and the suspension is lyophilized and milled before being weighted and packaged under vacuum.[7] A limitation of lyophilization as a step in formulation is the difficulty that can be encountered in dispersal of the occlusion bodies in the tank mixtures since some clumping of the lyophilized material occurs. Ignoffo reduced clumping by lyophilization of the virus in a lactose paste.[11] The powder was stable during lengthy periods of refrigeration. Coprecipitation of a concentrated suspension of occlusion bodies in lactose (4 to 6%) with acetone is an alternate method that has been used to obtain a concentrated wettable powder formulation.[12] These formulations are somewhat more easily resuspended in a tank mixture. In the early stages of *H. zea* NPV development, an experimental product, Viron H (International Minerals Corp., Libertyville, Ill.), was formulated by this method. Although formulations prepared by acetone precipitation are more easily resuspended in water and a significant loss in activity does not occur during the formulation process, the method has been little used because these preparations lack stability and rapidly lose activity in storage, especially at high temperatures (35°C).[4,13,14]

Spray drying has proven to be the most successful approach for producing stable wettable powder NPV formulations. To obtain spray dry formulations, the NPV is mixed with attapulgite clay and other selected diluents. The mixture is sprayed and dried, and yields a microencapsulated-type of formulation.[15] *H. zea* NPV formulated by spray dry techniques (Sandoz, Inc., San Diego, Calif.) was more active and exhibited increased stability when compared to previous formulations of this virus.[16] The most efficacious *H. zea* NPV spray dry formulation was registered by Sandoz, Inc. under the trade name Elcar®. Other NPVs, *Trichoplusia ni* NPV (Sandoz® 405) and *Autographa californica* NPV (Sandoz® 404) have also been successfully formulated by this method. However, when *Choristoneura fumiferana* NPV was formulated by spray drying, significant activity was lost during the process.[17]

The spray dry formulation technique may not be appropriate for granulosis virus (GV). A spray dry formulation of *Laspeyresia pomonella* GV (Sandoz® 406) has been less stable than NPV formulations, even with refrigeration. It may be necessary to consider flowable formulation for GV, although flowable NPV formulations have been generally less stable in storage than wettable powders. Sandoz® Inc. formulated *H. zea* NPV as a flowable, SAN 240-WDC, but some difficulty was encountered with spoilage by microbial contamination.[15]

H. zea NPV formulations have also been prepared by microencapsulation. Acetone precipitates of *H. zea* NPV alone or mixed with sunlight protectants were encapsulated with ethyl cellulose, gelatin, or polymer by National Cash Register Company, Dayton, Ohio. the average diameter of the capsules ranged from 10 to 100 μm for the various formulations. Bioassays revealed that carbon and the solvents used in the microencapsulating process reduced activity of the virus by 14 to 20 × and 1.7 to 8.0 ×, respectively.[18] Southwest Research Institute, San Antonio, Tex., also encapsulated an acetone precipitate of *H. zea* NPV. A series of encapsulated formulations were made by dispersing the materials with a high-shear stirrer in an ethyl acetate solution of polymer formed from a half ester of styrene maleic anhydride. The materials were metered onto a high-velocity rotating disk that fractionated the mixture. The ethyl alcohol evaporated in flight and the resulting dry capsules were collected on a shield. Encapsulation reduced viral activity by threefold in the two most promising formulations selected for field trials.[19]

Only in a few instances have baculoviruses been formulated for application other than as sprays. Lack of efforts to formulate for application as solids is due primarily to the scarcity of application equipment and the failure of such formulation tested to date to offer significant advantages over formulations designed for application as sprays. Montoya and Ignoffo[20] formulated the *H. zea* NPV as a dust. A virus suspension was mixed with attaclay and blended, and the slurry was then lyophilized, triturated, blended, sieved, and reblended into a fine powder containing 970 to 4890 polyhedral inclusion bodies (PIB) per milligram. Stacey et al.[21] lyophilized a *H. zea* NPV suspension and mixed it with powdered cellulose. This powder was combined with either corn meal, pulverized wheats, or cottonseed meal for a final concentration of 5.3×10^{10} PIB per kilogram. For the final formulation the combination was placed in a Nalgene drum, and rolled on a hammer mill for 3 hr. McGaughey prepared a dust of *Plodia interpunctella* GV by mixing a lactose precipitate of the virus with wheat flour.[13] Granular formulations of *H. zea* NPV were also prepared that consisted of 8 g of Coax®, either 2% gelatin or 1% Dacagin®, 0.2 mg Elcar®, 135 mℓ water, and 100 g of wheat bran, corn bran, corn hominy, oat, or rice hull filler. The mixtures were dried, ground, and sieved through a 40 mesh screen.[22]

Due in part to their instability, nonoccluded baculoviruses have not been successfully formulated for application as viral insecticides. For example, Zelazny[23] found that the *Oryctes rhinoceros* nonoccluded baculovirus retained little activity after 1 week when stored mixed with sawdust at 26°C. Inactivation was more rapid when the mixture was dried. If this is typical of nonoccluded baculoviruses, it would appear that they will be extremely difficult to stabilize in formulations.

C. Storage Stability

Commercial development of viral insecticides cannot be accomplished without formulations that are physically and biologically stable in storage and distribution. Stability during application and on the treated surface after deposition is also required. Couch and Ignoffo[2] stated that a formulation of a pathogen product with a shelf life in excess of 18 months is critical for industrialization. Stability for this period of time is very difficult to obtain with baculoviruses when both physical and biological stability are required. Studies to determine suitable additives for stabilization of virus formulations have been limited. However, the carriers and surface active agents found suitable for *B. thuringiensis* basic formulations should be comparable to those needed for viral formulations.[2]

Flowable formulations of baculoviruses that possess the desired characteristics for stability are lacking. Due to the particulate nature of viruses it has proven difficult to maintain physical stability in flowables. Furthermore, the conditions required to maintain viral activity have greatly restricted the choice of carriers. A water-based carrier with pH near neutrality is most desirable, but it is difficult to eliminate growth of microbial contaminants. As a result, fermentation in flowable formulations stored at room temperature may increase the pressure within the storage containers and result in rupture and spillage.[2,15]

With the exception of formulations prepared by acetone precipitation, stability of dry formulations has been better than that experienced with flowables. This has particularly been the case with spray dry formulations prepared as wettable powders by Sandoz®, Inc. Provided that the virus is compatible with the carrier and formulation method, surface-active agents often give a formulation its necessary physical performance qualities. These materials and their concentrations are proprietary in nature and the particulars of formulation studies by industry are seldom available. Basic formulations, other than those by Sandoz,® Inc. often do not include surface-active agents, with these subsequently added in tank mixtures. Wettable powders, however, always require the use of surface-active agents when applied. The surface-active agent must be such that atomization and deposition of the virus will be desirable for specific application conditions.

1. Temperature

Although nonoccluded virus particles are not stable in lengthy storage, polyhedra are relatively stable when stored refrigerated or frozen. Significant activity in polyhedral preparations may be present after freezing for 20 years.[24-26] Martignoni reported that the shelf life of TM-Biocontrol-1 stored in a cool dry place was at least 5 years.[7] Baculoviruses are less stable at ambient temperature but most polyhedra remain infective for several years at room temperature.[25] Lewis and Rollinson[27] found, however, that the polyhedra of *Lymantria dispar* NPV retained potency for up to 2 years depending upon the method of storage. Activity of *Diprion hercyniae* NPV decreased after 2 years of storage.[28] The GV may be less stable then NPV. David[29] reviewed studies on stability and reported that GV of *Pieris brassicae* and *P. rapae* retained little activity after storage from 1 to 5 years. Temperatures above ambient result in rapidly decreased stability of baculoviruses. At 38° to 42°C a significant loss in activity occurs within a few months or even weeks.[27,30,31] At temperatures of 50°C or above, viral activity is lost in a matter of hours or minutes.[26]

A few studies on the effect of moisture on inactivation of baculoviruses have been conducted. Couch and Ignoffo[2] and Jaques[26] summarized findings of studies and reported that stability of viral preparations generally increased as moisture content decreased. There is, however, a notable exception to this generalization. David[29] reported that *P. brassicae* and *P. rapae* GV were more stable when stored wet.

2. Light

Baculoviruses are rapidly inactivated by short- and long wavelength (254 to 310 nm) UV light and should not be exposed to such during or after formulation.[26,32-35] David[36] reported that when exposed to UV light the *P. brassicae* GV was more stable in wet films.

3. Chemicals

The effect of hydrogen-ion concentration on long-term stability of baculovirus formulations has not been reported. Baculoviruses are inactivated by extremes in hydrogen-ion concentration.[37,38] Although baculoviruses are stable for short periods at pHs varying from 4 to 9, the pH of a formulation should be near neutrality. Chemicals of extreme pH physically destroy the integrity of the inclusion body.[37-39] Some disinfectants also rapidly inactivate occluded baculoviruses. The effects of these and other chemicals on baculoviruses have been reviewed by Jaques[26] and David.[29]

Stability of the virus in chemicals used in the basic formulation process and tank mixtures is of particular importance in formulation. Information on stability of baculoviruses in formulation chemicals, however, is fragmentary. With the exception of chemicals used in dust and microencapsulated formulations of viruses, little information is available. Compatibility of baculoviruses with some chemicals used in the basic formulation process was presented earlier in this chapter.

D. Field Stability

Baculoviruses are rapidly inactivated by the UV spectrum of sunlight with most having a half-life of form two to several hours.[40-46] Loss of activity is not due to germicidal activity of the short wavelength UV spectrum, since only UV wavelengths longer than 290 nm typically are present in natural sunlight. David found that UV light of 290 to 320 nm quickly reduced activity of the *P. brassicae* GV.[36] Similar results have been reported for *H. zea* NPV[47] and *Galleria mellonella* NPV.[34] Exposure of baculoviruses to longer wavelength irradiation (about 360 mu) has little effect on viral activity.[34,44,47]

Activity of baculoviruses on plants is brief when applied to the upper surface of foliage. Several studies have shown that activity of *H. zea* NPV on the upper surface of cotton leaves exposed to sunlight was mostly lost by 48 hr.[18,48,49] Persistence of the *H. zea* NPV on the upper surface of soybean leaves and corn silks was similar to that on cotton,[49-51] but the virus persisted slightly longer on tomato leaves.[49] Improved persistence of the virus on tomato foliage was attributed to shading from sunlight due to the pilose nature and curvature of the tomato leaves.

Persistence of *H. zea* NPV on foliage of cotton, soybean, and tomato was shown to be primarily related to sunlight. When plants were covered at night but exposed to sunlight during the day, virus inactivation was as rapid as on plants uncovered for 24 hr. When the plants were mechanically shielded from sunlight during the day, however, most of the viral activity remained after 4 days.[49] *H. zea* NPV has also been shown to persist significantly longer on more shaded cotton plant parts, i.e., under surface of leaves, square bracts, and square calyces.[49]

On cabbage, *T. ni* NPV and *P. brassicae* GV lose much of their activity on the upper surface of leaves within 2 days.[43,52] Jaques[52] reported, however, that approximately 15% of the virus activity on cabbage has been retained for 10 days and that activity on the undersurface of leaves persisted much better than on the upper surface.[52] Persistence of *P. brassicae* GV applied to cabbage during the winter months when UV radiation from sunlight is reduced, was significantly improved with some activity remaining for at least 4 months.

Viral activity is also quickly lost on foliage in the forest canopy,[53,54] but persistence appears to be superior to that on row crops. Persistence of *L. dispar* NPV on foliage of red oak and red maple was from 3 to 15 days.[54] Elgee[55] reported that some *O. leucostigma* NPV activity remained on foliage of Balsam fir throughout the winter. Most of the activity of *O. pseudotsugata* NPV produced by virus-killed cadavers on fir, however, is inactivated on the foliage and only a small percentage of NPV is incorporated into duff.[56] *Gilpinia hercynaie* NPV has been shown to persist on Norway spruce and Sitka spruce throughout the winter, and this virus is considered to be a strong contributor to the primary inoculum each spring. Half-life for purified *G. hercyniae* NPV was 22.6 days in the summer and 38.4 days in the winter months.[57,58] Activity of *L. dispar* NPV on bark of red oak and red maple was much more persistent than on foliage, and some activity remained after 1 year.[54]

The contribution of other factors to inactivation of baculoviruses on plants under field conditions appears to be insignificant when compared to sunlight. Few data are available on the effect of temperature on field persistence of virus on host plants. Laboratory tests discussed earlier in the chapter suggest that only high temperatures during summer months and in climates where temperatures reach 40°C or higher for extended periods inactivate baculoviruses. Jaques[26] noted that baculoviruses should withstand maximum temperatures normally encountered in the field environment, at least for short periods. Although temperature seldom appears to have a direct effect on loss of viral activity in the field, McLeod et al.[35] found that inactivation of *H. zea* NPV exposed to shortwave UV light increased significantly as temperatures increased from 15 to 45°C.

Studies on cabbage have shown that little virus is removed from the foliage by rainfall. David[29] stated that virus is virtually impossible to wash from cabbage leaves, with most *P. brassicae* GV remaining on the plants following 254 cm of simulated rainfall. Scrubbing leaves with detergent during the washing process also failed to remove the virus. After washing cabbage leaves continuously for 4 days, 40% of the *T. ni* NPV that had been applied to the surface remained. Although studies on removal of baculoviruses from other host plants by rainfall are lacking, a cytoplasmic polyhedrosis virus persisted well on foliage of pine after simulated rainfall at 0.5 cm.[60]

The removal of virus from plants by wind has not been quantified. It is generally recognized, however, that polyhedra from virus-killed cadavers can be spread by wind to other plants or to the soil, and this is no doubt a factor in loss of viral activity on plants by weathering.

In addition to their physical properties, other characteristics of certain plants influence the persistence of baculoviruses. Some genera of the Malvaceae have alkaline leaf surfaces due to secretion of high concentrations of cations by glandular trichomes.[61] Cotton may have a leaf surface pH in excess of 10.0.[62-64] Andrews and Sikorowski[62] observed brownian movement of virus particles in *H. zea* NPV polyhedra when the matrix was solubilized by dew from cotton. When *H. zea* NPV was placed in dew collected from cotton foliage and the dew was dried and resuspended daily, viral activity declined steadily.[49] When dew with a neutral pH was collected from soybean and treated similarly, the virus was not inactivated. However, when *H. zea* NPV was applied to cotton, soybean and tomato plants in the field and shaded during the daylight hours to prevent sunlight inactivation, activity of the virus after 4 days did not differ significantly between plants uncovered at night to allow dew formation and plants covered continuously to prevent formation of dew. Also, activity of virus on cotton after 4 days did not differ significantly from that on the other hosts.

Baculoviruses are inactivated slowly in the soil and some activity persists for years. Jaques[65] applied *T. ni* NPV to soil in cabbage fields and found that much of the virus remained active in the upper 10 cm 2 years later. Assays of samples collected after 5 years revealed that 25% of the activity remained in the soil.[66] Thomas et al.[67] also reported that much of the *T. ni* NPV in cabbage fields remained in soil through the winter. *Pieris brassicae* GV showed little inactivation in soil after 2 years,[3] and *T. ni* GV remained active in soil 4 years after applications.[66] Persistence of other viruses in the soil does not appear to be as great. Application of *H. armigera* NPV to sorghum resulted in an epizootic in *H. zea* populations, but viral activity in soils fell to one third of its highest activity by winter.[69] *Pseudoplusia includens* NPV applied to soybean persisted from 1 year to the next in cultivated fields, but only a small percentage of the activity persisted through the year.[70,71]

Baculoviruses also persist for extended periods in forest soils. *Hyphantria cunea* NPV activity was found to be high in soils following epizootics in *H. cunea* during previous years.[72] *L. dispar* NPV activity in litter and soil under trees was high following application of NPV to trees infested with *L. dispar*.[54] Evans and Entwistle[58] reported that the *G. hercyniae* NPV persisted well in soil of spruce forests. *Orgyia pseudotsugata* NPV has been found in duff and soil of forests in which *O. pseudotsugata* populations had not been reported for 11 to 30 years.[56,73]

Baculoviruses are tightly bound to soil particles and are only slowly leached from soil.[74-76] Soil containing *Pieris brassicae* GV was washed with up to 121 cm of simulated rain and most of the activity in upper layers of soil remained.[30] Most of the active virus remaining following an epizootic is in the upper 4.5 cm of soils.[30,77,78] Although tightly bound to soil, Thomas et al.[67] found that the *T. ni* NPV was slowly inactivated by soil with an acidic pH.

E. Tank Mixtures

The need for improved performance specifications of viral formulations has led to the development of a variety of adjuvants for addition at application to tank mixtures containing viral sprays. Major efforts have been made in this area by the private and public sectors. Adjuvants have been developed for a variety of purposes, i.e., increasing stability of viruses on crops, gustatory stimulation of larval hosts, improved droplet characteristics and coverage, and evaporation retardation. These will be subsequently discussed in conjunction with application of baculoviruses.

III. APPLICATION

Application of viral insecticides has typically been with methods utilized for conventional chemical pesticides.[79,80] The potential for increased efficacy of viral insecticides through improved methods of application is generally recognized,[3,80-83] but few studies on the relation of application methodology to efficacy of viral insecticides have been conducted. Since the method of infection for viruses is per os, applications should be designed to provide optimum deposits at the feeding site(s) of the target pest. Feeding sites may differ with insect pest, host-plant species, stage of development of insect host, etc. Further, when viruses are applied to rapidly growing plants, the new growth will move away from the virus deposits, necessitating repeated application at short intervals in order to maintain adequate plant coverage. Thus, the problems encountered in application of viral insecticides are complex. Most conventional spray apparati, however, are not designed to selectively target droplets. Equipment capable of producing droplet sizes considered optimal for conventional contact insecticides, 50 μm, may not be the most desirable for particulate materials such as viruses.[84] Although development of equipment for application of microbial agents formulations has been lacking, Smith and Bodde[83] question if advantage has been taken of the delivery systems currently available.

A. Equipment

1. Forest

Conditions under which pesticides are applied to forested areas are complex and varied. The area requiring treatment may be several thousand hectares in size, and isolated in rough terrain with inverted air currents. Alternately, the area may be restricted to a limited number of trees in an urban situation. Equipment utilized for application of viral formulations to trees reflects these different conditions. The application of viruses to forested areas is further complicated by the extensive leaf area that must be covered, and a premium is placed on coverage, often in a canopy many layers thick. Additional components may need to be tank mixed with the formulation to increase persistence and/or stimulate insect feeding on the viral deposits. Despite these complexities, economics require that pest control often be accomplished with a single application at a spray volume that is often several-fold less than that used on agricultural crops.

Due to reluctance to utilize conventional pesticides in forest situations and the potential of microbial agents, particularly viruses, to suppress a forest pest population over a period of years, research on use of microbial agents in forest pest management has received considerable attention. Technology utilized for application of microbial agents in forests has generally been that developed for conventional chemical sprays. When technology has been developed specifically for microbial agents, it has most often been with *B. thuringiensis*. This technology in turn is often applied to viruses. Both aerial and ground equipment have been utilized to apply viruses as concentrated foliar sprays for control of numerous lepidopterous pest species in a variety of forest ecosystems. Aerial application of baculoviruses in forests often offers advantages over ground application, i.e., areas inaccessible to group equipment can be treated, large areas can be treated rapidly, low volume spray can be used, applications can be made to tall trees, etc. Both fixed wing aircraft and helicopters have been used in spray programs. Fixed wing aircraft utilized to apply viral pesticides are most often equipped with hydraulic nozzles on a boom. Flat fat nozzles have most often been used, although a variety of nozzles are available. More recently, rotating disk atomizers have been used more extensively, particularly with helicopters. The two rotating disk systems most used are Beecomist and Micronair models. The tank mix formulations have varied, In some instances the viral preparation alone has been mixed with water, but a common spray mixture consists of the virus molasses up to 25% total volume, Sandoz Shade®, a commercial spreader sticker, and water. Spray volumes have been as high as 100 ℓ/ha, but most range from 4.7 to 18.8 ℓ/ha. For most large aerial spray programs, volumes greater that 4.7 ℓ/ha are not economical or operationally feasible.

Although viral formulations have often been applied aerially in forest situations, very few studies have been conducted that definitively evaluate spray performance specifications of different equipment, formulations, and volumes. The influence of spray volume on the effectiveness of NPV against Douglas-fir tussock moth was examined using fixed wing aircraft equipped with flat-fan nozzles.[85] The virus was applied at 2.5×10^{11} PIB per hectare in spray volumes ranging from 9.6 to 19.2 ℓ/ha. Spray deposits were monitored with aluminum plates and white Kromekote® cards placed in openings adjacent to sample trees and by removing foliage. Volumetric deposits were determined by a fluorescent method,[86] and droplet sizes and densities on Kromekote® cards were assessed using a Quartimet® 720 particle analyzer. All three spray deposit assessment methods showed that NPV deposit doubled as the volume doubled but reduction in larval population level and defoliation were not altered by the increase in volume.

Smirnoff et al.[87] compared water and oil formulations of NPV at 2×10^6 PIB per milliliter on Swain's Jack pine sawfly on Jack pine using fixed wing aircraft with boom and swirl-jet 1/8 B 2 nozzles. The virus was applied at volumes of 4.8 and 38.0 ℓ/ha, and droplet cards were used to assess deposits. Volumetric deposits were better with an oil-based formulation (Span 80 and No. 2 fuel oil) than an aqueous formulation (water, latex, and dried blood). Both formulations were effective, however. Based on drift deposit data, they concluded that the oil-based formulations atomized more readily and are preferable at low application rates. Since aqueous sprays evaporate rapidly, application at low volumes above forest canopy may result in very small droplets at the target and increased drift. Oil-based sprays should minimize this problem and may allow viruses to be applied at lower volumes than those typically tested in forest situations without a reduction in deposits.

A series of tests has been conducted to compare the effectiveness of 52-8006 Flat Fan® tee jet nozzles and Beecomist nozzles with perforated sleeves for application of Gypcheck,[88] In 1977 Gypcheck was applied at 64×10^6 gypsy moth potency units in 4.8 ℓ molasses, 1.1 kg Shade®, 10.4 ℓ Chevron®, and 14.3 ℓ water per hectare. Mean droplet sizes produced by both types of nozzles ranged from 300 to 320 μm.[89] The reduction in egg masses was 77 and 65% for flat fan and Beecomist nozzles, respectively, compared to a 37% increase in the control plots. Defoliation was 49 and 55% for flat fan and Beecomist nozzles, respectively, compared to 80% in the control plots. Lewis[89] concluded that both types of nozzles were effective and not significantly different in performance. In 1978 dosages of Gypcheck of 64 to 128×10^6 gypsy moth potency per hectare units were evaluated in three tank mixtures: (1, 2) 0.55 or 1.1 kg Shade and 4.7 ℓ molasses, 0.4 ℓ Chevron, and 18.7 ℓ water per hectare, and (3) 2.3 ℓ Pro-tec and 16.4 ℓ water per hectare. The efficiency of nozzle depended upon the tank mixture. Larval reduction with Beecomist nozzles was two times greater than with flat fan nozzles with molasses mixtures, but the opposite was true for the Pro-tec® mixture.[88]

Desaulniers and Cunningham conducted a series of tests of NPV efficacy against European pine sawfly on 37 red pine plantations.[90] Fixed wing aircraft with Micronair units, a helicopter with Beecomist nozzles or ground application with a back pack mist blower was used. The dosages were 5×10^9 PIB per hectare for aerial applications and 1×10^{10} PIB per hectare for ground applications with both at a volume of 9.4 ℓ per hectare and applied in either water or 25% molasses plus 20 g/ℓ Sandoz Shade.® Larvae were in developmental stages L1 in the aerial tests and L3 and L4 in the ground tests. In the aerial tests, larval mortality was high, 97.7 and 72.5%, when plantations received either total or partial coverage, respectively, for the fixed wing aircraft and 99.4 and 90.8%, respectively, for the helicopter, compared to 29.5% mortality in untreated plantations. The reduction in foliage loss was acceptable in all aerial trials. Mortality resulting from ground applications averaged 98.4%, but the reduction in foliage loss was not acceptable. In the ground tests, they concluded that the application rate was too low and insect development too advanced at the time of treatment. In a similar test in 1979, Bordeleau (cited by Cunningham) compared helicopters with Micronair unit or boom and flat fan at a volume of 9.4 ℓ/ha with a back pack mist blower at 18.8 ℓ/ha.[90] Larval mortality 35 days posttreatment averaged 87, 98, and 94%, respectively. Check mortality was 7%.

Ground spray application of baculoviruses is usually not applicable to most commercial or recreational forest situations. Ground equipment can, however, be effectively utilized to treat forest insect pests in plantations of small trees or along roadways. Pressurized hand-held sprayers equipped with one or more hydraulic nozzles or backpack mist blowers are often used to treat small trees over a limited area. Vehicle mounted mist-blowers or high pressure, high volume hydraulic sprayers are most often utilized for treatment of larger trees or sizeable acreages of small trees. Most ground applications of baculoviruses have not been of an operational nature, instead, they have usually been preliminary small-scale evaluations of the efficacy of a given virus, formulation, or tank mixture. Should results of these tests prove favorable, additional evaluations are most often conducted with aerial equipment. The aim of tests utilizing ground application has not been evaluation or comparison of equipment.

2. Agriculture

As in forests, baculoviruses have been applied to agricultural crops with both aerial and ground equipment. However, unlike the situation with forest applications, which are mostly made by air, ground equipment has been predominantly used for agricultural crops. The selection of ground equipment for applications on agricultural crops has been primarily due to its widespread availability and the ability to use it on small plots, thus facilitating replication. In agricultural situations, baculoviruses have typically been applied as aqueous foliar sprays, although limited evaluations of dust or other formulations for dry applications have been conducted.

Aerial applications of viruses to agricultural crops have mostly been used in large-scale efficacy evaluations and, to a lesser degree, in integrated pest management programs. Definitive evaluations of aerial applications have not been conducted, and comparisons of different application equipment have been lacking. Aerial applications have been primarily with fixed wing aircraft equipped with boom and flat fan nozzles. The volume of spray in aerial treatment has typically varied from 19 to 47 ℓ/ha.

Ground equipment for row crops has usually been a high clearance sprayer with boom and nozzle. A variety of nozzles has been used, but most have been hollow cone or flat fan. Spray volumes usually vary from 19 to 114 ℓ/ha. Virus applications in fruit orchards and many vegetable crops have been with a variety of sprayers, i.e., mist blowers, high pressure-high volume equipment etc., typically used for the crop and locality. In small plot efficacy tests, viruses have often been applied with a variety of conveniently used sprayers, i.e., hand-held or back pack compressed air or knapsack sprayers equipped with various nozzles.

Research on application of viruses to agricultural crops has been primarily with *H. zea* NPV against *Heliothis* spp. on cotton. Andrews et al.[91] compared the effect of direction and volume of spray on efficacy of *H. zea* NPV (Viron H) on cotton. The virus, 253 larval equivalents (LE) per hectare was tank mixed in water with a cottonseed oil bait and sprayed at 9.6 or 47 ℓ/ha.[92] The spray was either directed downward from nozzles located above the plant to give primary coverage on upper leaf surfaces or from nozzles lowered between rows and directed upward to give primary coverage on the lower leaf surface. The 47ℓ/ha application volume resulted in better boll protection than the lower volume. Chapman and Ignoffo[93] had previously found that doubling the volume of spray from 94 to 188 ℓ/ha provided control equal to doubling the rate of *H. zea* NPV from 102 to 204 LE per hectare. Although Andrews et al.[91] found that sprays directed upward did not provide good plant coverage, the direction of spray did not alter effectiveness of the virus. Stacey et al.[94] also examined the effect of directional sprays on efficacy of *H. zea* NPV on cotton using hollow-cone nozzles. The virus was applied at 97 LE per hectare in 96 ℓ/ha with 1, 3, or 5 nozzles per row to cotton plants over the top, from the sides, from underneath, and all permutations thereof. Nozzle arrangement and orientation did not influence the efficacy of the virus.

Stacey et al.[95] measured the droplet spot diameter and location following application of *H. zea* NPV with a series of gustatory adjuvants using a conventional boom and hollow-cone nozzles. They found that the most efficaceous treatments were those with the largest drop spots (235 μm) and greatest deposits on the upper area of plants. However, droplet characteristics were confounded by the effects of the adjuvants in some virus treatments. In later tests, Stacey et al.[96] examined the effect of spray volume and droplet spot diameter (VMD) on efficacy of the virus. Droplet size was regulated between VMDs of 115-402 μm and volumes of spray between 9.4 and 93.5 ℓ/ha using either a "Model 141 Span Spray" system or conventional boom with hollow-cone nozzles. No significant differences in virus efficacy were detected between the various volumes and droplet spot diameters. In a subsequent test, the application of *H. zea* NPV with Raindrop® nozzles, VMD = >600 μm, to maximize droplet size failed to significantly improve the efficacy of the virus over that obtained with the conventional hollow-cone TX-4 nozzles.

The most extensive effort to develop spray performance specifications for applications of NPVs has been by Smith and co-workers (USDA/ARS, Columbia, Md.). They used a spinning disk droplet generator to obtain a variety of droplet sizes and densities for *H. zea* NPV sprays and found that application rate was more important than droplet size, density, or concentration. There was no significant interaction, however, between the concentration of virus deposited and droplet size. The combination of small droplet size, high droplet density, and high virus concentration gave highest mortality of *T. ni* and *H. zea* larvae in bioassays of virus deposits.[97] Smith et al.[98] evaluated TX-1 nozzles at 552 kPa and TX-4 nozzles at 373 kPa for application of a series of *H. zea* NPV formulations on soybean foliage. The TX-4 nozzles produced higher mortality of *H. zea* larvae in bioassays than did TX-1 nozzles for a given volumetric deposit. However, certain adjuvants that increased viscosity of the spray increased effectiveness of the TX-1 nozzles. In a following test, Smith et al.[99] compared deposits of a *H. zea* NPV formulation, with increased viscosity due to the addition of Shade® (2%) and polyvinyl alcohol (0.5%) sprayed through TX-1, TX-4, or TX-6 nozzles operating at 552 kPa. The TX-1 nozzles produced higher deposits of virus and higher *H. zea* larval mortalities than either TX-4 or TX-6 nozzles operating at 380 or 276 kPa, respectively.

Application of viruses as dust formulations has received little attention. Montoya and Ignoffo[20] described an apparatus and technique for assay of dust concentrations of the *H. zea* NPV. They applied *H. zea* NPV as a dust on cotton and found it was less effective for *Heliothis* spp. control than a spray of 50 μm VMD.[100] Stacey et al.[21] applied *H. zea* NPV at 97 LE per hectare as a granule in 11.4 kg/ha of dry baits using a Casaron granule applicator. The baits were less efficacious than aqueous sprays against *Heliothis* spp. on cotton.

Falcon et al.[101] performed a series of tests to examine the potential of cold aerosal generators for application of NPV in finely atomized sprays. A microgen fog generator equipped with Belvoir nozzles that delivered 1.0 to 1.5 ℓ/min effectively disseminated the *H. zea* NPV up to 150 m downwind of the application path. Bioassay of cotton fruiting structures collected at this distance resulted in 70 to 100% mortality of *Heliothis* larvae. In this test, cottonseed oil was found to be superior to skimmed milk in water as a carrier. In other tests, the effect of droplet size on coverage was examined.[101,102] When droplets of 10 to 40 μm VMD (produced by a cold aerosol generator with Belvoir nozzles) or 30 to 90 μm VMD (produced by an aerosol applicator equipped with a Calblower) were sprayed under a temperature inversion and low wind velocities, coverage of wide swaths was obtained. Bioassays detected virus 1 mi downwind following application with either machine. When *T. ni* NPV was drift sprayed over cotton with a cold aerosol generator, an epizootic occurred in *T. ni* larvae, and the degree of pest control obtained by use of the aerosol generator compared favorably with aerial application.[81]

B. Tank Mixtures

In direct comparison, viral insecticides have generally been found less effective than conventional chemical insecticides. Lack of success with baculoviruses has often been attributed to inadequate coverage, lack of persistence due to UV degradation and/or behavioral characteristics of target larvae that minimize consumption of a lethal dose, i.e., feeding within fruiting structures or other plant tissues. As a result, a variety of tank mixture adjuvants have been utilized in an effort to circumvent these apparent problems.

The simplest and most often used adjuvants for tank mixtures with viral pesticides have been surface-active agents, i.e., spreader-stickers, emulsifiers, wetting agents, etc. A variety of surfactants, most having similar properties, have been utilized and appear to be compatible with baculoviruses. Ignoffo and Montoya[103] found that Multifilm Buffer X, Triton® X-100, Triton® X-152, Triton® X-172, and Triton® B-1956 were compatible with *H. zea* NPV. Smith et al.[98] tank mixed a variety of surfactants including Triton ® CS-7, polyvinyl alcohol, mineral oils, and spreader-penetrates in laboratory tests with *H. zea* NPV and reported that they were compatible. Bivert (an emulsifiable crop oil) and Nalcotrol® (a spray thickener) have been mixed with sprays of baculoviruses and other microbial agents to reduce drift and retard evaporation.[2] Molasses and other sugar solutions have often been added to tank mixtures with viral insecticides. In addition to their sticking properties, sugar solutions also serve as spray thickeners and evaporation retardants and appear to increase on target deposits in row crops and forests. Wolfenbarger[104] tested the *H. zea* NPV with surfactants and suggested efficacy could be improved with their use. Although commercial surfactants, especially spreader-stickers are relatively inexpensive, often used and considered desirable, there have been few definitive tests to evaluate their impact on efficacy.[2,79] However, the added effectiveness of molasses (25%) to TM Biological-1 in field tests has led to a statement on the label of this product that stickers may enhance its performance.

A variety of materials with UV-absorbing properties have been mixed with baculoviruses in foliar sprays. Finely ground charcoal, such as carbon black, has been the most commonly used UV screen. Shade®, developed by IMC for use with microbial agent formulations, has generally been mixed at concentrations from 1 to 6% for use primarily against pests of forest and cotton. Other materials with UV screening properties that have been used in tank mixtures include lignin sulfate, gelatin, hemolymph, and Uval (Miles Laboratory, Inc.)[29] In addition, Jaques[105] found that a variety of potential UV screening materials, including proteins, stains, and particulates were compatible with the *T. ni* NPV and several of these protected the virus well in the laboratory and in the field. Jaques reported that a combination of egg albumin and India ink was the most effective mixture tested.

Additives containing components with gustatory-stimulant properties have been utilized, particularly with *H. zea* NPV on cotton. The sugars in molasses appear to make it act as a gustatory stimulant as well as a surfactant. Sucrose, glucose, and fructose have also been used at concentrations of 3.25%. Wheast, a cottage cheese byproduct, and powdered milk have also been added to spray mixtures.[95,106] Aqueous extracts of host plants including corn silk and seed of corn, cotton, and crimson clover have been mixed with the *H. zea* NPV in either dusts or sprays.[95,100,107,108]

More complex gustatory adjuvants consisting of two or more components and possessing properties of a UV screen, gustatory stimulant, and surfactant have been developed. The first of these was cottonseed oil mixed with sucrose, hydroxycellulose, and Dacagin®.[92] This formulation, evaluated by Andrews et al.[91] had several desirable properties, but spray qualities were poor. Bell and Kanavel[109,110] modified this formulation to consist of 5% cottonseed flour, 1% cottonseed oil, and 2% sucrose. A concentrated version of this bait with approximately 0.4% Tween® 80 added as a surfactant has been sold commercially under the trade name of Coax® (Wilbur-Ellis Co., Ft. Worth, Tex.) A soybean meal-based adjuvant with similar properties was developed by Sandoz, Inc. for use with Elcar®.[16] This formulation was later altered to improve spray qualities and marketed under the trade name Gustol®. Smith et al.[108] attempted to develop adjuvants that were biologically superior to Coax® and Gustol®. Of the numerous formulations evaluated in the laboratory, formulations containing corn flour and oil or soybean flour and oil were superior to formulations that contained only oil or corresponding flour.

1. Forest

Virus spray adjuvants used for forest applications have been primarily sunlight protectants and surface-active agents. Sawflies have been effectively controlled with NPV in a series of tests in which adjuvants were included in the spray mixture. Excellent control of *Neodiprion sertifer* was obtained at 1.2×10^{10} PIB of NPV in 2.4 ℓ/ha containing 5 g/ℓ of skim milk,[111] and at 5×10^{10} PIB in 9.4 ℓ/ha containing 3% IMC 90-001 UV screen and 0.13% commercial sticker.[112] *N. swainei* NPV at 1×10^6 PIB per milliliter in 4.7 or 37.6

ℓ/ha as an aqueous spray containing latex and dried blood or a fuel oil, magmabentonite, and span emulsion was effective when used against small larvae but not against older larvae.[113] In a series of tests on *N. lecontei* in Canada, NPV sprays containing Shade® (3 to 6%), molasses (25%) and a commercial spreader-sticker (0.1%) were found to be effective.[112,114,115] However, deGroot and Cunningham[116] compared *N. lecontei* NPV at 5 × 10^9 PIB per hectare in water alone or water containing 25% molasses and 3% Shade® at volumes of 2.4, 4.7, and 9.4 ℓ/ha and suggested that the *N. lecontei* NPV applied in water alone was as effective as with adjuvants.

Spray additives used with viruses of lepidopterous pests in forests have been similar to those used with NPV of sawflies. *Choristoneura fumiferana* NPV at 7.6 × 10^{11} PIB per hectare in 29 ℓ/ha with 2.5% Shade® was effective against eastern spruce budworm. Cunningham et al.[17] found that *C. fumiferana* NPV at 2.5 to 7.5 × 10^{11} PIB per hectare in 9.4 ℓ/ha containing 25% molasses, 6% Shade,® and 0.1% spreader-sticker was superior to a commercial formulation of the virus, Sandoz 285 WP.

Stelzer et al.[117] and Stelzer and Neisess[118] evaluated *O. leucostigma* NPV at 2.5 × 10^{11} PIB per hectare in 19 ℓ/ha containing 25% molasses, 25% molasses + 1.1 kg/ha IMC 90-001, or 1.1 kg/ha IMC 90-001. They found that the addition of the sunlight protectant to IMC 90-001 increased population reduction at 7 and 21 days posttreatment. In a similar test, Stelzer et al.[85] reaffirmed the importance of protecting the virus against UV inactivation. Ilnytzky et al.[119] also obtained satisfactory control with *O. leucostigma* NPV at 2.5 × 10^{10} PIB per hectare in 9.4 ℓ/ha with 25% molasses and a sunlight screen.

In two tests *Lymantria dispar* was controlled on hardwood trees treated with NPV at rates of either 2 or 3.1 × 10^7 PIB per milliliter containing 2% methyl cellulose or 1.8% skimmed milk powder.[120,121] Gypsy moth larvae were effectively controlled with NPV at 2.5 × 10^{13} PIB in 750 ℓ/ha with a commercial sticker and 0.15 g/ℓ of IMC 90-001 in one test and 6% Shade®, 2.5 mℓ Chevron sticker, and 25% molasses in another.[122] Wollam et al.[123] compared *L. dispar* NPV tank mixes of 6% Shade®, 12.5% molasses, and 4.7% Chevron sticker with or without 50% Sandoz adjuvant. Spray deposits were better with Sandoz adjuvant, but population reduction and foliage protection were similar for both formulations. Following a series of tests with these tank mix formulations, Lewis and Yendel[88] reported similar results.

Abrahamson and Harper[124] added 25% molasses to an NPV (6 × 10^{10} PIB in 28 ℓ/ha) applied aerially against L3 and L4 *Malacosoma disstria,* but failed to obtain foliage protection. Ives and Muldrew[125] applied the virus at 10^7 to 10^8 PIB per milliliter with 2.5% Shade® and obtained high larval mortality and excellent foliage protection when applications were timed against the egg stage 10 days prior to hatch.

Pritchett et al.[106] improved effectiveness of *Hyphantria cunea* NPV, *H. cunea* GV, and *A. californica* NPV against *H. cunea* by the addition of several adjuvants, including Coax®, Sandoz Adjuvants 1976 or 1977, Shade®, molasses, or powdered milk.

2. Agriculture

As with application in forests, viral sprays on agricultural crops typically include a spray additive, usually a spreader-sticker, although these do not appear to significantly influence efficacy. Most of the work with tank mix formulations in agricultural crops has been with those that provide better persistence against UV degradation and exhibit gustatory stimulant properties. Ignoffo et al.[126] reported on a series of field efficacy evaluations of *H. zea* NPV applied at 6 × 10^{11} PIB per hectare plus sunlight protectants on cotton against *Heliothis* spp. Virus applications in water with either activated charcoal or IMC 90-001, increased virus persistence on the plants and increased yields over virus alone.

Microencapsulation has also been used to increase sunlight protection of *H. zea* NPV formulations. Ignoffo and Batzer[18] found that nonencapsulated combinations were as effective as encapsulated combinations of virus plus carbon. Bull et al.[19] reported that microencapsulated *H. zea* NPV in titanium dioxide or carbon black had excellent persistence in sunlight and was superior to nonencapsulated mixtures of virus and protectant. In field tests, however, against moderate *Heliothis* spp. populations on cotton, yields from plots treated with encapsulated formulations were similar to plots treated with a nonprotected commercial formulation, Sandoz 270-WP. Although a number of materials and formulations have been shown to markedly increase the persistence of NPV on cotton, this increase in persistence is not reflected in increased virus efficacy.[3]

Molasses and sugar solutions have been tested in tank mixtures with viruses for use on agricultural crops. Results from these tests have been variable and inconclusive. Roome[127] reported that a spray of 100 LE per hectare of NPV to which 0.6% molasses had been added was as effective against *H. armigera* on cotton as 200 LE per hectare without molasses. Other workers have observed increased yields, on cotton by combining NPV on *Heliothis* NPV with molasses or other sources of sugars. Stacey et al.[95,96] increased yields by the addition of 3.3% sugar to the NPV suspension. The cottonseed oil bait used by Andrews et al.[91] that increased efficacy of an NPV formulation also contained 10% sugar. However, the others failed to obtain a yield increase when sugars were added to the Sandoz 240 WP formulation of *H. Zea* NPV.[19,128] Although sugars increase droplet density, appear to improve coverage, and act as weak gustatory stimulants and sunlight screens, it has been suggested that the apparent increase in NPV efficacy resulting from their use may have been due to physiological effects on plants that resulted in increased cotton yields.[95]

Adjuvants which contain gustatory components have also been added to *H. zea* NPV sprays in an attempt to increase virus consumption by *Heliothis* spp. larvae. Aqueous extracts from a variety of host plants and cottonseed oil have been demonstrated to stimulate feeding by larvae of *Heliothis* spp. and have been used as spray additives with NPV against *Heliothis* spp. on cotton.[100,129] *H. zea* NPV sprays containing aqueous extracts of fresh corn were more efficacious than sprays containing the virus alone.[100,107] However, Stacey et al.[95] failed to obtain a significant yield increase over the virus alone, with aqueous extracts of corn meal, corn seed, cottonseed, or clover seed in tank mixes. The addition of 3.3% wheast to the virus spray also failed to elicit an increased yield of cotton. Stacey et al.[21] also applied these gustatory stimulants in dry baits containing *H. zea* NPV on cotton without any increase in yields.

The inconsistent performance of spray additives consisting of one or two components eventually led to the use of multicomponent additives with several properties. Andrews et al.[91] improved the efficacy of *H. zea* NPV on cotton with a cottonseed oil-based preparation consisting of 22% cottonseed oil, 1.4% Thixen®, 0.5% Dacagin®, 1.1% hydroxycellulose, and 1.8% invertase sugars.

In a series of field evaluations of NPV with either Coax® or Gustol® against *Heliothis* spp. on cotton, efficacy was typically superior to that obtained by use of the virus alone but often less than that obtained with the conventional chemical insecticide standard.[95,128,130-133] The addition of Coax® also improved the performance of *A. californica* NPV against *P. gossypiella* and *Buccalatrix thurberiella* (Busck) on cotton.[110,134] Studies by Smith et al.[98,108] on a series of spray additives for *H. zea* NPV indicated that addition of soybean flour, cotton flour, citrus pulp, or sugars to virus sprays could substantially increase efficacy. They suggested use of an adjuvant consisting of 8% Nutrisoy® 7B soybean flour, 0.5% crude soybean oil, 1% sucrose, and 0.01% Trition CS-7. Hostetter and Pinnell[22] reported that granular formulations of *H. zea* NPV (8 × 10⁵ PIB) with 2% gelatin, 1% Dacagin®, 8 g Coax®, and 100 g filler (corn bran, corn hominy, reground oat hulls, rice mill feed, or ground rice hulls) were less effective against *H. zea* than an aqueous suspension of NPV plus Coax® when applied to soybean foliage.

C. Compatibility with Other Pesticides

Crops often require an array of pesticides during a growing season for control of diseases, weeds, insects, etc. It is often desirable to treat two or more of these pest problems simultaneously by applying a mixture of pesticides in a single application. It is thus advantageous for viral insecticides to be compatible in tank mixtures with as wide a range of pesticides as possible. The compatibility of viruses and other pesticides has been reported in previous reviews.[135,136] Viral insecticides have been found to be compatible in tank mixtures with most conventional synthetic organic insecticides, i.e., organochlorine, organophosphorus, carbamate, and pyrethyroid insecticides. Baculoviruses found compatible with some of these insecticides are *Pieris rapae* GV,[137] *Peridroma saucia* NPV,[138] *T. ni* NPV,[137,139-141] and *H. zea* NPV.[103,139,141] *H. zea* NPV activity was reduced with the addition of methyl parathion or EPN-methyl parathion mixtures.[142,143] Other pesticides with which *H. zea* NPV has been found compatible in sprays are chlordimeform and cyhexatin (acaracides), thiabenzadole and benomyl (fungicides), and methoprene and diflubenzuron (insect growth regulators). *H. zea* NPV did not appear to be compatible with fentin hydroxide and chlorothalonil (fungicides).[144] *Plodia interpunctella* GV was compatible with the fumigants phosphine, carbon tetrachloride, carbon bisulfide, or ethylene dichloride, but not methyl bromide.[145]

Jaques[137,146] found that mixtures of either *T. ni* NPV or *Pieris rapae* NPV with either methomyl or endosulfan were more efficacious than the virus alone. Also, mixtures of low concentrations of *A. californica* NPV or *P. rapae* GV with low dosages of chlordimeform were as effective or more effective than either material alone against *T. ni* and *P. rapae*, respectively, on cabbage.[147] On cotton, *H. zea* NPV-chlordimeform mixtures produced yields greater than or similar to either material alone.[148,149] Application of reduced rates of permethrin or methomyl in combination with a recommended rate of *H. zea* NPV for *Heliothis* spp. produced yields of cotton equal to those of *H. zea* NPV alone.[143] A mixture of EPN-methyl parathion-*H. zea* NPV, however, reduced yields compared to the *H. zea* NPV alone. Also, *H. zea* NPV-fenvalerate was more effective than *H. zea* NPV-methomyl and both were much superior to the *H. zea* NPV alone.[128]

Combinations of baculoviruses with *Bacillus thuringiensis* have also been tested in field trials. *T. ni* NPV or *P. rapae* NPV-*B. thuringiensis* mixtures were more effective than either pathogen alone against *T. ni* or *P. rapae*, respectively, in cabbage.[26] *A. californica* NPV-*B. thuringiensis* resulted in increased yields of cotton over either material alone against *Buccalatrix thurberiella* and *H. virescens*.[130,134] Johnson,[133] however, reported that *H. zea* NPV-*Bacillus thuringiensis* mixtures with Gustol® or Coax® were not superior in effectiveness to either pathogen alone with the adjuvant.

D. Application Timing

Proper timing of baculovirus applications is necessary for their efficacious utilization in pest management systems. Applications must be timed against the susceptible stages of the insect pest so that population levels will be effectively suppressed prior to development of the most damaging larval stages. Since baculoviruses must be ingested to bring about an infection and the infected individual continues to feed until it is close to death, the slow action of a baculovirus necessitates proper timing of application.

Baculoviruses evaluated to date as pesticides are those of lepidopterous and hymenopterous larvae. Since the pest status of these insects is due to damage resulting from larval feeding, viral insecticide applications must be directed against early larval stages so that death occurs before the larvae reach the large, more damaging developmental stages. For example, Alam[150] noted that more than 80% of the food consumption by *Pseudoplusia includens* was during the last two developmental stages. Thus, severe damage will occur if applications are delayed even though most of the larvae in the population succumb to the disease.[151,152] Normally the smaller the larvae at treatment the more effectively controlled will be the target generation.

Younger larvae are more susceptible to baculoviruses than older larvae and after treatment die earlier in development thus reducing food consumption. When treated such that they become infected in the L1 or L2 stages, death usually occurs prior to completion of the L4 stage. Since most food consumption by larvae occurs during the last two stages, infection of small larvae results in little feeding damage between virus ingestion and cessation of feeding.[153-155] Treatment of larvae older than L1 or L2 stage however, will often result in considerable damage even if all larvae are infected and eventually die.

Ideally, virus application should be timed against newly hatched larvae, but timing may vary depending on the feeding habits of the pest and the crop. Larvae of many species feed concealed during much of their developmental period and these must be treated during an exposed stage in which they feed at sites of virus deposits. For example, L1 *L. pomonella* larvae feed externally on apples but thereafter bore into the apples and feed concealed. Thus, application must be timed against newly hatched larvae.[156] *Choristoneura fumiferana* feed as needle miners during L1 and L2 stages, then bore into buds until L4, when they are exposed by flushing of the buds. Treatment has been directed against the small larvae, but control is much better when timed against L4 larvae.[17,157] Treatment of such large larvae did not provide foliage protection in the year of treatment but did provide protection in the following year. Alam[150] found that treatment of *P. includens* on soybean with NPV resulted in better control of L2 or L3 larvae than of L1 larvae. Alam suggested that this was due to a difference in feeding habits between L1 and older larvae. Young larvae feed low on the plant on the undersurface of leaves, but older larvae feed higher on the plant and chew through leaves to the upper surface. Thus, when L1 larvae are treated, virus deposited on the upper surface of leaves is not available to them until later instars, and by then much of the virus will have been inactivated by sunlight.

Timing of application for a virus-host complex may also vary with plant host. Larvae of *Heliothis* spp. on cotton feed concealed much of the time, and application of virus must be directed against very young larvae that feed on new terminal growth before moving down into fruiting structures. On some crops such as grain sorghum, larvae feed exposed on the open fruiting heads and timing is less critical.

Since viruses are rapidly inactivated by sunlight, it has often been suggested that they be applied at dusk rather than earlier in the day to delay this rapid loss in activity. Although this would often appear advantageous, data are not available from field tests to verify the value of this approach. The time interval between treatments with viral insecticides may need to be shorter than with chemical insecticides. This is due to the more critical timing of application against specific-sized larvae and rapid inactivation of virus on crops. Also, small larvae of many species such as *H. zea* feed on new growth such as in plant terminals where growth is often rapid. The plant quickly grows away from the virus deposits and the virus must again be applied to maintain plant coverage.[15]

Timing of application may differ when efficacy evaluations are directed against a series of generations, during succeeding years, etc., rather than only against the target generation. In this situation, the objective is not necessarily to treat for maximum mortality and minimal damage in the target generation, but rather to sustain an epizootic and minimize damage through succeeding generations. This may require treatment at reduced rates or in later instars.

Improper timing of viral insecticide applications is often considered a primary reason for their failure to provide efficacious control. When variable results are obtained with a series of field trials, much of the variability can often be attributed to improper timing of the less effective treatments. Proper timing requires a knowledge of insect behavior and the presence of an effective scouting system. Dependable monitoring systems are seldom available in many cropping systems and when in place are typically designed for chemical insecticides, which require less indepth scouting than viral insecticides. Efficacious application of viral pesticides can only be accomplished with proper timing, and on many crops, consistent control with viruses will await the development of adequate monitoring systems.

REFERENCES

1. **Van Valkenburg, W.,** in *Pesticide Formulations,* Van Valkenburg, W., Ed., Marcel Dekker, New York, 1973, 93.
2. **Couch, T. L. and Ignoffo, C. M.,** Formulation of insect pathogens, in *Microbial Control of Pests and Plant Diseases, 1970—1980,* Burges, H. D., Ed., Academic Press, New York, 1981, 621.
3. **Yearian, W. C. and Young, S. Y.,** Control of insect pests of agricultural importance by viral insecticides, in *Microbial and Viral Pesticides,* Kurstak, E., Ed., Marcel Dekker, New York, 1982, 387.
4. **Ignoffo, C. M. and Shapiro, M.,** Characteristics of baculovirus preparations processed from living and dead larve, *J. Econ. Entomol.,* 71, 186, 1978.
5. **Ignoffo, C. M.,** Standardization of products containing insect viruses, *J. Invertebr. Pathol.,* 8, 547, 1966.
6. **Martignoni, M. E. and Iwai, P. J.,** Sanitation program for diet preparation, in Standard Operating Procedures, NWV-2203, USDA For. Sci. Lab., Corvallis, Ore., 1977.
7. **Martignoni, M. E.,** Virus in biological control: production activity and safety, in *The Douglas Fir Tussock Moth: A Synthesis,* Brookes, M. H., Stark, R. W., and Campbell, R. W., Eds., U.S. Department of Agriculture, Washington, D.C., 1978, 140.
8. **Hughes, P. R. and Wood, H. A.,** A synchronous peroral technique for the bioassay of insect viruses, *J. Invertebr. Pathol.,* 37, 154, 1981.
9. **Dulmage, H. and Burgerjon, A.,** Industrial and international standardization of microbial pesticides. II. Insect viruses, *Entomophaga,* 22, 131, 1977.
10. **Shapiro, M.,** *In vivo* mass production of insect viruses, in *Microbial and Viral Pesticides,* Kurstak, E., Ed., Marcel Dekker, New York, 1982, 463.
11. **Ignoffo, C. M.,** Production and virulence of a nuclear polyhedrosis virus from larve of *Tricholplusia ni* (Hubner) reared on a semisynthetic diet, *J. Insect Pathol.,* 6, 318, 1964.
12. **Dulmage, H. T., Martinez, A. J., and Correa, J. A.,** Recovery of the nuclear polyhedrosis virus of the cabbage looper, *Trichoplusia ni,* by coprecipitation with lactose, *J. Invertebr. Pathol.,* 16, 80, 1970.
13. **McGaughey, W. H.,** A granulosis virus for Indian meal moth control in stored wheat and corn, *J. Econ. Entomol.,* 68, 346, 1975.
14. **Hunter, D. K., Collier, S. S., and Hoffman, D. F.,** Granulosis virus of the Indian meal moth as a protectant for stored in-shell almonds, *J. Econ. Entomol.,* 70, 493, 1977.
15. **Bull, D. L.,** Formulations of microbial insecticides: microencapsulation and adjuvants, *Misc. Publ. Entomol. Soc. Am.,* 10, 11, 1978.
16. **Ignoffo, C. M., Hostetter, D. L., and Smith, D. B.,** Gustatory stimulant, sunlight protectant, evaporation retardant: three characteristics of a microbial insecticidal adjuvant, *J. Econ. Entomol.,* 69, 207, 1976.
17. **Cunningham, J. C., Kaupp, W. J., House, G. M., McPhee, J. R., and deGroot, P.,** Aerial application of spruce budworm baculovirus: tests of virus strains, dosages and formulations in 1977, *Can. For. Serv. Inf. Rep.,* FPM-X-3, 1978.
18. **Ignoffo, C. M. and Batzer, O. F.,** Microencapsulation and ultraviolet protectants to increase sunlight stability of an insect virus, *J. Econ. Entomol.,* 64, 850, 1971.
19. **Bull, D. L., Ridgway, R. L., House, V. S., and Pryon, N. W.,** Improved formulations of the *Heliothis* nuclear polyhedrosis virus, *J. Econ. Entomol.,* 69, 731, 1976.
20. **Montoya, E. L. and Ignoffo, C. M.,** Laboratory technique and apparatus for testing viruses applied as dusts, *J. Invertebr. Pathol.,* 8, 251, 1966.
21. **Stacey A. L., Yearian, W. C., and Young, S. Y.,** Efficacy of *Baculovirus heliothis* and *Bacillus thuringiensis* applied in dry baits, *Ark. Farm. Res.,* 26, 3, 1977.
22. **Hostetter, D. L. and Pinnell, R. E.,** Laboratory evaluation of plant-derived granules for bollworm control with a virus, *J. Ga. Entomol. Soc.,* 18, 155, 1983.
23. **Zelazny, B.,** Studies on *Rhabdiovirus oryctes.* I. Effect on larvae of *Oryctes rhinoceros* and inactivation of the virus, *J. Invertebr. Pathol.,* 20, 235, 1972.
24. **Aizawa, K.,** The nature of infections caused by nuclear-polyhedrosis viruses, in *Insect Pathology: An Advanced Treatise,* Vol. 1, Steinhaus, E. A., Ed., Academic Press, New York, 1963, 383.
25. **Huger, A.,** Granulosis of insects, in *Insect Pathology: An Advanced Treatise,* Vol. 1, Steinhaus, E. A., Ed., Academic Press, New York, 1963, 531.
26. **Jaques, R. P.,** Stability of entomopathogenic viruses, *Misc. Publ. Entomol. Soc. Am.,* 10, 99, 1977.

27. **Lewis, F. B. and Rollinson, W. D.,** Effect of storage on the virulence of gypsy moth nucleopolyhedrosis inclusion bodies, *J. Econ. Entomol.,* 71, 719, 1978.
28. **Neilson, M. M. and Elgee, D. E.,** The effect of storage on the virulence of polyhedrosis virus, *J. Insect Pathol.,* 2, 165, 1960.
29. **David, W. A. L.,** The granulosis virus of *Pieris brassicae* L. and relationship with its host, in *Advances in Virus Research,* Lauffer, M. A., Bang, F. R., Maramorosch, K., and Smith, K. M., Eds., Academic Press, New York, 1975, 111.
30. **David, W. A. L. and Gardiner, B. O. C.,** The effect of heat, cold, and prolonged storage on a granulosis virus of *Pieris rapae, J. Invertebr. Pathol.,* 9, 555, 1967.
31. **Hunter, D. K., Hoffman, D. F., and Collier, S. J.,** Pathogenicity of a nuclear polyhedrosis virus of the almond moth, *Cadra cautella, J. Invertebr. Pathol.,* 21, 282, 1973.
32. **Aizawa, K.,** Dissolving curve and the virus activity of the polyhedral bodies of *Bombyx mori, Sanshi Kenkyu,* 8, 52, 1954.
33. **Jaques, R. P.,** The inactivation of the nuclear-polyhedrosis virus of *Trichoplusia ni* by gamma and ultraviolet radiation, *Can. J. Microbiol.,* 14, 1161, 1968.
34. **Witt, D. J. and Stairs, G. R.,** The effects of ultraviolet irradiation on a baculovirus infecting *Galleria mellonella, J. Invertebr. Pathol.,* 26, 321, 1975.
35. **McLeod, P. J., Yearian, W. C., and Young, S. Y.,** Inactivation of *Baculovirus heliothis* by ultraviolet irradiation, dew and temperature, *J. Invertebr. Pathol.,* 30, 237, 1977.
36. **David, W. A. L.,** The effect of ultraviolet irradiation of known wavelengths on a granulosis virus of *Pieris brassicae, J. Invertebr. Pathol.,* 14, 336, 1969.
37. **Ignoffo, C. M. and Garcia, C.,** The relation of pH to the activity of inclusion bodies of a *Heliothis* nuclear polyhedrosis, *J. Invertebr. Pathol.,* 8, 426, 1966.
38. **Gudauskas, R. T. and Canerday, D.,** The effect of heat, buffer salt and H-ion concentration, and ultraviolet light on the infectivity of *Heliothis* and *Trichoplusia* nuclear-polyhedrosis viruses, *J. Invertebr. Pathol.,* 12, 405, 1968.
39. **Harpaz, I. and Raccah, B.,** Nucleopolyhedrosis virus (NPV) of the Egyptian cottonworm, *Spodoptera littoralis* (Lepitoptera:Noctuidae): temperature and pH relations, host range and synergism, *J. Invertebr. Pathol.,* 32, 368, 1974.
40. **David, W. A. L.,** The granulosis virus of *Pieris brassicae* L. in relation to natural limitation and biological control, *Ann. Appl. Biol.,* 56, 331, 1965.
41. **Cantwell, G. E.,** Inactivation of biological insecticides by irradiation, *J. Invertebr. Pathol.,* 9, 138, 1967.
42. **Jaques, R. P.,** The persistence of a nuclear polyhedrosis virus in the habitat of the host insect, *Trichoplusia ni*. I. Polyhedra deposited on foliage, *Can. Entomol.,* 99, 785, 1967.
43. **David, W. A. L., Gardiner, B. O. C., and Woolner, M.,** The effects of sunlight on a purified granulosis virus of *Pieris brassicae* applied to cabbage leaves, *J. Invertebr. Pathol.,* 11, 496, 1968.
44. **Morris, O. N.,** The effect of sunlight, ultraviolet and gamma radiations and temperature on the infectivity of nuclear polyhedrosis virus, *J. Invertebr. Pathol.,* 18, 292, 1971.
45. **Smirnoff, W. A.,** The effect of sunlight on the nuclear-polyhedrosis virus of *Neodiprion swainei* with measurement of the solar energy received, *J. Invertebr. Pathol.,* 19, 179, 1972.
46. **Nordin, G. L.,** Inactivation of a baculovirus of *Hyphantria cunea* by sunlight and ultraviolet irradiation (Lepidoptera:Arctiidae), *J. Kan. Entomol. Soc.,* 50, 18, 1977.
47. **Bullock, H. R., Hollingsworth, J. P., and Hartstack, A. W.,** Virulence of *Heliothis* nuclear polyhedrosis virus exposed to monochromatic ultraviolet irradiation, *J. Invertebr. Pathol.,* 16, 419, 1970.
48. **Bullock, H. R.,** Persistence of *Heliothis* nuclear polyhedrosis virus on cotton foliage, *J. Invertebr. Pathol.,* 9, 434, 1967.
49. **Young, S. Y. and Yearian, W. C.,** Persistence of *Heliothis* NPV on foliage of cotton, soybean, and tomato, *Environ. Entomol.,* 3, 253, 1974.
50. **Ignoffo, C. M., Parker, F. D., Boening, O. P., Pinnell, R. E., and Hostetter, D. L.,** Field stability of the *Heliothis* nucleopolyhedrosis virus on corn silks, *Environ. Entomol.,* 99, 785, 1973.
51. **Ignoffo, C. M.,** Production and virulence of a nuclear polyhedrosis virus from larvae of *Trichoplusia ni* reared on a semisynthetic diet, *J. Insect Pathol.,* 6, 318, 1974.
52. **Jaques, R. P.,** The inactivation of foliar deposits of viruses of *Trichoplusia ni* and *Pieris rapae* and tests on protectant activities, *Can. Entomol.,* 104, 1985, 1972.
53. **Bird, F. T.,** Transmission of some insect viruses with particular reference to ovarian transmission and its importance in the development of epizootics, *J. Invertebr. Pathol.,* 3, 352, 1961.
54. **Podgwaite, J. D., Shields, K. S., Zerillo, R. T., and Bruen, R. B.,** Environmental persistence of the nucleopolyhedrosis virus of the gypsy moth, *Lymantria dispar, Environ. Entomol.,* 8, 528, 1979.
55. **Elgee, E.,** Persistence of a virus of the white-marked tussock moth on Balsam fir foliage, *Environ. Can. For. Serv. Bimon. Res. Notes,* 31, 33, 1975.

56. **Thompson, C. G. and Scott, D. W.,** Production and persistence of the nulcear polyhedrosis virus of the Douglas-fir-tussock moth, *Orgyia Pseudotsugata* (Lepidoptera:Lymantriidae) in the forest ecosystem, *J. Invertebr. Pathol.,* 33, 57, 1979.
57. **Entwistle, P. F. and Adams, P. H. W.,** Prolonged retention of infectivity in the nuclear polyhedrosis virus of *Gilpinia hercyniae* (Hymenoptera: Diprionidae) on foliage of spruce species, *J. Invertebr. Pathol.,* 29, 392, 1977.
58. **Evans, H. F. and Entwistle, P. F.,** Epizootiology of the nuclear polyhedrosis virus of European spruce sawfly with emphasis on persistence of virus outside the host, in *Microbial Viral Pesticides,* Kurstak, E., Ed., Marcel Dekker, New York, 1982, 449.
59. **David, W. A. L. and Gardiner, B. O. C.,** Persistence of a granulosis virus of *Pieris brassicae* on cabbage leaves, *J. Invertebr. Pathol.,* 8, 180, 1966.
60. **Burgerjon, A. and Grison, P.,** Adhesiveness of preparations of *Smithiavirus pitocompae vago* on pine foliage, *J. Invertebr. Pathol.,* 7, 281, 1965.
61. **Smith, C. M.,** Excretions from leaves as a factor in arsenical injury to plants, *J. Agric. Res.,* 26, 291, 1923.
62. **Andrews, G. L. and Sikorowski, P. O.,** Effects of cotton leaf surfaces on the nuclear polyhedrosis virus of *Heliothis zea* and *Heliothis virescens* (Lepidoptera:Noctuidae), *J. Invertebr. Pathol.,* 22, 290, 1973.
63. **Young, S. Y., Yearian, W. C., and Kim, K. S.,** Effect of dew from cotton and soybean foliage on activity of *Heliothis* nuclear polyhedrosis virus, *J. Invertebr. Pathol.,* 29, 105, 1977.
64. **Harr, J., Guggenheim, R., Boller, T., and Oertli, J. J.,** High pH-values on the leaf surface of commercial cotton varieties, *Cotton Fibres Trop.,* 35, 379, 1980.
65. **Jaques, R. P.,** The persistence of a nuclear polyhedrosis virus in soil, *J. Insect Pathol.,* 6, 251, 1964.
66. **Jaques, R. P.,** The persistence of a nuclear polyhedrosis virus in the habitat of the host insect, *Trichoplusia ni.* II. Polyhedra in soil, *Can. Entomol.,* 99, 820, 1967.
67. **Thomas, E. C., Reichelderfer, C. G., and Heimpel, A. M.,** The effect of soil pH on the persistence of cabbage looper nuclear-polyhedrosis virus, *J. Invertebr. Pathol.,* 21, 21, 1972.
68. **Jaques, R. P.,** Occurrence and accumulation of viruses of *Trichoplusia ni* in treated field plots, *J. Invertebr. Pathol.,* 23, 140, 1974.
69. **Roome, R. E. and Daoust, R. A.,** Survival of the nuclear polyhedrosis virus of *Heliothis armigera* on crops and in soil in Botswana, *J. Invertebr. Pathol.,* 27, 7, 1976.
70. **Young, S. Y. and Yearian, W. C.,** Soil applications of *Pseudoplusia* NPV: persistence and incidence of infection in soybean looper caged on soybean, *Environ. Entomol.,* 8, 860, 1979.
71. **McLeod, P. J., Young, S. Y., and Yearian, W. C.,** Application of a baculovirus of *Pseudoplusia includens* to soybean: efficacy and seasonal persistence, *Environ. Entomol.,* 11, 412, 1982.
72. **Hukuhara, T. and Namura, H.,** Distribution of a nuclear polyhedrosis virus of the fall webworm, *Hyphantriacunea,* in soil, *J. Invertebr. Pathol.,* 19, 308, 1972.
73. **Thompson, C. G., Scott, D. W., and Wickman, B. E.,** Long-term persistence of the nuclear polyhedrosis virus of the Douglas-fir tussock moth, *Orgyia pseudotsugata* (Lepidoptera:Lymantriidae), in forest soil, *Environ. Entomol.,* 10, 254, 1981.
74. **Hukuhara, T. and Namura, H.,** Microscopic demonstration of polyhedra in soil, *J. Invertebr. Pathol.,* 18, 162, 1971.
75. **Hukuhara, T.,** Demonstration of polyhedra and capsules in soil with scanning electron microscope, *J. Invertebr. Pathol.,* 20, 375, 1972.
76. **Hukuhara, T. and Wada, H.,** Adsorption of polyhedra of a cytoplasmic virus by soil particles, *J. Invertebr. Pathol.,* 20, 309, 1973.
77. **Jaques, R. P.,** Leaching of the nuclear polyhedrosis virus of *Trichoplusia ni* from soil, *J. Invertebr. Pathol.,* 13, 256, 1969.
78. **Jaques, R. P.,** Occurrence and accumulation of the granulosis virus of *Pieris rapae* in treated field plots, *J. Invertebr. Pathol.,* 23, 351, 1974.
79. **Ignoffo, C. M.,** Microbial Insecticides: No-Yes; Now-When!, in *Proc. Tall Timbers Conf. on Ecological Annual Control and Habitat Management,* Tall Timbers Research Station, Tallahassee, Fla., February 1970, 41.
80. **Yearian, W. C. and Young, S. Y.,** Application of microbial insecticides, *Misc. Publ. Entomol. Soc. Am.,* 10, 21, 1978.
81. **Falcon, L. A.,** Application technology: improving coverage with microdroplet applicator, in *Proc. NSF-USDA-Univ. Fla. Workshop; Microbial Control of Insect Pests: Future Strategies in Pest Management Systems,* Allen, G. E., Ignoffo, C. M., and Jaques, R. P., Eds., University of Florida, Gainesville, 1978, 113.
82. **Falcon, L. A., Sorensen, A., and Akesson, N. B.,** Pioneering research on aerosol application of insect pathogens, *Calif. Ag.,* 28, 11, 1974.

83. **Smith, D. B. and Bodde, L. F.,** Machinery and factors that affect the application of pathogens, in *Microbial Control of Pests and Plant Diseases, 1970—1980,* Burges, H. D., Ed., Academic Press, New York, 1981, 635.

84. **Himel, C. M. and Moore, A. D.,** Spray droplet size in the control of spruce budworm, boll weevil, bollworm, and cabbage looper, *J. Econ. Entomol.,* 62, 916, 1969.

85. **Stelzer, M. J., Neisess, J., Cunningham, J., and McPhee, J.,** Field evaluation of *Baculovirus* stocks against Douglas-fir tussock moth in British Columbia, *J. Econ. Entomol.,* 70, 243, 1977.

86. **Maksymiuh, B. and Orchard, R. D.,** Techniques for evaluating *Bacillus thuringiensis* and spray equipment for aerial application against forest defoliating insects, *USDA For. Serv. Res. Pap.,* No. PNW-183, 13pp., 1974.

87. **Smirnoff, W. A., Fettes, J. J., and Haliburton, W.,** A virus disease of Swain's Jack pine sawfly, *Neodiprion swainei* Midd. sprayed from an aircraft, *Can. Entomol.,* 94, 477, 1962.

88. **Lewis, F. B. and Yendol, W. G.,** Gypsy moth nuclear polyhedrosis virus: efficacy, *U.S. Dept. Agric. Tech. Bull.,* No. 1584, 503, 1981.

89. **Lewis, F. B.,** Control of the gypsy moth by a baculovirus, in *Microbial Control of Pests and Plant Diseases, 1970—1980,* Burges, H. D., Ed., Academic Press, New York, 1981, 363.

90. **Cunningham, J. C.,** Field trials with baculoviruses. Control of forest insect pests, in *Microbial and Viral Pesticides,* Kurstak, E., Ed., Marcel Dekker, New York, 1982, 335.

91. **Andrews, G. L., Harris, F. A., Sikorowski, P. P., and McLaughlin, R. E.,** Evaluation of *Heliothis* nuclear polyhedrosis virus in a cottonseed oil bait for control of *Heliothis virescens* and *Heliothis zea* on cotton, *J. Econ. Entomol.,* 68, 87, 1975.

92. **McLaughlin, R. E., Andrews, G., and Bell, M. R.,** Field tests for control of *Heliothis* spp. with a nuclear polyhedrosis virus included in a boll weevil bait, *J. Invertebr. Pathol.,* 18, 304, 1971.

93. **Chapman, A. J. and Ignoffo, C. M.,** Influence of rate and spray volume of a nuclear polyhedrosis virus on control of *Heliothis* in cotton, *J. Invertebr. Pathol.,* 20, 183, 1972.

94. **Stacey, A. L., Young, S. Y., and Yearian, W. C.,** *Baculovirus heliothis:* effect of selective placement of *Heliothis* on mortality and efficacy in directed sprays on cotton, *J. Ga. Entomol. Soc.,* 12, 167, 1977.

95. **Stacey, A. L., Yearian, W. C., and Young, S. Y.,** Evaluation of *Baculovirus heliothis* with feeding stimulants for control of *Heliothis* larvae on cotton, *J. Econ. Entomol.,* 70, 779, 1977.

96. **Stacey, A. L., Luttell, R., Yearian, W. C., Matthews, E. J., and Young, S. Y.,** Field evaluation of *Baculovirus heliothis* on cotton using selected application methods, *J. Ga. Entomol. Soc.,* 15, 365, 1980.

97. **Smith, D. B., Hostetter, D. L., and Ignoffo, C. M.,** Laboratory performance specifications for a bacterial *(Bacillus thuringiensis)* and a viral *(Baculovirus heliothis)* insecticide, *J. Econ. Entomol.,* 70, 437, 1977.

98. **Smith, D. B., Hostetter, D. L., and Ignoffo, C. M.,** Formulation effects on application of a viral insecticide *(Baculovirus heliothis), J. Econ. Entomol.,* 71, 814, 1978.

99. **Smith, D. B., Hostetter, D. L., and Ignoffo, C. M.,** Nozzle size-pressure and concentration combinations for *Heliothis zea* control with an aqueous suspension of polyvinyl alcohol and *Baculovirus heliothis, J. Econ. Entomol.,* 72, 920, 1978.

100. **Montoya, E. L., Ignoffo, C. M., and McGarr, R. L.,** A feeding stimulant to increase effectiveness of, and a field test with a nuclear polyhedrosis virus of *Heliothis, J. Invertebr. Pathol.,* 8, 320, 1966.

101. **Falcon, L. A.,** Insect pathogens: integration into a pest management system, in *Proc. Summer Institute on Biological Control of Plant Insects and Diseases,* Maxwell, F. G. and Harris, F. A., Eds., University Press of Mississippi, Jackson, 1974, 618.

102. **Falcon, L. A.,** Problems associated with the use of arthropod viruses in pest control, *Annu. Rev. Entomol.,* 21, 305, 1976.

103. **Ignoffo, C. M. and Montoya, E. L.,** The effects of chemical insecticides and insecticidal adjuvants on a *Heliothis* nuclear polyhedrosis virus, *J. Invertebr. Pathol.,* 8, 409, 1966.

104. **Wolfenbarger, D. A.,** Polyhedrosis virus — surfactant and insecticide combinations and *Bacillus thuringiensis* surfactant combinations for cabbage looper control, *J. Invertebr. Pathol.,* 7, 33, 1965.

105. **Jaques, R. P.,** Protectants for foliar deposits of a polyhedrosis virus, *J. Invertebr. Pathol.,* 17, 9, 1971.

106. **Pritchett, D. W., Young, S. Y., and Yearian, W. C.,** Efficacy of baculoviruses against field populations of fall webworm, *Hyphantria cunea* (Drury), *J. Ga. Entomol. Soc.,* 15, 333, 1980.

107. **Allen, G. E. and Pate, T. L.,** The potential role of a feeding stimulus in connection with the nuclear polyhedrosis virus of *Heliothis, J. Invertebr. Pathol.,* 6, 129, 1966.

108. **Smith, D. B., Hostetter, D. L., Pinnell, R. E., and Ignoffo, C. M.,** Laboratory studies of viral adjuvants:formulation development, *J. Econ. Entomol.,* 75, 16, 1982.

109. **Bell, M. R. and Kanavel, R. F.,** Potential of bait formulations to increase effectiveness of nuclear polyhedrosis virus against the pink bollworm, *J. Econ. Entomol.,* 68, 389, 1975.

110. **Bell, M. R. and Kanavel, R. F.,** Tobacco budworm: development of a spray adjuvant to increase effectiveness of a nuclear polyhedrosis virus, *J. Econ. Entomol.,* 71, 350, 1978.

111. **Bird, F. T.,** The use of a virus disease in the biological control of the European pine sawfly, *Neodiprion sertifer* (Geoff.), *Can. Entomol.,* 85, 437, 1953.

112. **Cunningham, J. C., Kaupp, W. J., McPhee, J. R., Sippell, W. L., and Barnes, C. A.,** Aerial applications of nuclear polyhedrosis virus to control European pine sawfly, *Neodoprion sertifer* (Geoff.), at Sandbanks Provincial Park, Quinte Island, Ontario in 1975, *Can. For. Serv. Inf. Rep.,* IP-X-7, 1975.

113. **Smirnoff, W. A.,** The biological control of *Neodiprion swainei* Midd. with a nuclear-polyhedrosis virus, *Can. For. Serv. Bimon. Res. Notes,* 23, 35, 1967.

114. **Knapp, W. J. and Cunningham, J. C.,** Aerial application of a nuclear polyhedrosis virus against the red-headed pine sawfly, *Neodiprion lecontei* (Fitch), *Can. For. Serv. Inf. Rep.,* IP-X-14, 1977.

115. **deGroot, P., Cunningham, J. C., and McPhee, J. R.,** Control of red-headed pine sawfly with a baculovirus in Ontario in 1978, and a survey of areas treated in previous years, *Can. For. Serv. Inf. Rep.,* FPM-X-20, 1979.

116. **deGroot, P. and Cunningham, J. C.,** Aerial spray trials with a baculovirus to control red-headed pine sawfly in Ontario in 1979 and 1980, *Can. For. Serv. Inf. Rep.,* FPM-X-63, 1983.

117. **Stelzer, M. J., Neisess, J., and Thompson, C. G.,** Aerial application of nucleopolyhedrosis virus and *Bacillus thuringiensis* against the Douglas-fir tussock moth, *J. Econ. Entomol.,* 68, 269, 1975.

118. **Stelzer, M. J. and Neisess, J.,** Field efficacy tests, *U.S.D.A. For. Serv. Tech. Bull.,* No. 1585, 140, 1978.

119. **Ilnytzky, S., McPhee, J. R., and Cunningham, J. C.,** Comparison of field-propagated nuclear polyhedrosis virus from Douglas-fir tussock moth with laboratory produced virus, *Can. For. Serv. Bimon. Res. Notes,* 33, 5, 1977.

120. **Magnolier, A.,** The differing effectiveness of purified and non-purified suspensions of the nuclear polyhedrosis virus of *Porthetria dispar, J. Econ. Entomol.,* 71, 186, 1968.

121. **Magolier, A.,** Field dissemination of a nucleopolyhedrosis virus against the gypsy moth, *Lymantria dispar* L., *Z. Pflanzenbr.,* 9, 497, 1974.

122. **Yendol, W. G., Heldung, R. C., and Lewis, F. B.,** Field investigation of a baculovirus of the gypsy moth, *J. Econ. Entomol.,* 70, 598, 1977.

123. **Wollam, J. D., Yendol, W. G., and Lewis, F. B.,** Evaluation of aerially applied nuclear polyhedrosis virus for suppression of the gypsy moth, *Lymnatria dispar,* L., *U.S.D.A. N. E. For. Serv. Exp. Stn. Res. Pap.,* 396, 1978.

124. **Abrahamson, L. P. and Harper, J. D.,** Microbial insecticides control forest tent caterpillar in southwestern Alabama, *U.S.D.A. For. Serv. Res. Note,* SO—157, 1973.

125. **Ives, W. G. H. and Muldrew, J. A.,** *Can. For. Serv. Inf. Rep.,* NOR-X-24, 1978.

126. **Ignoffo, C. M., Bradley, J. R., Gilliland, F. R., Harris, F. A., Falcon, L. A., McGarr, R. L., Sikorowski, P. P., Watson, T. F., and Yearian, W. C.,** Field studies on stability of the *Heliothis* nucleopolyhedrosis virus at various sites throughout the cotton belt, *Environ. Entomol.,* 1, 388, 1972.

127. **Roome, R. E.,** Field trial with a NPV and *Bacillus thuringiensis* against *Heliothis armigera* larvae on sorghum and cotton in Botswana, *Bull. Int. Res.,* 65, 507, 1975.

128. **Pfrimmer, T. R.,** *Heliothis* spp. control on cotton with pyrethroids, carbamates, organophosphates, and biological insecticides, *J. Econ. Entomol.,* 72, 593, 1979.

129. **Guerra, A. A. and Shaver, T. N.,** Feeding stimulants from plants for larvae of the tobacco budworm and bollworm, *J. Econ. Entomol.,* 62, 98, 1968.

130. **Bell, M. R. and Romine, C. L.,** Tobacco budworm field evaluation of microbial control in cotton using *Bacillus thuringiensis* and a nuclear polyhedrosis virus with a feeding adjuvant, *J. Econ. Entomol.,* 73, 427, 1980.

131. **Luttrell, R. G., Yearian, W. C., and Young, S. Y.,** Effects of Elcar (*Heliothis zea* nuclear polyhedrosis virus) treatments on *Heliothis* spp, *J. Ga. Entomol. Soc.,* 17, 211, 1982.

132. **Luttrell, R. G., Yearian, W. C., and Young, S. Y.,** Effect of spray adjuvants on *Heliothis zea* (Lepidoptera:Noctuidae) nuclear polyhedrosis virus efficacy, *J. Econ. Entomol.,* 72, 162, 1983.

133. **Johnson, D. R.,** Suppression of *Heliothis* spp. on cotton by using *Bacillus thuringiensis, Baculovirus heliothis* and two feeding adjuvants, *J. Econ. Entomol.,* 75, 207, 1982.

134. **Bell, M. R. and Romine, C. L.,** Cotton leaf perforator (Lepidoptera:Lyonetiidae): Effect of two microbial insecticides on field populations, *J. Econ. Entomol.,* 75, 1140, 1982.

135. **Benz, G.,** Synergism of micro-organisms and chemical insecticides. in *Microbial Control of Insects and Mites,* Burges, H. D. and Hussey, N. W., Eds., Academic Press, New York, 1971, 327.

136. **Jaques, R. P. and Morris, O. N.,** Compatibility of pathogens with other methods of pest control and with different crops, in *Microbial Control of Pests and Plant Diseases, 1970—1980,* Burges, H. D., Ed., Academic Press, New York, 1981, 695.

137. **Jaques, R. P.,** Tests on microbial and chemical insecticides for control of *Trichoplusia ni* and *Pieris rapae* on cabbage, *Can. Entomol.,* 105, 21, 1973.

138. **Harper, J. D. and Thompson, C. G.,** Mortality in *Peridroma saucia* following single and combined challenge with DDT and nuclear polyhedrosis virus, in *Proc 4th Int. Colloq. Insect Pathogens,* College Park, Md., 1970, 307.

139. **Genung, W. G.,** Comparison of insecticides, insect pathogens and an insecticide-pathogen combination for control of cabbage looper, *Trichoplusia ni, Fla. Entomol.,* 43, 65, 1960.

140. **Getzin, L. W.,** The effectiveness of polyhedrosis virus for control of cabbage looper, *Trichoplusia ni, J. Econ. Entomol.,* 55, 442, 1962.

141. **Creighton, C. S., McFadden, T. L., and Bell, J. V.,** Pathogens and chemicals tested against caterpillars on cabbage, *Prod. Res. Rep.,* 114, 10, 1970.

142. **McGarr, R. L. and Ignoffo, C. M.,** Control of *Heliothis* spp with a nuclear polyhedrosis virus, EPN and two newer insecticides, *J. Econ. Entomol.,* 59, 1284, 1966.

143. **Luttrell, R. G., Yearian, W. C., and Young, S. Y.,** Laboratory and field studies on the efficacy of selected chemical insecticide-Elcar *(Baculovirus heliothis)* combinations against *Heliothis* spp. *J. Econ. Entomol.,* 72, 57, 1979.

144. **Mohamed, A. I., Young, S. Y., and Yearian, W. C.,** Susceptibility of *Heliothis virescens* (F) (Lepidoptera:Noctuidae) larvae to microbial agent-chemical pesticide mixtures on cotton foliage, *Environ. Entomol.,* 12, 1403, 1983.

145. **McGaughey, W. H.,** Compatibility of *Bacillus thuringiensis* and granulosis virus treatments of stored grain with four grain fumigants, *J. Invertebr. Pathol.,* 26, 247, 1975.

146. **Jaques, R. P.,** Control of the cabbage looper and the imported cabbage-worm by viruses and bacteria, *J. Econ. Entomol.,* 65, 757, 1972.

147. **Jaques, R. P. and Laing, D. R.,** Efficacy of mixtures of *Bacillus thuringiensis,* viruses, and chlordimeform against insects on cabbage, *Can. Entomol.,* 110, 443, 1978.

148. **Pieters, E. P., Young, S. Y., Yearian, W. C., Sterling, W. L., Melville, A. R., and Gilliland, R.,** Efficacy of Elcar-chlordimeform and Dipel-chlordimeform mixtures for control of *Heliothis* spp on cotton, *Southwest Entomol.,* 3, 237, 1978.

149. **Yearian, W. C., Luttrell, R. G., Stacey, A. L., and Young, S. Y.,** Efficacy of *Bacillus thuringeinsis* and *Baculovirus heliothis*-chlordimeform spray mixtures against *Heliothis* spp on cotton, *J. Ga. Entomol. Soc.,* 3, 260, 1980.

150. **Alam, M. Z.,** A Nuclear Polyhedrosis Virus (NPV) of *Pseudoplusia includens* (Walker) (Lepidoptera:Noctuidae): Effects on Larval Feeding and Biological Control Potential, Ph.D. thesis, University of Arkansas, Fayetteville, 1974.

151. **Chamberlain, F. S. and Dutley, R. S.,** Test of pathogens for the control of tobacco insect, *J. Econ. Entomol.,* 51, 560, 1958.

152. **Glass, E. H.,** Laboratory and field tests with the granulosis of the red-banded leaf roller, *J. Econ. Entomol.,* 51, 454, 1958.

153. **Harper, J. D.,** Food consumption by cabbage loopers infected with nuclear polyhedrosis virus, *J. Invertebr. Pathol.,* 21, 191, 1973.

154. **Tatchell, G. M.,** The effect of granulosis virus infection and temperature on the food consumption of *Pieris rapae* (Lepidoptera:Noctuidae), *Entomophaga,* 26, 291, 1981.

155. **Ramambrisman, N. and Chaudhari, S.,** Effect of nuclear polyhedrosis disease on consumption, digestion, and utilization of food by the tobacco caterpillar, *Spodoptera litura* (Fabricus), *Indian J. Entomol.,* 36, 93, 1974.

156. **Tanada, Y.,** A granulosis virus of the codling moth, *Carpocapsa pomonella* (L). (Olethreutidae: Lepidoptera), *J. Insect Pathol.,* 6, 378, 1964.

157. **Cunningham, J. C., Knapp, W. J., McPhee, J. R., Howse, G. M., and Harnden, A. A.,** Aerial application of nuclear polyhedrosis virus on spruce budworm on Manitoulin Island, Ontario in 1975, *Can. For. Serv. Inf. Rep.,* IP-X-9, 1975.

Chapter 7

USE OF BACULOVIRUSES IN PEST MANAGEMENT PROGRAMS

J. Huber

TABLE OF CONTENTS

I. INTRODUCTION

The use of baculoviruses in insect pest management can be traced back to the 19th century. One of the first reported attempts to use such a virus on an operational scale is the supposed introduction of a nuclear polyhedrosis virus (NPV) into populations of the nun moth, *Lymantria monacha,* during a mass outbreak of this severe pine pest in Germany in 1892.[1] At that time, the so-called "Wipfelkrankheit" of the nun moth was believed to be caused by a microorganism referred to as Hofmann's bacterium. Attempts were made to propagate this germ in horse meat mixed with potatoes. Pieces of the contaminated substrate were then suspended on trees in the threatened parts of the forests. It is reported that shortly afterward larvae of the nun moth with typical disease symptoms could be found, and finally the whole population collapsed. With our present knowledge, it is obvious that this success was merely due to the natural outbreak of an epizootic among the larvae. Since there was no untreated and no further examination, this was not recognized. There was much enthusiasm about this novel method of forest protection, and it was widely used by the German forest service for control of the nun moth. Results were not always very good for reasons which are obvious today.[2]

Since these early stages in the utilization of baculoviruses in microbial control, considerable progress has been made. Cunningham,[3] in one of the most recent reviews on the use of baculoviruses in the forest lists 24 species of forest insect pests which have been used in field tests of these viruses. Some viruses have been applied aerially to several thousand hectares and, in general, with good to excellent results. The use of viruses is economically and ecologically more advantageous in forests than in agricultural crops, because in the latter the damage thresholds are usually very low. Nevertheless, there is a great potential for the use of these biological control agents in pest management programs in agriculture. A partial list by Yearian and Young[4] enumerates 29 baculoviruses whose usefulness against agricultural pests has been evaluated. Falcon[5] estimated that in the Western Hemisphere about 30% of the current insect pest problems in agricultural crop production are amenable to control by insect viruses, especially baculoviruses. He calculated that in Nicaragua, for example, biological control agents, including viruses, could reduce expenditures for pesticides by 50%.

Unfortunately, commercialization of baculoviruses to be used for pest control has not progressed at the same rate as the cognition of their potential. Between 1960 and 1975, 17 different virus preparations were produced and marketed commercially in the U.S. or were near commercialization (Table 1). Many of these operations were small-scale, individually owned businesses, aimed only at local markets.[9] To date (March 1984), only four of these preparations have been registered by the U.S. Environmental Protection Agency (see Chapter 9). Only one product, Elcar®, has been developed by private industry and is being sold on the retail market. It is also the only viral pesticide registered for use in agriculture in the U.S. The other three preparations, Gypcheck, TM BioControl-1, and Neochek-S were registered by the U.S. Forest Service and are not for sale to the public.

Even though the usefulness and potential of insect viruses in pest control are well documented and generally recognized, there seems to be astonishingly little practical realization of their considerable potential; in particular, private industry shows very little interest in this novel groups of biopesticides. In the following sections, some reasons will be given for this paradox which is encountered in the U.S. as well as in most other industrialized nations (see also Chapter 10). The advantages of using baculoviruses in pest management programs will also be summarized.

II. INCENTIVES FOR INTEGRATING BACULOVIRUSES INTO PEST MANAGEMENT PROGRAMS

The most important attribute of baculoviruses for pest control is their host-specificity. In

Table 1
COMMERCIAL AND SEMICOMMERCIAL NPV-PREPARATIONS PRODUCED IN THE U.S. BETWEEN 1960 AND 1975[6-8]

Target insect	Trade name	Manufacturer
Autographa californica	—	USDA, Insect Pathol. Lab., Beltsville, Md.
Heliothis zea	Biotrol VHZ	Nutrilite Products, Inc.
	Virex	Hayes-Sammons
	Viron/H	International Minerals and Chemical Corp.
	Elcar®	Sandoz Inc.
Lymantria dispar	—	U.S. Forest Service, Hamden, Conn.
Neodiprion sertifer	Polyvirocide	Indiana Farm Bur. Co-op Assoc.
	—	U.S. Forest Service, Hamden. Conn.
Orygia pseudotsugata	TM BioControl-1	U.S Forest Service, Corvallis, Ore.
Prodenia sp.	Biotrol VPO	Nutrilite Products Inc.
	Viron/P	International Minerals and Chemical Corp.
Spodoptera sp.	Biotrol VSE	Nutrilite Products Inc.
	Viron/S	International Minerals and Chemical Corp.
Trichoplusia ni	Biotrol VIN	Nutrilite Products Inc.
	Viron/T	International Minerals and Chemical Corp.
	Trichoplusia virus	Rudd Associates
	Trichoplusia virus	Biological Control Supplies

most cases, only a few insect species, belonging all to the same family or even the same genus, are susceptible to a given virus.[10] No member of the baculovirus family has ever been isolated from a host other than an arthropod. Their specificity makes them ideal for use in integrated pest management (IPM) programs, since one of the main concepts in such programs is the selective control of the few pest species in a given crop that exceed the economic damage threshold, leaving the rest of the fauna in the crop undisturbed. Thus, the whole potential of the beneficial arthropods present in an intact ecosystem can be exploited. Many secondary pests are kept below damage level and the necessity for additional plant protection measures is greatly reduced. Consequently, a treadmill situation, which often results from the use of broad-spectrum chemical insecticides, can be avoided. In European apple orchards, control of European red mite, *Panonychus ulmi,* and wooly apple aphid, *Eriosoma lanigerum,* is often superfluous if the chemical control of codling moth, *Cydia pomonella,* is replaced by more selective control measures. When the effect of codling moth control by Guthion was compared to that of the highly selective granulosis virus (GV) of this moth, European red mite infestation was more than 100 times lower in the virus-treated plots and well below this economic damage level. In chemically treated trees, the foliage was severely damaged and additional applications of acaricides would have been inevitable for a commercial apple grower.[11]

In contrast to chemicals, viruses have the ability to multiply in their target host. Artificially disseminated for pest control they may persist or even spread in the population of their host initiating real epizootics which keep the target pest at a low level for several years and make further control measures unnecessary. A classical example is the termination of a severe outbreak of the introduced European spruce sawfly, *Gilpinia hercyniae,* in eastern Canada and northeastern U.S. by the dissemination of an NPV imported from Europe. Originally, the virus was probably introduced accidentally with imported parasites, and it spread rapidly through the infested forests.[12] Subsequently there were planned transfers of the virus in order to hasten the spread of the epizootic in Canada. Later, the virus was introduced into New-foundland where the disease became established and successfully controlled the sawfly populations.[13]

Evidence gathered to date indicates that viral insecticides do not create resistance problems in either target or nontarget species. There are several reports of induced resistance to baculoviruses in laboratory-reared insects, and field-collected insects from different locations have been found to show different levels of susceptibility to the same virus. However, in both cases the degree of resistance was small and not comparable to that which can result from the use of chemical insecticides (see Chapter 11). Significantly, some insect viruses, such as the NPVs of several sawflies, have been used in pest management for more than 50 years without showing any signs of loss of efficacy. In a recent study of the susceptibility of codling moth to GV, diapausing larvae were collected in the spring of 1983 from an orchard where the codling moth GV had been applied least four times every year since 1974, and examined for resistance to the virus. The larvae were reared in the laboratory to the adult stage, and the first instars of the next generation were checked for susceptibility to the GV by bioassay. A laboratory strain of codling moth which had not changed in its susceptibility since 1973 was tested in the same way. No difference in slope and LD_{50} between the two strains was detected.[14] (The bioassay system used in this test is sensitive enough to show a twofold difference in LD_{50} with statistical significance.)

Another important consideration is that viral pesticides do not show cross resistance with chemical compounds. A virus preparation can provide relief in situations where a whole group of chemical insecticides has become ineffective due to pest resistance as, e.g., in the case of tobacco budworm in cotton which, in the late 1960s, could no longer be controlled by organochlorines, carbamates, or OP-compounds, (see Section IV.A.1) Of course, there is even less danger of cross resistance to viral pesticides in nontarget pests. The opposite can be true of chemicals. For example, mosquito larvae living in irrigation ditches near cotton fields can become insensitive to the chemicals used in their own control programs due to exposure to broad-spectrum insecticides applied against the cotton pests.[117]

An advantage of using virus pesticides over other biological control methods is that conventional techniques can be used for the application of viruses. For the future, however, it would be desirable to develop novel application methods particularly suited for use with microbials.[15] Aerial applications of chemical insecticides and use of ULV-equipment are becoming more and more controversial due to uncontrolled spray drift. In Norway, aerial spraying of pesticides has been banned, and this situation may arise in other countries, too.[3] In West Germany, the only pesticides registered for application with ULV-equipment in the field are preparations of *Bacillus thuringiensis*, ULV-applications of chemicals are allowed only in greenhouses. Viral pesticides could relieve situations where the use of effective chemicals is restricted because of possible undesirable secondary effects. They are entirely acceptable both ecologically and environmentally.

A further advantage of using viruses is that they do not create residue problems. Though viruses are rather persistent in soil, spray deposits on fruit and leaves are quickly inactivated by the UV radiation of sunlight.[16] They do not accumulate in food chains, interact with other pesticide residues, or degrade into secondary products with undesirable attributes. In short, they do not create the problems associated with the use of many chemical insecticides.[17]

Insect viruses are ideally suited for integration with most other plant protection measures used in IPM. They can be combined with other entomopathogens, parasites, fungicides, and even reduced concentrations of chemical insecticides (see Chapter 5).

With regard to production, viral insecticides have the advantage that they require only low energy input technology. Since little mechanization is possible in their production, they are ideally suited for a cottage-type industry.

III. LIMITATIONS TO THE USE OF BACULOVIRUSES IN PEST MANAGEMENT

In view of the positive aspects listed above, insect control by viruses seems to be the ideal method for plant protection. However, some attributes of viral insecticides have prevented their widespread use in forestry and agriculture.[18] The first "negative" attribute of insect viruses is the same feature which headed the list of positive qualities: host-specificity. Even though a highly selective pesticide is very desirable for ecological and environmental reasons, it is not necessarily so from a commercial standpoint. An insecticide that can be used against a single, though sometimes regionally important, pest species is often unattractive to industrial management; its market potential is very limited, and the economic return from the sales of such a product is low.[19] The cost for developing a pesticide with a narrow insect host range is nearly the same as developing one with a broad-spectrum. Even if developmental costs for a viral insecticide were estimated to be two to five times below those for a chemical pesticide,[20] the only preparations which have a chance to be commercially successful are those against key pests of main agricultural crops, such as corn and cotton. The sales potential is further limited because integrated pest management programs require less frequent pesticide applications than conventional spraying programs.

Specificity can also be negative for the user of the pesticide. If pests other than the target insect are present, which are not kept below economic threshold by natural control factors, additional specific control methods must be applied. If these are not available and conventional spraying of chemicals has to be used, the benefit from the selective control of the first pest species is lost. In orchards where codling moth is controlled by the use of the specific GV, damage by leafrollers sometimes degrades the market value of fruit considerably. When pest control programs using conventional chemical insecticides were in effect, these species had passed unnoticed since they were controlled by the insecticide treatments against codling moth.[11] The situation is further complicated by the fact that there is not only one leafroller, but a complex of several potentially damaging species with fluctuating populations.[21] In some cases, it may be possible to use a mixture of different viruses;[22] however, a farmer faced with such a situation will, out of economic considerations, most probably fall back upon a nonselective chemical insecticide which controls all pests. In general, the farmer does not realize that by doing so, he is running the risk of creating secondary pest problems which will force him to use additional pesticides. Through education, farmers must be made aware that it can be economical to replace one broad-spectrum insecticide with two or three specific ones.

Even if there is only one key pest to control, education of the user is a most important point for the effective use of viral pesticides. Insect viruses require time to produce a response in their target, and farmers conditioned to the knockdown effect of most chemical insecticides need to be properly informed. Insect viruses must be ingested to become effective; they do not work by contact or fumigation action as many chemicals do. Bait formulations and gustatory stimulations might help to overcome this problem,[23] otherwise, good coverage by the spray deposits is most important.

Another handicap of insect viruses is that they can attack only certain stages of their host. They have no effect on the insect eggs, very seldom act on the adults, and in the larval stages susceptibility usually decreases with age. Evans[24] found a 34,000-fold increase in LD_{50} values from first to fifth instars of *Mamestra brassicae*. Even though the larger larvae of the later stages consume more food thereby ingesting more inoculum, it is better to direct virus application against neonate or very young larvae which are often difficult to monitor. The farmer must be taught to look for these early stages of the pest; he can not afford to wait until damage becomes visible as he might do with chemical insecticides. Considerable care must be used in the timing of a virus application.[25] The spraying of the virus-suspension has to be done exactly in phase with the development of the target insect and its host plant. If application is even a few days too early, the success of the treatment can be jeopardized by the rapid degradation of the virus deposit and the possible growth of plant foliage or fruit away from the spray cover. On the other hand, if the virus is applied too late, control is inadequate if substantial portions of the pest population are in advanced, less susceptible, developmental stages.

Another peculiarity of viral and microbial insecticides is the extreme flatness of their rate-effect curve. If efficiency (e.g., percent reduction of damage or population size) is plotted on a probit-scale against the logarithm of the virus rate or concentration, the slope of the corresponding line will seldom exceed a value of one. In a field trial using three different application rates of Viron/H, an experimental preparation of *Heliothis* NPV for the control of *H. zea* and *H. virescens* on cotton, the average slope of the probit lines for reduction of number of worms and reduction of injuries to squares and bolls was 0.8.[26] Similar trials designed to evaluate the dose-efficacy relationship for a GV to be used for control of codling moth, *Cydia pomonella,* gave a slope of about 0.6.[27,28] Using economically feasible amounts of virus preparations, these shallow slopes usually make it impossible to obtain efficacy values which are comparable to chemical insecticides which show a quite different rate effect relationship. In order to augment the efficacy of the codling moth GV from 90 to 99%, it would be necessary to increase the dosage rate 55 times. A 100-fold dilution of the rate giving 90% control, on the other hand, will still result in 50% protection. This is a novel situation for farmers accustomed to the application of conventional insecticides. From their experience with pesticides, they know that doubling the application rate usually has substantial effect on the success of the treatment.

Baculoviruses have some attributes which are important to potential producers of insecticides involving these pathogens. Occlusion bodies of NPVs and GVs make them rather stable in comparison to most other viruses. They can be stored frozen ($-10°C$ or lower) for many years without loss of activity. They even maintain an acceptable level of activity if stored for several years in a refrigerator. Preparations held at average temperatures will lose activity slowly; at elevated temperatures, they will degrade very rapidly.[29] Dried powder preparations of the GV codling moth had a half-life of 5, 1.5, and 0.5 months at room temperature, 26 and 35°C, respectively.[30] This seems very short, compared to the half-life of conventional pesticides. However, if the user of viral insecticides can be educated to refrigerate these preparations, storage performance should be of little importance for the introduction of viral pesticides on the market. Biological control by baculoviruses competes favorably with the use of predators or parasites which cannot be stored or produced in stock.

Since viruses are naturally occurring pathogens, they are not patentable as yet. A private company which has developed a virus preparation is not legally protected against competitors using the same virus isolate in their products.[18] It seems that the importance of this fact has been overestimated in the past. Lists of candidate baculoviruses with promising potential for use in pest control have been elaborated by scientists from universities or governmental agencies all over the world,[31] and virus isolates can be obtained from the corresponding laboratories. This saves private industry time and money by eliminating screening for compounds suited for further development to insecticides. Since no money had to be invested, there is also no need for legal protection of this first step in the development of a viral pesticide. Further steps, such as the establishment of production technology, are mainly the responsibility of the commercial producer and have patent potential. A simple and perhaps even more effective method for protecting innovations in virus processing is to simply keep them secret from possible competitors.

To date, all the virus material produced for commercial preparations has been propagated in vivo, in living larvae of a host insect. This whole-organism technique is fairly labor-intensive and can only partly be facilitated by automation. However, relatively little money has to be invested to mass rear insects when compared to preparations produced in cell culture. The cell culture technique has not yet advanced to the point where it can be used in commercial virus production (see Chapter 3). This fact and the small market for virus preparations tend to place small, versatile firms in more of a position to process insect viruses than big industrial companies. Small biological control companies which have experience and equipment for insect rearing from their production of beneficial insects would be ideal candidates. These laboratories could switch to rearing host larvae for virus propagation in fall and winter when few parasites or predators can be sold and only a maintenance rearing is necessary.

There is one major obstacle which continues to impair this approach: the cost for registration. The procurement of safety and efficacy data needed for registration of a preparation is almost as expensive for a product destined for a small market as for one which will be used on a much broader scale. Since it is not the virus being registered but the virus preparation, a considerable amount of data about the final product has to be submitted even though the literature is full of evidence for the safety of baculoviruses and only few additional data are required to demonstrate the safety of a particular member of this virus family.[32] Only a large potential market can guarantee enough economic return to cover the cost of the tests required. A similar problem is faced regarding conventional insecticides which are often only registered for use on a few major crops but are also needed for plant protection in minor crops. Due to the low market potential, registration for use in these crops is not very attractive to the manufacturer. Some moves have been made by governmental agencies to overcome this problem. In the U.S., the IR-4 program has been established to help registration of minor-use pesticides and in West Germany special procedures to promote registration for pesticides to be used on minor crops have been developed.[33] In 1983, IR-4 expanded its scope to include biorational pesticides such as baculoviruses.

IV. EXAMPLES OF THE SUCCESSFUL USE OF BACULOVIRUSES IN PEST MANAGEMENT PROGRAMS

It has been estimated that the baculoviruses can be used against approximately 30% of the major pest species of the world interfering with food and fiber production, and furthermore that insect viruses could reduce pesticide consumption by 60% in California and by 80% in Central America.[31] The actual use of insect viruses is rather disappointing compared to this apparent potential. Maksymiuk[34] estimated that, from the beginning of insect virus research to 1974, baculoviruses have been applied to a total area of 5,900 ha in Canada and 104,400 ha in the U.S. The total number of baculoviruses applied in North America during that time is assumed to be about 2.7×10^{15} inclusion bodies or less than 2 kg of pure virus. In comparison, agriculture in California alone consumed 38 million kg of pesticides on 6.1 million ha in a single year (1976).[31] In order to appreciate the present extent and future possibilities of baculovirus applications in pest control, an attempt is made below to summarize some data on the most successfully used viruses in the Western Hemisphere.*

A. Field Crops
1. The Nuclear Polyhedrosis Virus of Heliothis Species

The *Heliothis* NPV was the first viral insecticide to be registered in the U.S. Until today, it remains the only virus preparation registered for use in agriculture. It is effective against several *Heliothis* species on cotton, corn, sorghum, soybeans, tobacco, and tomato.[20] The history of its industrialization and commercial development began in 1961.[35,36] At this time, cotton growers were faced with increasing resistance of cotton pests to most of the available pesticides, and new control techniques were most desperately needed.[118] Growing public awareness of the benefits of integrated pest management in the early 1970s contributed further to the great interest of industry in this novel type of insecticide. Several companies, including Hayes-Sammons, International Minerals and Chemical Corp. (IMC), and Nutrilite

---

* The author thanks all the colleagues who have contributed to this survey by supplying unpublished data.

Products Inc. (NPI), started producing experimental preparations of the virus (Table 1). In 1973, the Environmental Protection Agency (EPA) granted exemption from requirement of a tolerance for Viron/H, the preparation from IMC, which was already registered for sale in Spain. Finally, after Sandoz Inc. had further developed the product under the experimental number SAN 240 I,[37] a label was approved under the trade name Elcar® in December 1975. Subsequently, the preparation was also registered in Australia. In 1979, Sandoz opened an Elcar® production facility in Wasco, Calif. This is the first such facility designed expressly for commercial production of a viral pesticide. Considerable efforts have been made to familiarize growers with the properties of this novel pesticide and to assist them before, during and after its application. It had been recognized that education was most important during the introduction of an insecticide with unprecedented qualities.

In 1975 and 1976, when the U.S. Department of Agriculture (USDA) and the EPA were promoting integrated pest management, the use of Elcar® increased. Then, a new group of insecticides, the pyrethroids, appeared on the market. They were highly effective, showed no cross resistance to the insecticides used previously, and their sales soared while the popularity of the biorationals decreased. By 1979, the use of Elcar® had dropped to a level one seventh that of the previous year.[36] Between 1980 and 1982, only 87,000 ha were sprayed with Elcar® in spite of decreases in its price.[38] The economic return from Elcar® was too low for a large company like Sandoz, which spent on its development a considerable portion of the estimated overall cost of $2 million (U.S.).[20] Developmental costs have been greatly influenced by the fact that the *Heliothis* NPV was the first virus preparation ever developed into a commercial product. Much pioneering work had to be done, particularly with regard to registration. Additional viral pesticides are expected to have lower commercialization costs. In contrast, the costs for development leading to the commercialization of a chemical insecticide have been estimated to average $20 million in 1977.[39]

On the whole, development of the *Heliothis* virus to a commercial product was successful technically, but not economically. None of the difficulties, however, are a feature of the virus itself. Cotton growers are rather reluctant to use new methods in pest control. For obvious reasons, the farmer will be interested in a new product when he sees a chance to reduce his cost of crop production. Biological pesticides, being more expensive and requiring greater care in application, are therefore at a disadvantage. They are valued only in special cases, such as combatting resistance to the traditional insecticides. At present, the pyrethroids are already beginning to face this problem,[119] and it will be only a matter of time before the value of the *Heliothis* virus is fully appreciated.

B. Plantation and Orchard Crops
1. The Granulosis Virus of the Codling Moth

The codling moth, *Cydia pomonella*, is one of the key pests in commercial apple and pear orchards all over the world. It is often the limiting factor in integrated control programs because natural enemies of this pest are incapable of keeping it below economic tolerance. In warmer areas, it can produce several overlapping generations a year making regular insecticide application necessary throughout the whole growing season. A fruit infested with a larva of codling moth is not only unmarketable for the commercial grower, but also does not meet the more lenient quality standards of home gardeners and consumers of organic food. In a time of growing public concern about pesticide residues in food, an effective but environmentally safe insecticide for control of this orchard pest is desired.

The granulosis virus of the codling moth fulfills these requirements. The virus is highly effective against young larvae of codling moth and is very specific. Only a few tortricids which are closely related to the codling moth are susceptible to it, but they are not orchard fauna.[40] The codling moth GV is one of the most virulent baculoviruses known. The LD_{50}, determined by measuring the surface area of virus-contaminated apple leaf discs consumed by neonate larvae, was calculated to be 1.2 virus granules per larva.[41]

Table 2
FIELD TRIALS WITH THE CODLING MOTH GV

Country	Period	Estimated area treated (ha)	Maximum number of orchards in 1 year	Ref.
Australia	1969—1972	0.9	2	44
Canada	1974—1983	1.0	5	45,46
Europe	1970—1971	0.6	2	47
	1973—1980	8.7	13	40,48
	1980—1983	26.3	29	40
U.S.	1966—1968	0.6	2	43
	1980—1983	15.0	7	49

The first field trials using the virus, which was originally isolated from a few codling moth larvae collected in Mexico,[42] were conducted by Falcon and collaborators[43] in California in 1966. Because of the promising results of these early studies, field tests with the virus were conducted in most of the apple-growing areas of the world. In addition to the trials listed in Table 2, the virus has also been tested on a smaller scale in Chile, New Zealand, and South Africa. From 1975 to 1980, research on the virus in Europe was partly coordinated by a working group of the International Organization for Biological Control, West Palearctic Regional Section (IOBC/WPRS). This working group was originally established for exchange of information on use of sterile insect technique for control of codling moth, but shifted its interest to other control methods later on.[50] In 1979, the Commission of the European Communities (CEC) started a program on integrated and biological control which supported research on use of baculoviruses for control of tortricid larvae in orchards.[51] These activities culminated in 1982 with field trials in ten European countries in 29 different orchards.

In the 18 years of research with the codling moth virus a total of about 1.5×10^{16} virus granules have been applied to nearly 100 different orchards world-wide. About one million codling moth larvae had to be reared to produce this amount of virus. Although the total of 50 to 60 ha of apple and pear orchards (including some walnut plantations) treated with the virus may seem extremely small compared to the areas treated with other insect viruses, it must be noted that apple is a high value crop, and in the warmer apple growing areas up to ten applications of chemical or viral insecticides are necessary to give adequate coverage through the entire season.

A limiting factor in all these trials was the availability of virus material. In spite of the good efficacy of the GV, industry was reluctant to develop it into a commercial product for a long time because it was thought that the market potential of an insecticide specific for codling moth was too small to make such a product profitable. It was only in 1980, that Sandoz Inc. started research on the virus and produced a preparation with the experimental designation SAN 406 I. For 2 years, the number and size of field tests were greatly increased, particularly in Europe and the U.S. In most trials results were excellent, and it was for the first time, that a biological control agent by itself was capable of effectively reducing codling moth populations below economical threshold levels. At this moment, Sandoz, disappointed by the decreasing sales of Elcar®, stopped all research with insect viruses, not realizing the fundamental differences between pest management in cotton and in orchard crops. In the latter, more and more farmers feel uneasy about the growing public concern about pesticide residues on food and are inclined to reduce the use of chemical pesticides. Further incentive comes from the fact that most consumers are willing to pay better prizes for "untreated" fruit.

At present, possibilities for industrial production of the codling moth GV are being evaluated in three countries. In France and West Germany, two large chemical companies began to produce virus material for extensive field testing, partly in collaboration with the ministry for agriculture and the extension service of the corresponding countries. In the U.S., a new biotechnology firm, MicroGeneSys, has started commercial production of the codling moth virus. The tradename DECYDE has been given to this new GV product. This year (1985), the Environmental Protection Agency (EPA) granted the company an experimental use permit for DECYDE on apples, pears, walnuts, and plums/prunes. Limited sales to selected growers have already started. The company hopes to get full EPA registration in 1986 or 1987. So one can be confident that this novel insecticide, which may revolutionize pest control in apple, pear, and walnut, will be available to the growers in the near future.

2. The Baculovirus of the Coconut Palm Rhinoceros Beetle

Control of the coconut palm rhinoceros beetle, *Oryctes rhinoceros,* by a nonoccluded baculovirus is a prime example of classical biological control.[52] (Classical biological control is understood as the regulation of a pest population by exotic natural enemies, imported for this purpose.) *O. rhinoceros* is rated the most serious coconut pest in the world. The beetle, which also attacks oil palms, is native to Southeast Asia and was accidentally introduced into a number of South Pacific countries and into Mauritius.[53] Due to its biology, chemical control of the beetle is not feasible. After several fruitless attempts to find an effective natural control agent of this pest, a survey conducted by Huger[54] in 1963, in Malaysia, Fiji, and Western Samoa resulted in the discovery of a baculovirus disease commonly referred to as "Malaya disease".[55] In contrast to most other baculoviruses, the *Oryctes* virus infects not only larval stages, but also adults.[56] In the latter, the virus multiplies mainly in the nuclei of midgut cells. The infection induces a severe cell proliferation in the regeneration crypts of the midgut and fills its lumen with dense aggregations of cells. Active virus reproduction takes place in the hypertrophied nuclei of this cell mass.[57] The beetles lives for an average of 30 days following virus infection and are still capable of flying. They defecate infective virus material in species-specific habitats (feeding, mating, and breeding sites) and are ideal distributors of the virus.[58] It is estimated that up to 0.3 mg virus per day may be produced and disseminated by an infected adult.[59] The disease can be introduced effectively into a virus-free pest population by release of a relatively small number of virus-infected beetles. A small virus inoculum from an infected adult midgut is sufficient to start an epizootic in a beetle population. The wave of infection has been estimated to spread from the initial center of virus release at a rate of 100 m/day.[60] Thus, within a relatively short time, the pest population on a whole island can be infected.

Using the Longworth and Kalmakoff classification system,[61] the *Oryctes* virus exemplifies the "limited release approach" in the use of a virus for pest control because very little virus material is needed to obtain effective control. It is obvious that such a virus is unattractive for commercial development by private industry. However, broader use and practical study of the virus was made possible by the "Rhinoceros Beetle Project" established by the U.N. and the South Pacific Commission (UN/SPC) and by some smaller projects supported by the governments of individual nations such as the programs of the German Agency for Technical Cooperation (GTZ). From 1970 to 1974, a total of 10,000 virus-infected beetles were released in Fiji alone.[62] The costs for producing a single beetle were calculated to be about U.S. $0.9 in 1974.[63] These expenditures compared favorably to the overall costs caused by the presence of *O. rhinoceros* (loss in copra, costs for control measures, quarantine procedures, and replanting), which in six South Pacific countries exceeded U.S. $1 million in 1968 (in Fiji: U.S. $114,000).[64] Since the discovery of the virus in Malaysia, it has been used for biological control of *O. rhinoceros* in several countries in the South Pacific and the Indian Ocean.[65] These include Western Samoa, Fiji, Wallis Island, Tonga, American Samoa, Takelau Islands, Mauritius, and Papua New Guinea.[66] Although the cogeneric species *O. monoceros* seems to be less susceptible to the virus, attempts have been made to use it against this beetle in the Ivory Coast and in the Seychelles, where the virus became established after its first introduction in 1973.[67]

The *Oryctes* virus has effectively controlled the pest population and drastically reduced damage to the coconut and oil palms on most islands where it has been introduced. The result has been a new impetus to the copra industry in these countries. Since establishment, the virus has persisted in the beetle populations, keeping the pest below economic damage level over several years.[68] To take full advantage of the potential of this virus and to preserve its control capacity, it has been recommended that the virus be included in an integrated control program[69] which comprises:

1. Rigorous field sanitation measures (removal of decaying palm trunks and other breeding sites).
2. Use of the fungus *Metarhizium anisopliae* for contamination of remaining breeding substrates.
3. Establishment of a dense ground cover vegetation in replantings to suppress breeding and beetle attack.
4. Regular surveys on palm damage and on incidence of virus disease.
5. Release of virus-infected beetles, when appropriate.

Properly used, the baculovirus of *O. rhinoceros* is a very effective pathogen for biological control of this pest, especially in developing countries. It requires almost no input of capital, little labor, and not costly equipment.[70] Once established, it gives long-term control without further expenditures.

C. Forest
1. The Nuclear Polyhedrosis Virus of the Douglas-Fir Tussock Moth

The Douglas-fir tussock moth, *Orgyia pseudotsugata*, is a severe pest of the Douglas-fir and other coniferous trees in western North America from British Columbia in Canada to Mexico. During periods of mass outbreaks, which occur irregularly, the insect is capable of defoliating and eventually killing trees over large areas of the forest. The devastating effects of such an outbreak in the northwestern U.S. in 1973 resulted in the formation of the USDA "Combined Forest Pest Research and Development Program" (CFPP), which included the tussock moth, gypsy moth, and southern pine beetle, and was supported by four agricultural agencies. The "Expanded Douglas-Fir Tussock Moth Research and Development Program", a part of the overall program, greatly augmented existing research and development projects,[71] particularly the research on the use of a NPV for control of tussock moth by the Forestry Sciences Laboratory, Pacific Northwest Forest and Range Experiment Station in Corvallis, Ore.

Two distinct NPVs have been isolated from *Orgyia pseudotsugata*, a single-embedded NPV (SNPV) and a multiple-embedded NPV (MNPV).[72] The MNPV, being slightly more virulent, was further developed for use in a technical grade viral preparation registered by the EPA under the name TM BioControl-1 in 1976. The preparation was and is being produced by various private companies under contract of the U.S. Forest Service. It is used only by or under the supervision of the Forest Service and is not sold on the retail market. The virus is propagated in the laboratory in larvae of *O. pseudotsugata* and is further processed to a wettable powder by lyophilization.[73] In 1983, another preparation of the Douglas-fir tussock moth MNPV, named Virtuss, was granted temporary registration status by Agriculture Canada.[74] This product is also restricted for use by or under the supervision of Federal or Provincial Forest Service employees. Registration has to be renewed annually, and further safety testing and environmental studies have been requested for full registration. Virtuss was developed by and is being produced at the Forest Pest Management Institute (FPMI) at Sault Ste. Marie, Ontario. It consists of virus-diseased larvae of the white-marked tussock moth, *O. leucostigma*, which are freeze-dried and ground to a fine powder.

Since its registration in 1976, TM BioControl-1 has only been used operationally two

Table 3
AERIAL APPLICATIONS OF *ORYGIA PSEUDOTSUGATA* MNPV IN THE U.S. AND IN CANADA

Year	Location	Virus preparation	Total area treated (ha)	Total virus amount applied ($\times 10^{12}$ PIB)	Ref.
1971	Californica	TM-BioControl-1	32	8	75
1973	Oregon	TM BioControl-1	48	48	75, 76
1974	British Columbia	Virtuss	10	5	77
1975	British Columbia	Virtuss	105	26	77, 78
		TM BioControl-1	618	153	
1978	New Mexico	TM BioControl-1	380	95	79
1979	New Mexico	TM BioControl-1	430	105	80
1981	British Columbia	Virtuss	20	4	81
1982	British Columbia	Virtuss	40	6	82

times in the U.S with the total area treated amounting to 810 ha (Table 3). A total of about 4.5×10^{14} polyhedral inclusion bodies (PIBs), produced in about one million larvae, have been applied on 1700 ha in the U.S. and Canada. Most of the treatments resulted in exceptionally good control. The reduction of the pest population obtained in some of the trials was close to 100%. Aerial applications of the virus, the most common method employed, gave excellent foliage protection, and even when minor defoliation occurred, the trees recovered the year after treatment. Thus, the NPV is an efficient and environmentally safe alternative to chemical agents in the regulation of the Douglas-fir tussock moth.

One of the major problems preventing extensive use of the virus is the high cost of producing the viral preparation. *O. pseudotsugata,* the host in which the virus is propagated, has considerable disadvantages for commercial production. Its eggs have an obligatory diapause of 3 to 4 months, and larval growth is relatively slow. It costs U.S. $40 to produce the 2.5×10^{11} PIB needed to treat 1 ha,[83] a cost which is very high for an insecticide used against a forest pest. Attempts have been made to eliminate the expensive laboratory rearing of the host larvae by propagating the virus in field populations of tussock moth,[84] but due to naturally occurring virus and other pathogens already present in the populations, the purity of field-produced virus cannot be guaranteed. As a result, viruses from other insects which are easier to produce, have been considered for use in the control of the tussock moth.[83] Additionally, it may be possible in the near future, to reduce costs by using lower rates of application. Recent field studies have demonstrated that the amount of virus applied can be reduced to 8.3×10^{10} PIB per hectare without considerable loss of efficacy.[82] Since other microbial insecticides, such as *Bacillus thuringiensis,* are not as effective against the tussock moth as the virus, there is still hope that the latter, with its good efficacy and high specificity, will obtain the rank it merits in the management of this insect pest.

2. The Nuclear Polyhedrosis Virus of the Gypsy Moth

The gypsy moth, *Lymantria dispar,* is a significant defoliator of forest, shade, and fruit trees in many parts of the world. It causes serious problems in the northeastern U.S., Central and Western Europe, the Mediterranean region, and Japan. In 1971, 800,000 ha of forest were defoliated by this pest in the U.S. alone.[85] The gypsy moth is the only forest insect regulated by a USDA Federal Domestic quarantine. The larvae are not only forest pests; they are a great nuisance in the urban environment, where the use of chemical insecticides for their control is confronted with rising suspicion by the public.

Since 1960, attempts have been made in several countries to control *L. dispar* with a NPV which often causes natural epizootics in the pest population. Field trials have been conducted in Italy, West Germany, Romania, and Yugoslavia on a total area of 150 ha.[86]

Table 4
USE OF GYPCHECK FOR CONTROL OF *LYMANTRIA DISPAR* IN NORTH AMERICA

Year	Location	Total area treated (ha)	Total virus amount applied ($\times 10^{12}$ PIB)	Ref.
1976	Pennsylvania	170	275	97
1977	Pennsylvania	250	140	97
1978	Pennsylvania	405	250	98
1979	Wisconsin	130	65	99
	Michigan	90	45	
	Massachusetts	160	60	
	Vermont	120	60	
	New York	200	100	
	Pennsylvania	2020	1000	
1980	—	—	—	99
1981	Connecticut	180	210	99
1982	Ontario, Canada	70	35	82

A commercial preparation of the virus, Virin-ENS, is used in the U.S.S.R., but details on its production and efficacy have not been published. It is reported that in 1978 alone, Virin-ENS was used on 53.000 ha.[87]

As with the tussock moth NPV, research on the gypsy moth virus in the U.S. was intensified by an ''Expanded Gypsy Moth Research, Development, and Application Program'' which was part of the ''Combined Forest Pest Research and Development Program'' of the USDA.[85] Studies coordinated by the Forest Insect and Disease Laboratory, Northeastern Forest Experiment Station, in Hamden, Conn. culminated in the registration of a virus product, Gypchek, by the EPA in 1978.[88,89] Since production of the preparation by private companies under contract of the Forest Service (FS) proved unsatisfactory, Gypchek is being produced at Otis Air Force Base, Mass. in a joint program of the FS, the Science and Education Administration (SEA), and the Animal and Plant Health Inspection Service (APHIS).[90] In the initial production of the viral preparation, larval homogenate was purified by isopycnic centrifugation in a zonal ''K''-rotor to remove bacteria and gypsy moth hair and scales which presented a potential hazard.[91] This was quite effective, but it brought the cost for the production of material to be used on 1 ha to an intolerable price of U.S. $120 to 150.[88] Subsequently, it was concluded that zonal centrifugation would not be required for Gypchek as long as the final product met the safety requirements prescribed for registration.[92] Thus, the purification step was eliminated in mass production of the NPV.[93] At the APHIS Gypsy Moth Methods Development Laboratory, Otis Air Force Base, sophisticated rearing facilities for the gypsy moth were developed which permitted a daily production of up to 50,000 larvae for less than U.S. $20 per thousand.[94] Ten days after being inoculated with the virus, containers of moribund and virus-killed larvae were deep-frozen. The caterpillars were collected by hand and then lyophilized, dehaired, and ground to a fine powder.[95] By applying this improved technology and optimizing all parameters in the production process, the product cost for the field use of the NPV was decreased to U.S. $1.50 to 3.70/ha.[96]

Since the first use of the unpurified NPV preparation in 1976, a total of about 3800 ha in the U.S. and Canada have been treated with Gypchek (Table 4). To obtain acceptable foliage protection, most trials required two aerial applications of 2.5×10^{11} PIB per hectare, 10 days apart. When applied properly, Gypchek should reduce the number of new egg masses by 75% or more. Typically, defoliation could be expected to stay below 55 to 60%, sometimes even below 30%, on virus-treated plots.[25] Though it could be argued that conventional insecticides are more effective, Gypchek guarantees sufficient protection against

Table 5
NEODIPRION SERTIFER **NPV SOLD BY
KEMIRA OY, HELSINKI**[101,102]

Year	Country	Total area treated (ha)	Virus supplied ($\times 10^{12}$ PIB)
1974	Finland	15	0.12
	Norway	300	2.40
1975	Finland	326	2.61
1976	Finland	6	0.04
	Norway	1	0.01
1977	Finland	4	0.03
1978	Finland	3	0.02
	Bulgaria	1	0.01
1979	Finland	17	0.14
	Canada	10	0.08
1980	Finland	1363	10.90
1981	Finland	2451	19.61
	Norway	10	0.08
1982	Finland	4068	32.54
	Great Britain	15	0.12
1983	Finland	51	0.41

gypsy moth damage and has the invaluable advantage of being ecologically and environmentally acceptable. Its development and subsequent registration in the U.S. is a substantial contribution to the overall reduction of pesticide loads in the environment.

3. The Nuclear Polyhedrosis Viruses of Sawflies

NPVs have been reported from 25 sawfly species in the forest habitat.[100] They are highly host-specific and replicate, in contrast to most other NPVs, only in the epithelial cells of the larval midgut. Larval feeding decreases and finally stops long before death. Many economically important species of sawflies are gregarious, which makes them particularly vulnerable to viral epizootics. If one individual of a colony of sawfly larvae becomes infected with a virus, the whole colony is doomed to death because the diseased larva will contaminate the foliage and infect the other members of the group. This is advantageous in practical control programs because very low dosages, capable of infecting only a small percentage of larvae, are sufficient to wipe out the entire pest population.

The feasibility of sawfly control by NPVs has been tested in field trials with nine different sawfly species.[100] In 1983, the NPV of the European pine sawfly, *Neodiprion sertifer,* was registered by the EPA for use by the U.S. Forest Service. Pertinent data for registration of the virus preparation, named Neochek-S, had been compiled and were submitted to the EPA by the Forest Insect and Disease Laboratory in Hamden, Conn. In the same year, a preparation of the NPV of the redheaded pine sawfly, *N. lecontei,* was granted temporary registration status by Agriculture Canada following a petition by the Canadian Forest Service, Forest Pest Management Institute in Sault Ste. Marie.[74] The product, named Lecontvirus, carries a restricted label and can only be used by, or under the supervision of, the Forest Service. Additionally, a commercial preparation of the *N. sertifer* NPV is produced and marketed by a Finnish state-owned company, Kemira Oy. Up to 1982, enough virus material for the treatment of about 9000 ha had been sold by Kemira, most of which was used in Scandinavian countries (Table 5).

The NPV of *N. sertifer* is the most widely used virus in forestry. At least 20,000 ha have been treated during the last 30 years. Trials have been conducted in nine European countries, the U.S.S.R. and North America.[86,100] In Canada and the U.S., the virus had been suc-

Table 6
AERIAL AND GROUND APPLICATIONS OF LECONTVIRUS IN CANADA

Year	Province	Plots	Total area treated (ha)[a]	Total virus amount applied ($\times 10^{12}$ PIB)	Ref.
1976	Ontario	3	43	0.2	103
1977	Ontario	3	48	0.3	104
1978	Ontario	2	26	0.1	105
	Quebec	37	700	3.9	106
1979	Ontario	4	34	0.2	106
	Quebec	43	330	1.7	106
1980	Ontario	96	540	4	106
	Quebec	12	21	0.1	106
1981	Ontario	65	759	4.5	107
1982	Ontario	49	374	0.6	108

[a] Partly strip or spot-application.

cessfully used by foresters, Christmas tree growers, and private landowners before authorities became concerned with viral registration around 1970. With the registration of Neochek-S, more than 10 years later, the virus is again available for use against the European pine sawfly in the U.S.

The NPV of the red-headed pine sawfly has mainly been used in eastern Canadian provinces, for control of this serious pest of red pine, *Pinus resinosa*. Up to 1983, Lecontvirus has been experimentally applied to nearly 3000 ha (Table 6). The virus preparation is propagated in *N. lecontei* larvae in the field. Selected plantations with high sawfly populations are sprayed with virus. Several days later, diseased and moribund larvae are then picked from the foliage and frozen. Subsequently, they are lyophilized and ground to a powder.[08] The registration of Lecontvirus has demonstrated that field-propagated virus can fulfill the safety requirements prescribed by health authorities. Since no expenditures are needed for rearing the host larva, such a virus preparation can be produced at much lower costs. For approximately U.S. $2 enough Lecontvirus can be manufactured to treat 1 ha. The recommended rate for aerial application is 5×10^9 PIB or about 50 virus-killed larvae per hectare. Since the virus readily spreads throughout the pest population, it is often sufficient to use spot or strip applications of the preparations. This is particularly appropriate if the virus is distributed from the ground with, for example a mistblower or ULV-equipment. If applied early in larval development (compared to a chemical insecticide the virus is slow acting), the NPV gives good control and foliage protection. One application is sufficient to free plantations of sawfly populations for several years.[105]

Lecontvirus and Neochek-S are economic, highly effective, and environmentally safe biocontrol agents. They are sufficient replacements for the chemical insecticides which have been used against the two sawfly pests.

In addition to the viruses discussed above, several other baculoviruses have been extensively field-tested for use in biological control.[3,4] They have definite potential, but most of them are still far from being a technical success. In Canada, the NPV of the eastern spruce budworm, *Choristoneura fumiferana,* has been sprayed on more than 2000 ha since 1971,[109] but the use of this baculovirus is still at an experimental stage. The efficacy of the virus applications did not sufficiently justify the high production costs. The virus is now being tested for use against the western spruce budworm, *C. occidentalis,* which is considered to be more susceptible to the NPV than the eastern species.[82] It is hoped that the NPV will provide a carryover effect in *C. occidentalis* populations. If this occurs, a single virus application can regulate the pest for a number of years and the use of the virus will be economically justified.[110] In the U.S.S.R., four viral insecticides are produced commercially. They are the GV of *Hyphantria cunea* and the NPVs of *Mamestra brassicae, Malacosoma neustria,* and *Lymantria dispar.*[87] Detailed information about the extent of use of these preparations and their efficacy has not been published.

V. FUTURE ROLE OF BACULOVIRUSES IN PEST CONTROL

A. Industrialized Countries

After reviewing the history of the virus preparations in Section IV, it is evident that private industry is very reluctant to develop baculoviruses as microbial pesticides. Most of the research and development of these viruses had been conducted through publicly funded programs. It is not surprising that private companies see little profit in the production of viral pesticides. As discussed above, one of the major obstacles is the market potential of a virus product which, in most cases, will be much smaller than that of a conventional insecticide and will give very little economic return (if the price of the product stays within reasonable limits). This aspect of the cost balance will be difficult to alter. However, it should be possible to decrease the financial contribution of the manufacturing company by covering part of the developmental cost with public funds. To some extent, this has already been done; most of the basic research on baculoviruses has been performed by public institutions, and the results of these studies are available to any potential producer of a virus preparation. If relief from the registration cost of the product could be provided, even small companies might become interested in the development of viral pesticides. Baculovirus preparations could then have the same economic incentives as the commercially produced parasites and predators already reared and sold by numerous small companies, mainly for local markets. Considerable progress in lowering registration costs for the manufacturers of viral pesticides in the U.S. was made when the IR-4 support program for minor use pesticides expanded to include biorationals. Overall cost of registration could also be reduced if authorities would agree on protocols for registration which are suitable for several different countries at the same time. In Europe, efforts are being made to harmonize the registration guidelines applied to the countries of the European community.

Viral insecticides used only as ecologically safe replacement for chemical compounds are not competitive with conventional pesticides. More emphasis should be placed on the integration of virus applications into specially designed pest management programs which make use of the particular attributes of the viruses. Only by comparing the overall cost of these programs to the expenditures of conventional pest management in a given crop it is possible to demonstrate that the use of relatively expensive viral preparations can be economically interesting. Consideration of the compatibility of viral control with most other control methods greatly enhances the efficacy of pest management programs. By combining the virus treatment, which is mainly larvicidal, with a completely independent method, such as, the release of *Trichogramma,* which parasitizes the egg stage, the efficacy of control is multiplied. Assuming that each method on its own is capable of killing 70% of the pest population, one can expect a 91% reduction by the combined effect of both. To achieve the same mortality with the virus treatment alone, the dosage rate would have to be increased more than tenfold. This would be more expensive than using two methods at a low efficacy level (especially since *Trichogramma* is also effective against other Lepidoptera pests in the same crop).

By applying this integrated concept in pest management and by making the production of viral pesticides more attractive for private industry, it should be possible to use baculoviruses in plant protection without having to pay for their uncontested environmental benefits with increased expenditures in agricultural production. The support required for the development of ecologically oriented pest strategies and for the development and registration of commercial virus preparations can only come from public funds through universities, governmental agencies and institutions, and from international organizations.[18]

B. Developing Countries

Baculoviruses have a great potential for use in pest control programs in developing nations.[111] Little energy input and equipment is needed for their propagation, whereas the required hand labor often is abundant in these countries. Viral pesticides can be produced within the country, in a cottage type industry, using locally available resources thus saving currency currently used to import insecticides. It can be feasible even for individual farmers to produce themselves the virus material for their personal needs. In Thailand, this strategy has been used for control of *Spodoptera exigua*. Small samples of *S. exigua* NPV, produced by the Department of Agriculture in Bangkok, are distributed through the agricultural extension service to individual farmers which use them as inoculum for their own virus propagation in field-collected larvae. The crude suspension of mascerated virus-killed larvae is stored in jars kept in water until it is used for spraying on crops threatened by *S. exigua*.[112] The method works most successfully; in 1982 more than 400 farmers took part in the program. Of course, uncontrolled contaminations of the crude preparations is a potential hazard.

A more advanced technology is used in China. The local supply of baculoviruses, such as the NPV of *Heliothis armigera,* is produced in laboratory-reared larvae by the members of individual communes.[113] In the last 3 years, a GV preparation was used to control *Pieris rapae* on 16,000 ha in more than 30 provinces in China.[114] In contrast to many other insect viruses used in developing countries, this pathogen was extensively safety tested at the Wuhan University before being introduced for practical use. One of the most successful examples of insect pest control by baculoviruses in the third world is the use of the NPV of *Trichoplusia ni* in Colombia. Massive application of the virus against *T. ni* on cotton has been so successful that it replaced all other control methods, and the insect ceased to be an important cotton pest in Colombia.[115]

In the topical climates of many developing countries, agricultural and forest ecosystems are particularly sensitive to any disturbances. Use of broad-spectrum pesticides can induce dramatic and unexpected secondary effects. Many potential tropical pests are kept in check by natural enemies but can increase disastrously when their parasites and predators are reduced or eliminated by the inconsiderate use of broad-spectrum insecticides.[116] In this situation, use of highly specific baculoviruses is attractive because it affords the least possible interference with the environment. More extensive use of these viruses in the third world is hindered mostly by the lack of information in these countries. Developing countries must obtain the considerable amount of know-how on production and use of insect viruses from industrialized countries. Furthermore, they should increase their own research activities in order to make these environmentally safe and nondisruptive pest control agents available for use by their farmers and to have a more self-supporting agriculture in the area of pesticide supply.

REFERENCES

1. **Gehren, U.,** Bekämpfung der Nonne durch Impfung mit dem Hofmannschen Bacillus, *Z. Forst. Jagdwesen,* 24, 499, 1892.
2. **Escherich, K.,** *Die Forstinsekten Mitteleuropas,* Vol. 1, Paul Parey, Berlin, 1914, 346.
3. **Cunningham, J. C.,** Field trials with baculoviruses: control of forest insect pests, in *Microbial and Viral Pesticides,* Kurstak, E., Ed., Marcel Dekker, New York, 1982, 335.
4. **Yearian, W. C. and Young, S. Y.,** Control of insect pests of agricultural importance by viral insecticides, in *Microbial and Viral Pesticides,* Kurstak, E., Ed., Marcel Dekker, New York, 1982, 387.

5. **Falcon, L. A.,** Viruses as alternatives to chemical pesticides in the Western Hemisphere, in *Proc. Symp. Viral Pesticides: Present Knowledge and Potential Effects on Public and Environmental Health, Myrtle Beach, 1977,* Summers, M. D. and Kawanishi, C. Y., Eds., U.S. Environmental Protection Agency, Washington, D.C., 1978, 11.

6. **Briggs, J. D.,** Commercial production of insect pathogens, in *Insect Pathology,* Vol. 2, Steinhaus, E. A., Ed., Academic Press, New York, 1963, 519.

7. **Ignoffo, C. M.,** Entomopathogens as insecticides, *Environ. Lett.,* 8, 23, 1975.

8. **Harrap, K. A. and Tinsely, T. W.,** The international scope of invertebrate virus research in controlling pests, in *Proc. Symp. Viral Pesticides: Present Knowledge and Potential Effects on Public and Environmental Health, Myrtle Beach, 1977,* Summers, M. D. and Kawanishi, C. Y., Eds., U.S. Environmental Protection Agency, Washington, D.C., 1978, 27.

9. **Harper, J. D.,** Major developmental and applied programs for microbial control of invertebrates in the United States, in *Proc. 1st Int. Colloq. Invertebr. Pathol.,* Kingston, 1976, 69.

10. **Ignoffo, C. M.,** Specificity of insect viruses, *Bull. Entomol. Soc. Am.,* 14, 265, 1968.

11. **Dickler, E.,** Mehrjährige Erfahrungen bei der Verwendung des spezifischen Apfelwickler-Granulosevirus: Einfluss auf den Zielorganismus und andere Schadarthropoden, *Mitt. Dtsch. Ges. Allg. Angew. Entomol.,* 4, 59, 1983.

12. **Balch, R. E. and Bird, F. T.,** A disease of the European spruce sawfly, *Gilpinia hercyniae* (Htg.), and its place in natural control, *Sci. Agric.,* 25, 65, 1944.

13. **Clark, R. C., Clarke, L. J., and Pardy, K. E.,** Biological control of the European spruce sawfly in Newfoundland, *Can. For. Serv. Bimon. Res. Notes,* 29, 2, 1973.

14. **Huber, J.,** unpublished data, 1983.

15. **Falcon, L. A.,** The use of microdroplet applicators and light traps to facilitate the introduction and colonization of insect pathogens, *Progress in Invertebrate Pathology, Proc. IInd Int. Colloq. Invertebr. Pathol.,* Weiser, J., Ed., Prague, 1979, 59.

16. **Jaques, R. P.,** Stability of entomopathogenic viruses, *Misc. Publ. Entomol. Soc. Am.,* 10(3), 1977, 99.

17. **Pimentel, D., Andow, D., Dyson-Hudson, R., Gallahan, D., Jacobson, S., Irish, M., Kroop, S., Moss, A., Schreiner, I., Shepard, M., Thompson, T., and Vinzant, B.,** Environmental and social costs of pesticides: a preliminary assessment, *OIKOS,* 34, 126, 1980.

18. **Falcon, L. A.,** Problems associated with the use of arthropod viruses in pest control, *Annu. Rev. Entomol.,* 21, 305, 1976.

19. **Dulmage, H. T.,** Economics of microbial control, in *Microbial Control of Insects and Mites,* Burges, H. D. and Hussey, N. W., Eds., Academic Press, New York, 1971, 581.

20. **Ignoffo, C. M. and Couch, T. L.,** The nucleopolyhedrosis virus of Heliothis species as a microbial insecticide, in *Microbial Control of Pests and Plant Diseases. 1970—1980,* Burges, H. D., Ed., Academic Press, New York, 1981, 329.

21. **Dickler, E. and Gehr, V.,** Untersuchungen über den Einfluss von Apfelwickler-Granuloseviren auf die Biozoenose in Apfelanlagen: Analyse des Schalenwickler-komplexes, *Mitt. Dtsch. Ges. Allg. Angew. Entomol.,* 2, 146, 1981.

22. **Jaques, R. P.,** Field efficacy of viruses infectious to the cabbage looper and imported cabbage worm on late cabbage, *J. Econ. Entomol.,* 70, 111, 1977.

23. **Couch, T. L. and Ignoffo, C. M.,** Formulation of insect pathogens, in *Microbial Control of Pests and Plant Diseases, 1970—1980,* Burges, H. D., Ed., Academic Press, New York, 1981, 621.

24. **Evans, H. F.,** Quantitative assessment of the relationships between dosage and response of the nuclear polyhedrosis virus of *Mamestra brassicae, J. Invertebr. Pathol.,* 37, 101, 1981.

25. **Lewis, F. B., McManus, M. L., and Schneeberger, N. F.,** Guidelines for the use of Gypchek to control the gypsy moth, *USDA For. Serv. Res. Pap.,* NE-441, 1979.

26. **Chapman, A. J. and Ignoffo, C. M.,** Influence of rate and spray volume of a nucleopolyhedrosis virus on control of *Heliothis* in cotton, *J. Invertebr. Pathol.,* 20, 183, 1972.

27. **Glen, D. M. and Cranham, J. E.,** L.A.R.S. and E.M.R.S.: field trials of codling moth granulosis virus in the United Kingdom (1979—1981), in *C.E.C. Programme on Integrated and Biological Control: Progress Report 1979/1981,* Cavalloro, R. and Piavaux, A., Eds., Luxembourg Office for Official Publications of the European Communities, Brussels, 1983, 41.

28. **Huber, J. and Dickler, E.,** Influence of the concentration on the efficacy of the granulosis virus in the field, in *The Use of Integrated Control and the Sterile Insect Technique for Control of the Codling Moth,* Dickler, E., Ed., Mitt. Biol. Bundesanst. Land- u. Forstwirtsch., Berlin-Dahlem, Heft 180, 1978, 75.

29. **Lewis, F. B. and Rollinson, W. D.,** Effect of storage on the virulence of gypsy moth nucleopolyhedrosis inclusion bodies, *J. Econ. Entomol.,* 71, 719, 1978.

30. **Huber, J.,** unpublished data, 1981.

31. **Falcon, L. A.,** Economical and biological importance of baculoviruses as alternates to chemical pesticides, in *Proc. Symp. Safety Aspects of Baculoviruses as Biological Insecticides, Jülich, 1978,* Miltenburger, H. G., Ed., Bundesministerium für Forschung und Technologie, Bonn, 1980, 27.

32. **Burges, H. D., Croizier, G., and Huber, J.,** A review of safety tests on baculoviruses, *Entomophaga,* 25, 329, 1980.
33. **Herfs, W.,** Prüfung und Zulassung von Pflanzenschutzmitteln in Fällen von Lückenindikationen:, *Nachrichtenbl. Deut. Pflanzenschutzd. (Braunschweig),* 35, 95, 1983.
34. **Maksymiuk, B.,** Pattern of use and safety aspects in application of insect viruses in agriculture and forestry, in *Proc. Symp. Baculoviruses for Insect Pest Control: Safety Considerations,* Summers, M., Engler, R., Falcon, L. A., and Vail, P., Eds., American Society of Microbiology, Washington, D.C., 1975, 123.
35. **Ignoffo, C. M.,** Development of a viral insecticide: concept to commercialization, *Exp. Parasitol.,* 33, 380, 1973.
36. **Bohmfalk, G. T.,** Progress with the nuclear polyhedrosis virus of *Heliothis zea* by commercialization of Elcar, in *Proc. IIIrd Int. Colloq. Invertebr. Pathol.,* Brighton, 1982, 113.
37. **Bassand, D. and Knutti, H. J.,** SAN 240 I, the first commercial virus based insecticide, in *Proc. British Crop Prot. Conf. — Pest and Diseases,* 1977, 547.
38. **Bassand, D.,** personal communication, 1983.
39. **Metcalf, R. L.,** Changing role of insecticides in crop protection, *Annu. Rev. Entomol.,* 25, 219, 1980.
40. **Huber, J.,** The baculoviruses of *Cydia pomonella* and other tortricids, in *Proc. IIIrd Int. Colloq. Invertebr. Pathol.,* Brighton, 1982, 119.
41. **Huber, J.,** unpublished data, 1977.
42. **Tanada, Y.,** A granulosis virus of the codling moth, *Carpocapsa pomonella,* (Linnaeus) (Olethreutidae, Lepidoptera), *J. Insect Pathol.,* 6, 378, 1964.
43. **Falcon, L. A., Kane, W. R., and Bethell, R. S.,** Preliminary evaluation of a granulosis virus for control of the codling moth, *J. Econ. Entomol.,* 61, 1208, 1968.
44. **Morris, D. S.,** A cooperative programme of research into the management of pome-fruit pests in southeastern Australia. III. Evaluation of a nuclear granulosis virus for control of codling moth, in *Abstr. 14th Int. Congr. Entomol.,* Canberra, 1972, 238.
45. **Jaques, R. P., MacLellan, C. R., Sanford, K. H., Proverbs, M. D., and Hagley, E. A. C.,** Preliminary orchard tests on control of codling moth larvae by a granulosis virus, *Can. Entomol.,* 109, 1079, 1977.
46. **Jaques, R. P., Laing, J. E., MacLellan, C. R., Proverbs, M. D., Sanford, K. H., and Trottier, R.,** Apple orchard tests on the efficacy of the granulosis virus of the codling moth, *Laspeyresia pomonella* (Lep: Olethreutidae), *Entomophaga,* 26, 111, 1981.
47. **Keller, S.,** Mikrobiologische Bekämpfung des Apfelwicklers *Laspeyresia pomonella* (L.) (=*Carpocapsa pomonella*) mit spezifischem Granulosisvirus, *Z. Angew. Entomol.,* 73, 137, 1973.
48. **Huber, J.,** 7 Jahre Freilandversuche mit dem Granulosevirus des Apfelwicklers in der Bundesrepublik Deutschland, in *Proc. Int. Symp. IOBC/WPRS Integrierter Pflanzenschutz in der Land- und Forstwirtschaft,* Vienna, 1979, 583.
49. **Falcon, L. A.,** The baculoviruses of *Autographa, Trichoplusia, Spodoptera,* and *Cydia,* in *Proc. IIIrd Int. Colloq. Invertebr. Pathol.,* Brighton, 1982, 125.
50. International Organization for Biological Control (IOBC), Biological control of pests in orchards/Biology and control of codling moth, *WPRS Bull.,* 3(6), 1980.
51. **Cavalloro, R. and Piavaux, A.,** *C.E.C. Programme on Integrated and Biological Control: Progress Report 1979/1981,* Luxembourg Office for Official Publications of the European Communities, Brussels, 1983.
52. **Caltagirone, L. E.,** Landmark examples in classical biological control, *Annu. Rev. Entomol.,* 26, 213, 1981.
53. **Bedford, G. O.,** Biology, ecology, and control of palm rhinoceros beetles, *Annu. Rev. Entomol.,* 25, 309, 1980.
54. **Huger, A. M.,** Untersuchungen über mikrobielle Begrenzungsfaktoren von Populationen des Indischen Nashornkäfers, *Oryctes rhinoceros,* (L.), in SO-Asien und in der Südsee, *Z. Angew. Entomol.,* 58, 89, 1966.
55. **Huger, A. M.,** A virus disease of the Indian rhinoceros beetle, *Oryctes rhinoceros* (Linnaeus), caused by a new type of insect virus, *Rhabdionvirus oryctes* gen n., sp. n., *J. Invertebr. Pathol.,* 8, 38, 1966.
56. **Zelazny, B.,** Studies on *Rhabdionvirus oryctes.* II. Effects on adults of *Oryctes rhinoceros, J. Invertebr. Pathol.,* 22, 122, 1973.
57. **Huger, A. M.,** Grundlagen zur biologischen Bekämpfung des Indischen Nashornkäfers, *Oryctes rhinoceros* (L.), mit *Rhabdionvirus oryctes:* Histopathologie der Virose bei Käfern, *Z. Angew. Entomol.,* 72, 309, 1973.
58. **Zelazny, B.,** Transmission of a baculovirus in populations of *Oryctes rhinoceros, J. Invertebr. Pathol.,* 27, 221, 1976.
59. **Monsarrat, P. and Veyrunes, J. C.,** Evidence of *Oryctes* virus in adult feces and new data for virus characterization, *J. Invertebr. Pathol.,* 27, 387, 1976.
60. **Young, E. C.,** The epizootiology of two pathogens of the coconut palm rhinoceros beetle, *J. Invertebr. Pathol.,* 24, 82, 1974.

61. **Longworth, J. F. and Kalmakoff, J.,** Insect virus for biological control: an ecological approach, *Intervirology,* 8, 68, 1977.

62. **Bedford, G. O.,** Use of a virus against the coconut palm rhinoceros beetle in Fiji, *PANS,* 22, 11, 1976.

63. **Bedford, G. O.,** Mass rearing of the coconut palm rhinoceros beetle for release of virus, *PANS,* 22, 5, 1976.

64. **Catley, A.,** The coconut rhinoceros beetle *Oryctes rhinoceros* (L.) (Coleoptera: Scarabaeidae: Dynastinae), *PANS,* 15, 18, 1969.

65. **Bedford, G. O.,** Control of the rhinoceros beetle by baculovirus, in *Microbial Control of Pest and Plants Diseases, 1970—1980,* Burges, H. D., Ed., Academic Press, New York, 1981, 409.

66. **Gorick, B. D.,** Release and establishment of the baculovirus disease of *Oryctes rhinoceros* (L.) (Coleoptera: Scarabaeidae) in Papua New Guinea, *Bull. Entomol. Res.,* 70, 445, 1980.

67. **Paul, W.-D.,** personal communication, 1983.

68. **Young, E. C. and Longworth, J. F.,** The epizootiology of the baculovirus of the coconut palm rhinoceros beetle *(Oryctes rhinoceros)* in Tonga, *J. Invertebr. Pathol.,* 38, 362, 1981.

69. **Huger, A. M.,** Virusanwendung als Komponente eines integrierten Programms zur Bekämpfung des Indischen Nashornkäfers, *Orcytes rhinoceros* (L.) (Col.), *Mitt. Dtsch. Ges. Allg. Angew. Entomol.,* 1, 246, 1978.

70. **Marschall, K. J.,** Progress in microbial control (1975—1980): developing integrated pest management programs: D. Coconut, in *Proc. Workshop Insect Pest Management with Microbial Agents: Recent Achievements, Deficiencies, and Innovations,* Ithaca, N.Y., 1980, 31.

71. **Brookes, M. H., Stark, R. W., and Campbell, R. W., Eds.,** *The Douglas-Fir Tussock Moth: A Synthesis,* USDA For. Serv. Tech. Bull., No. 1585, 1978.

72. **Hughes, K. M. and Addison, R. B.,** Two nuclear polyhedrosis viruses of the Douglas-fir tussock moth, *J. Invertebr. Pathol.,* 16, 196, 1970.

73. **Martignoni, M. E.,** Virus in biological control: production, activity, and safety, *USDA For. Serv. Tech. Bull.,* No. 1585, 140, 1978.

74. **Cunningham, J. C.,** Two viruses granted temporary registration in Canada, *Can. For. Serv. FPMI News.,* 2(2), 4, 1983.

75. **Stelzer, M. J. and Neisess, J.,** Virus in biological control: field efficacy tests, *USDA For. Serv. Tech. Bull.,* No. 1585, 149, 1978.

76. **Stelzer, M. J., Neisess, J., and Thompson, C. G.,** Aerial applications of a nucleopolyhedrosis virus and *Bacillus thuringiensis* against the Douglas-fir tussock moth, *J. Econ. Entomol.,* 68, 269, 1975.

77. **Cunningham, J. C. and Shepherd, R. F.,** Douglas-fir tussock moth, *Orgyia pseudotsugata* (McDunnough), in *Biological Control Programmes Against Insects and Weeds in Canada, 1969—1980,* Kelleher, J. S. and Hulme, M. A., Eds., Commonwealth Agricultural Bureaux, Ottawa, 1984, 363.

78. **Stelzer, M. J., Neisess, J., Cunningham, J. C., and McPhee, J. R.,** Field evaluation of baculovirus stocks against Douglas-fir tussock moth in British Columbia, *J. Econ. Entomol.,* 70, 243, 1977.

79. **Hofacker, T. H., Holland, D. G., and Smith, T.,** 1978 cooperative Douglas-fir tussock moth NPV pilot control project, *USDA For. Serv. S. W. Reg. Rep.,* R-3 79-6, 1979.

80. **Hofacker, T. H., Smith, T., Graham, D. P., Sandquist, R. E., Barry, J. W., and Blackwell, G.,** 1979 Douglas-fir tussock moth suppression project, *USDA For. Serv. S. W. Reg. Rep.,* R-3 80-2, 1980.

81. **Shepherd, R. F., Otvos, I. S., Chorney, R. J., Friskie, L., and Cunningham, J. C.,** unpublished data, 1981.

82. **Cunningham, J. C. and Kaupp, W. J.,** Aerial application of viruses in 1982, *Can. For. Serv. FPMI Newsl.,* 1(3), 2, 1982.

83. **Martignoni, M. E., Stelzer, M. J., and Iwai, P. J.,** Baculovirus of *Autographa californica* (Lepidoptera: Noctuidae): a candidate biological control agent for Douglas-fir tussock moth (Lepidoptera: Lymantriidae), *J. Econ. Entomol.,* 75, 1120, 1982.

84. **Ilnytzky, S., McPhee, J. R., and Cunningham, J. C.,** Comparison of field-propagated nuclear polyhedrosis virus from Douglas-fir tussock moth with laboratory-produced virus, *Can. For. Serv. Bimon. Res. Notes,* 33, 5, 1977.

85. **Doane, C. C. and McManus, M. L., Eds.,** The gypsy moth: research toward integrated management, *USDA For. Serv. Tech. Bull.,* No. 1584, 1981.

86. **Franz, J. M. and Huber, J.,** Feldversuche mit insektenpathogenen Viren in Europa, *Entomophaga,* 24, 333, 1979.

87. **Lipa, J. J.,** Progress in microbial control (1975—1980): world activites: B. Eastern Europe, in *Proc. Workshop Insect Pest Management with Microbial Agents: Recent Achievements, Deficiencies and Innovations,* Ithaca, N.Y., 1980, 9.

88. **Lewis, F. B.,** Control of the gypsy moth by a baculovirus, in *Microbial Control of Pests and Plant Diseases, 1970—1980,* Burges, H. D., Ed., Academic Press, New York, 1981, 363.

89. **Podgwaite, J. D. and Mazzone, H. M.,** Development of insect viruses as pesticides: the case of the gypsy moth *(Lymantria dispar,* L.) in North America, *Protect. Ecol.,* 3, 219, 1981.

90. **Shapiro, M.,** Gypsy moth nucleopolyhedrosis virus: In vivo production at Otis Air Base, Mass., *USDA For. Serv. Tech. Bull.,* No. 1584, 464, 1981.
91. **Breillatt, J. P., Brantley, J. N., Mazzone, H. M., Martignoni, M. E., Franklin, J. E., and Anderson, N. G.,** Mass purification of nucleopolyhedrosis virus inclusion bodies in the K-series centrifuge, *Appl. Microbiol.,* 23, 923, 1972.
92. **Podgwaite, J. D. and Bruen, R. B.,** Procedures for the microbial examination of production batch preparations of the nuclear polyhedrosis virus (baculovirus) of the gypsy moth, *Lymantria dispar* L., *USDA For. Serv. Gen. Tech. Rep.,* NE-38, 1978.
93. **Podgwaite, J. D.,** Gypsy moth nucleopolyhedrosis virus: large-scale productions, *USDA For. Serv. Tech. Bull.,* No. 1584, 462, 1981.
94. **Bell, R. A., Owens, C. D., Shapiro, M., and Tardif, J. R.,** Development of mass-rearing technology, *USDA For. Serv. Tech. Bull.,* No. 1584, 599, 1981.
95. **Shapiro, M., Bell, R. A., and Owens, C. D.,** In vivo mass production of gypsy moth nucleopolyhedrosis virus, *USDA For. Serv. Tech. Bull.,* No. 1584, 633, 1981.
96. **Shapiro, M., Owens, C. D., Bell, R. A., and Wood, H. A.,** Simplified, efficient system for in vivo mass production of gypsy moth nucleopolyhedrosis virus, *J. Econ. Entomol.,* 74, 341, 1981.
97. **Lewis, F. B. and Yendol, W. G.,** Gypsy moth nucleopolyhedrosis virus: efficacy, *USDA For. Serv. Tech. Bull.,* No. 1584, 503, 1981.
98. **Lewis, F. B., Reardon, R. C., Munson, A. S., Hubbard, H. B., Schneeberger, N. F., and White, W. B.,** Observations on the use of Gypchek, *USDA For. Serv. Res. Pap.,* NE-447, 1979.
99. **Podgwaite, J. D.,** personal communication, 1982.
100. **Cunningham, J. C. and Entwistle, P. F.,** Control of sawflies by baculoviruses, in *Microbial Control of Pests and Plant Diseases, 1970—1980,* Burges, H. D., Ed., Academic Press, New York, 1981, 379.
101. **Evers, A.-M.,** personal communication, 1983.
102. **Aapola, A.,** personal communication, 1983.
103. **Kaupp, W. J. and Cunningham, J. C.,** Aerial application of a nuclear polyhedrosis virus against red-headed pine sawfly, *Neodiprion lecontei* (Fitch), *Can. For. Serv. Inf. Rep.,* IP-X-14, 1977.
104. **Kaupp, W. J., Cunningham, J. C., and de Groot, P.,** Aerial application of baculovirus on red-headed pine sawfly, *Neodiprion lecontei,* (Fitch), in 1977, *Can. For. Serv. Res.,* FPM-X-1, 1978.
105. **de Groot, P., Cunningham, J. C., and McPhee, J. R.,** Control of red-headed pine sawfly with a baculovirus in Ontario in 1978, and a survey of areas treated in previous years, *Can. For. Serv. Inf. Rep.,* FPM-X-20, 1979.
106. **Cunningham, J. C. and de Groot, P.,** Red-headed pine sawfly, *Neodiprion lecontei* (Fitch), in *Biological Control Programmes Against Insects and Weeds in Canada, 1969—1980,* Kelleher, J. S. and Hulme, M. A., Eds., Commonwealth Agricultural Bureaux, Ottawa, 1984, 323.
107. **de Groot, P. and Cunningham, J. C.,** unpublished data, 1981.
108. **Cunningham, J. C. and de Groot, P.,** Ontario Ministry of Natural Resources staff use virus to control red-headed pine sawfly, *Can. For. Serv. FPMI Newsl.,* 1(4), 4, 1982.
109. **Cunningham, J. C. and Howse, G. M.,** Eastern spruce budworm viruses: application and assessment, in *Biological Control Programmes Against Insects and Weeds in Canada, 1969—1980,* Kelleher, J. S. and Hulme, M. A., Eds., Commonwealth Agricultural Bureaux, Ottawa, 1984, 323.
110. **Shepherd, R. F. and Cunningham, J. C.,** Western spruce budworm, *Choristoneura occidentalis* Freeman, in *Biological Control Programmes Against Insects and Weeds in Canada, 1969—1980,* Kelleher, J. S. and Hulme, M. A., Eds., Commmonwealth Agricultural Bureaux, Ottawa, 1984, 323.
111. **Leon, G. O.,** Potential of microbial pest control in developing nations: ecological and economic advantages, in *Proc. Workshop Insect Pest Management with Microbial Agents: Recent Achievements, Deficiencies, and Innovations,* Ithaca, N.Y., 1980, 52.
112. **Anon.,** Control of onion lesser armyworm, *Spodoptera exigua* (Hübner) by using nuclear polyhedrosis virus (NPV), in *Biological Control of Insect Pests,* Plant Prot. Serv., Dep. Agric. Extension, Thailand, 1981, 1.
113. **Franz, J. M. and Krieg, A.,** Mikrobiologische Schädlingsbekämpfung in China. Ein Reisebericht, *Forum Mikrobiol.,* 3, 163, 1980.
114. **Liu, N. C., Liang, D. R., and Zhang, Q. L.,** The isolation, purification, identification, ultrastructure and application of a granulosis virus of the cabbage butterfly, *Pieris rapae, Wuhandaxue Xuebao,* 2, 49, 1981.
115. **Bellotti, A. C. and Reyes, J. A.,** Progress in microbial control (1975—1980): world activities: H. South and Central America, in *Proc. Workshop Insect Pest Management with Microbial Agents: Recent Achievements, Deficiencies, and Innovations,* Ithaca, N.Y., 1980, 20.
116. **Wood, B. J.,** Integrated control: critical assessment of case histories in developing economies, in *Insects: Studies in Population Management,* Geier, P. W., Clark, L. R., Anderson, D. J., and Nix, H. A., Eds., Ecological Society of Australia, Canberra, 1973, 196.

117. **Georghion, G. P., Breeland, S. G., and Ariaratnam, V.,** Seasonal escalation of organophorphorus and carbamate resistance in *Anopheles albimanus* by agricultural sprays, *J. Environ. Entomol.,* 2, 369, 1973.
118. **Bottrell, D. G. and Adkisson, P. L.,** Cotton insect pest management, *Annu. Rev. Entomol.,* 22, 451, 1977.
119. **Martinez-Carillo, J. L. and Reynolds, H. T.,** Dosage-mortality studies with pyrethroids and other insecticides on the tobacco budworm (Lepidoptera: Noctuidae) from the Imperial Valley, California, *J. Econ. Entomol.,* 76, 983, 1983.

Chapter 8

REGISTRATION OF BACULOVIRUSES AS PESTICIDES

Frederick S. Betz

TABLE OF CONTENTS

I. INTRODUCTION

This chapter describes the pesticide registration process and summarizes the kinds of data and information normally required to support the registration of viral pesticides.

A. FIFRA

The Environmental Protection Agency (EPA) regulates the use of pesticides in the U.S. under the authority of the Federal Insecticide, Fungicide, and Rodenticide Act[1] (FIFRA). FIFRA gives the EPA the authority to require the submission of data and information to support the registration of each pesticide product. The Agency reviews and evaluates the supporting data and may then register the pesticide product for specified uses, provided it has determined that such uses will not pose an unreasonable risk to human health or to the environment.

The basic elements of pesticide registration are addressed in Sections 3 and 5 of FIFRA. Section 3, ''Registration of pesticides'', sets forth the requirement for pesticides to be registered, exemptions from registration, the procedures for registration, and the classification and reregistration of pesticides. Section 5 addresses the issuance of experimental use permits (EUP).

B. Registration Procedures

Part 162, Title 40, of the Code of Federal Regulations[2] contains the procedures for the registration of pesticide products, including products containing viruses. This regulation explains the principal statutory provisions of FIFRA pertaining to pesticide registration; it indicates who may apply for registration and what a registration application should contain. It defines terms such as ''pesticide'', ''hazard'', ''nontarget organisms'', ''outdoor application'', and ''pest''. It also sets forth labeling requirements and criteria for approval of registration applications, and the criteria for determination of unreasonable adverse effects. Part 162 outlines the major steps in the review process, provides detailed information concerning the mechanics of the registration process and submission of registration applications, includes a description of the relevant registration forms, and gives recommendations on preparation of an application package.

C. Experimental Use Permits

An experimental use permit (EUP) is required in order to perform the field testing necessary to support registration under Section 3 of FIFRA. The procedures and general requirements pertaining to EUPs are specified in Part 172 — Experimental Use Permits.[3] This regulation specifies when an EUP must be obtained, what kinds of information and data are needed prior to applying for a permit, and how experimental products should be labeled.

D. Data Requirements

Part 158, Data Requirements for Registration,[4] specifies the kinds of data required to support registration and indicates when these data are necessary to support registration. Part 158 also contains several important policies relevant to the registration of viral pesticides. These policies address the general conditions under which the Agency will consider waiving certain data requirements or requiring additional data, and discusses the Agency's regulation of minor use pesticides and genetically modified and nonindigenous microbial pesticides.

E. Testing Guidelines

The Pesticide Assessment Guidelines — Subdivision M — Biorational Pesticides[5] are EPA's recommendations concerning test methods. They are intended to guide and assist developers of microbial pesticides, including viruses, in producing the data necessary to

Table 1
VIRAL PESTICIDES REGISTERED IN THE U.S.

Name of microbial agent (species)	Year first registered	Use pattern
Heliothis NPV (inclusion bodies)	1975	Cotton bollworm and cotton budworm on cotton
Douglas-fir tussock moth NPV (inclusion bodies)	1976	Douglas-fir tussock moth on Douglas fir
Gypsy moth NPV (inclusion bodies)	1978	Gypsy moth on forest, shade, and ornamental trees
Neodiprion sertifer NPV (inclusion bodies)	1983	Pine sawfly

register their products. Each guideline includes the Agency's recommended test standards, examples of acceptable protocols, and guidance on evaluation and reporting of data. The protocols are set forth as examples and recommendations rather than strict requirements because the Agency recognizes that other protocols may also be acceptable, or even preferable, depending on the particular virus in question and its intended use. Applicants are encouraged to consult with the Agency concerning the use of new or alternative protocols before they are used to ensure that they will provide the appropriate information.

The Agency is continuing to support research on the development and validation of test methods for evaluating viral pesticides, and will update and revise the guidelines accordingly from time to time.

II. BACKGROUND AND HISTORY

EPA's experience with the registration of viral pesticides began in the early 1970s with an evaluation of the nuclear polyhedrosis virus (NPV) of *Heliothis zea*. The Agency, at that time, had no specific policies, data requirements, or procedures for reviewing, evaluating, and registering viral pesticides. As a result, EPA dealt with *Heliothis* virus and many of the viral pesticides that followed, on a case-by-case basis. This approach proved to be unsatisfactory because developers of viral pesticides, and microbial pesticides in general, did not have a clear idea of what data would normally be required to support registration. In addition, too much emphasis was placed on concerns relevant to the safety of conventional chemical pesticides (e.g., toxicity) and testing protocols for chemicals, and too little emphasis was put on the more relevant characteristics of these living entities such as infectivity and pathogenicity.

As a result, the nuclear polyhedrosis virus of *Heliothis* was subject to a wide variety of short- and long-term testing. No significant adverse effects were observed in these tests or were predicted to occur as a result of its use. Therefore *Heliothis* NPV was registered for use in 1973 to control the cotton budworm and tobacco budworm. Today, there are four registered viral pesticides (all NPVs) for use against a variety of pests in agriculture and forestry (Table 1). This brings the total number of registered microbial pesticides to 14 (6 bacteria, 4 viruses, 3 fungi, and 1 protozoan), compared to about 1400 registered chemical pesticides.

Recognizing that viral pesticides posed some potential hazards that were entirely different from conventional pesticides, and that the conventional approaches to chemical pesticide evaluation were inappropriate, the Agency initiated an effort to correct this deficiency in its registration program. During 1974 to 1979, EPA sponsored a variety of symposia, workshops, and expert panels[6-9] to evaluate the potential hazards associated with viral pesticides,

PEST CONTROL AGENTS

CONVENTIONAL CHEMICAL
PESTICIDES

BIOCHEMICAL AND BIOLOGICALLY
DERIVED PEST CONTROL AGENTS

BIOCHEMICAL AND MICROBIAL
PESTICIDES

ALL OTHER LIVING
PEST CONTROL AGENTS

● INSECT PREDATORS
● MACROSCOPIC PARASITES
● NEMATODES

BIOCHEMICAL PESTICIDES

MICROBIAL PESTICIDES

(NATURALLY OCCURRING AND
GENETICALLY ENGINEERED)

● SEMIOCHEMICALS
● HORMONES
● NATURAL PLANT
   REGULATORS
● ENZYMES

● VIRUSES
● BACTERIA
● FUNGI
● PROTOZOA

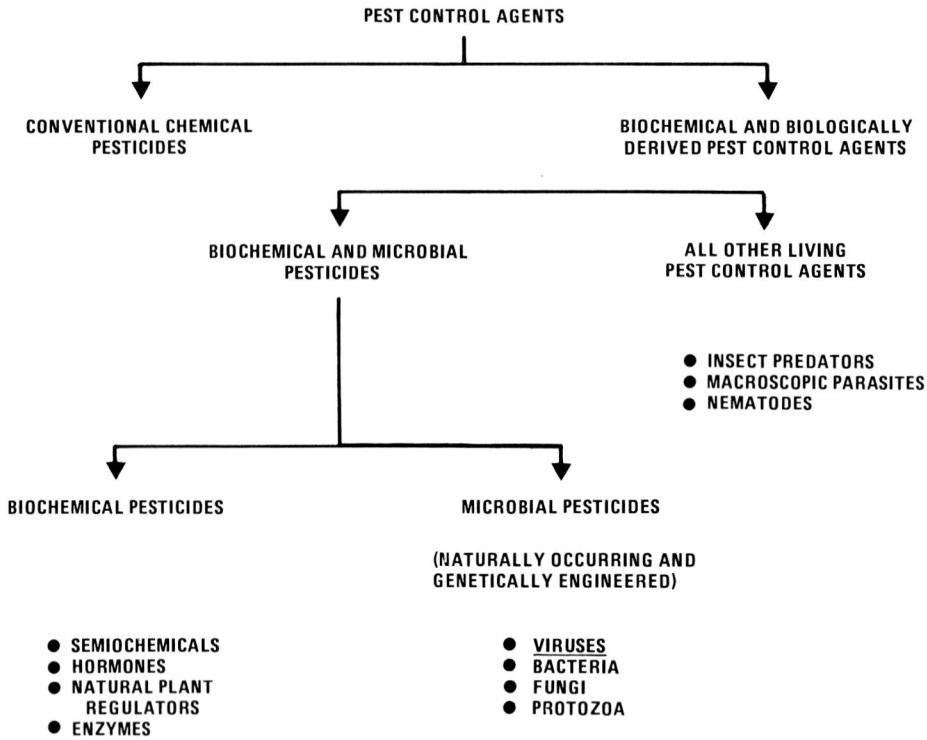

FIGURE 1.   Relationships between conventional and biological pesticides.

to identify relevant human health and environmental concerns, and to recommend appropriate approaches for testing and evaluating microbial pesticides.

By 1979, industry had demonstrated considerable interest in registering microbial pesticides. Therefore, EPA issued a policy statement in the *Federal Register*[10] outlining several steps it would take to improve the evaluation and registration process for microbial pesticides, including viruses. This policy statement included several elements that have had a significant effect on the subsequent registration of microbial pesticides.

First EPA recognized microbial pesticides as being inherently different from conventional chemical pesticides (Figure 1) and committed itself to implementing a comprehensive registration program for this kind of pesticide product. As part of this program EPA stated it would facilitate the registration of "environmentally integrated" pesticides by assuring that the registration requirements for microbial pesticides would be appropriate to their nature and not unduly burdensome. The Agency would also give priority in the registration process to innovative microbial pesticides, thereby reducing the time required to register such products. Finally, EPA stated it would develop guidelines for human and environmental safety testing of microbial pesticides within 24 months.

In 1980, EPA completed the first draft of the testing guidelines and beginning in February 1983, the Agency made the Pesticide Assessment Guidelines (Subdivision M) available through the National Technical Information Service (NTIS). The date required to support registration of viral pesticides are listed in the regulation titled "Data Requirements for Registration — Part 158".

III. PRODUCT SAFETY

A. Safety Concerns

Pesticides pose certain potential hazards by virtue of their application and use in the environment to ''control'' pest organisms. Historically, EPAs concern has focused on a wide range of acute, subacute, and chronic human health effects, including toxicity, allergenicity, reproductive, oncogenic and teratogenic effects; adverse ecological effects, including acute and reproductive effects (on nonhuman, nontarget species); and other environmental effects (e.g., ground water contamination). Viral pesticides pose similar concerns, although safety testing and use history thus far reveal no serious human health or environmental concerns. However, because viral agents are living entities capable of survival and reproduction, they pose some additional concerns, namely, pathogenicity and infectivity in nontarget hosts.

Tests to assess infectivity must determine whether the virus is able to survive, persist, and/or replicate in mammalian and other nontarget organism systems. Short-term survival of viruses, in the absence of adverse effects, may not be a cause for concern. However, because of the nature of the their parasitism and their potential for prolonged survival or replication as cellular nucleic acid entities, many additional concerns and potential hazards must be considered. In addition, the consequences of abortive infections without completion of the viral replication cycle can be more severe and longer lasting as in oncogenesis.[11] The Agency also recognizes that, while a viral pesticide may be unable to infect normal, healthy nontarget animals, animals with altered defenses may be susceptible. Thus, many of the infectivity tests are designed to address the Agency's concern regarding the ability of a virus to behave as an opportunistic pathogen.

Viruses, in particular, pose some unique concerns related to infectivity. Viruses are parasites at the cellular level, capable of inserting their genomes directly into the host cell. Consequently, more permanent effects could result with these agents. Many serious diseases caused by viruses are preceded by long periods of subclinical or latent infection. Viruses in latent states are often undetectable by conventional light or electron microscopy, or even by serological techniques. They can remain cell-associated in the form of nucleic acid molecules that are detectable only by nucleic acid hybridization procedures.[12]

Although viruses themselves might not be expected to elicit an allergic response, allergens may be present in the viral preparation. For example, one of the registered nuclear polyhedrosis viruses products contained allergenic larval setae as a residual from in vivo production. In this case, an alternative processing method was devised to remove the offending contaminant.[18] To date, the commercial viral preparations are produced by whole insect (in vivo) techniques. Such preparations may contain irritants resulting from the metabolic activities of the virus. Therefore, the potential for dermal and ocular irritancy must also be assessed to protect those people handling the product.

Application of viruses in various environmental situations (e.g., agricultural areas, forests, rangeland) also raises concerns beyond the potential for overt population reductions in nontarget organisms such as entomophagous predators and parasites, birds, mammals, fish, aquatic invertebrates, and plants. For example, the possibility of more subtle (nonlethal) effects on populations and resulting ecosystem disruption is also of concern, particularly if the virus is applied in habitats where it is not already part of the indigenous microbial population.

Finally, quality assurance considerations encompass such concerns as virus identification, genetic stability, contamination, and potency. These are variables that are all subject to change during production. Therefore the Agency is concerned that the virus identified and tested for safety prior to registration is the same as that which is eventually produced and marketed for use.

B. Approaches to Safety Testing

The Agency's approach to toxicology and ecological effects safety testing incorporates several unique features in order to yield maximum information on which to base a risk

Table 2
PRODUCT ANALYSIS
DATA REQUIREMENTS

- Product identity
- Manufacturing process
- Discussion of formation of un-
  intentional ingredients
- Analysis of samples
- Certification of limits
- Analytical methods
- Physical and chemical prop-
  erties of formulated product
- Submittal of samples

assessment. First, the data requirements and the recommended test protocols pay particular attention to the virus's potential to survive, replicate, infect, and cause disease. For example, an extended observation period and appropriate methods for detecting and quantifying the virus in the test organism are necessary in order to evaluate survival and infectivity.

Second, the required data are organized into separate tiers or groups of tests, and testing is conducted in a step-wise fashion. The first group of tests, Tier I, is mandatory and consists of a variety of relatively short-term (up to 30 days), inexpensive laboratory tests. Upon completion of Tier I testing, results are evaluated and a risk assessment is made based on the proposed use of the pesticide. If no problems are foreseen, then no additional testing is required. On the other hand, should a problem be identified or should additional questions arise, then further testing in Tier II would be required. The tier testing scheme provides for three levels of testing, with any potential problems identified in Tier I being examined more closely in subsequent tiers, in order to determine whether any adverse effects would be expected under actual conditions of use.

The third unique feature of the data requirements is use of a maximum challenge approach to testing in Tier I.[9] Using this approach, the recommended test methods were designed using the most sensitive test species or life stages, immunologically depressed test animals (mammals), high dose levels, various routes of exposure, and for viruses, the most infectious form of the agent. Having carried out Tier I testing under the maximum challenge conditions, there is greater assurance that any adverse effects associated with a viral pesticide will be identified. As previously indicated, any adverse effects noted in Tier I can then be more fully evaluated in subsequent tiers of testing.

IV. REGISTRATION DATA REQUIREMENTS

The minimum information and data requirements for the registration of a viral pesticide pertain to product analysis, toxicology (human health), and ecological effects.

A. Product Analysis

Information on product analysis (Table 2) is required to support the registration of all viral pesticide products, regardless of their intended use. The requirements include provisions for the identification, assay, and standardization of each virus in products proposed for registration.

All active and inert ingredients in the product must be identified and quantified to the extent possible. Each viral product submitted for registration is accompanied by a confidential statement of ingredients containing the identity and quantity of virus in the formulated product, as well as the identity and quantity of all inert ingredients such as UV screens, stickers, spreaders, and diluents. In addition to verifying product composition, this information is also used as an aid to determine appropriate dose levels in the toxicology and ecological effects tests, in making subsequent exposure and risk assessments, and in developing appropriate product labeling. For example, the amount of virus present in each formulated product is specified not only in the confidential statement of formula but also on the product label. Recognized units of potency, percentage of weight, units of viability or replication, or another appropriate expression of biological activity must be used on the label.

Each viral agent must be identified using the most sensitive and specific method available, and should be accompanied by the precise test procedures for such identification. To the extent possible, this would include the serotype and strain or other designated type. Data to support the designation should include morphologic, physiologic, nutritional, biochemical, and molecular biological data where appropriate. For viruses, the submission of restriction enzyme digest patterns for deoxyribonucleic acid (DNA) viruses and two-dimensional oligonucleotide fingerprints for ribonucleic acid (RNA) viruses is required. Careful and complete identification is essential in order to provide the necessary base line information against which postregistration monitoring results can be compared.

Information pertaining to the production process is required prior to registration. Such information includes a list of starting and intermediate materials. The steps taken to ensure the purity of these materials, and the steps taken to limit any extraneous chemical or biological contaminants in the product must also be identified. As a further safeguard against the introduction of unwanted contaminants or impurities, each manufacturer should indicate the procedures used to establish the identity and purity of the ''culture'' from which the unformulated virus is produced. The method of manufacture and techniques to ensure a uniform, standard product should be indicated. Information provided here includes an outline of the manufacturer's standard plant sanitation and cleanup operations, as well as any provision for the use of bactericidal or fungicidal additives to control the growth of unwanted microorganisms.

Quality assurance provisions consist of periodic monitoring during production to ensure that levels of impurities and contaminants continue to be within an acceptable range and to ensure that the species and strain of the microbe remains constant. The Agency also seeks information on the natural occurrence of each viral agent, its known geographic distribution, and its relationship to other viruses, particularly known pathogens.

The need for proper standardization of biological activity has been demonstrated by experience with several registered microbial agents. In the case of viruses, some of the original product labeling specified the application rate of product in terms of the number of polyhedral inclusion bodies (PIBs) per gram of product. Due to variations in production methodology and storage conditions, it was found that the number of PIBs per gram did not always yield a uniform insecticidal activity.[14] This in turn can lead to erratic and unpredictable results in the field — a situation that can hinder the success of a product.[15] As a result, most viral agents are now standardized by bioassay, and application rates are based on activity units per gram of product.

Finally, data on various chemical, physical, and biological properties of the product must be submitted in order to characterize the product and provide some of the necessary information for precautionary or other product label statements.

B. Toxicology

Toxicology testing addresses the potential for adverse effects in nontarget mammalian species in order to evaluate potential risks to humans and domestic animals. Tier I toxicology tests (Table 3) consists of short-term studies in several mammalian test species (e.g., rat, hamster, guinea pig) using oral, dermal, ocular, and injection routes of exposure. In addition, tests to assess irritancy, hypersensitivity, and effects on the immune system are also required.

Table 3
TOXICOLOGY DATA
REQUIREMENTS

Tier I
- Acute oral
- Actue dermal
- Acute inhalation
- I.V. and I.C. injection
- Infectivity studies
- Primary dermal
- Primary eye
- Hypersensitivity study
- Hypersensitivity incidents
- Immune response
- Tissue culture

Tier II
- Acute oral
- Acute inhalation
- Primary dermal
- Primary eye
- Immune response
- Teratogenicity
- Mammalian mutagenicity

Tier III
- Chronic feeding
- Oncogenicity
- Mutagenicity
- Teratogenicity

Tissue culture studies in a variety of mammalian cell lines are required as a further screen for the infective potential of viruses in nontarget mammalian hosts.

By requiring several routes of administration such as oral, dermal, ocular, and injection, different potential routes of human exposure are tested. Test animals are observed for at least 14 to 21 days posttreatment or until signs of reversible infectivity in survivors subside, whichever occurs later. All deaths, lesions, clinical signs of illness, and signs of recovery are reported. Surviving animals are sacrificed and subjected to a complete necropsy at test termination. Microorganism dissemination, replication, and survival in animal tissues and/or intestinal tract is reported. Survival and replication is determined using the appropriate qualitative and quantitative methods.

Primary dermal and primary ocular irritation studies are required in Tier I to determine the need for precautionary label statements. Test methodology is similar to that for chemical pesticides. Laboratory studies to assess allergenicity consist of repeated intradermal injection of the virus as a standard sensitizing treatment. Erythema, edema, and other lesions are scored according to the standard Draize method.[16] In addition, each manufacturer must report any hypersensitivity experiences in persons handling the product during development or production.

Immunological tests are included in Tier I to evaluate the ability of viral agents to survive and grow in vertebrates and/or impair their immune system. Recommended tests include a lymphocyte proliferation assay; an in vitro assay for cellular immunocompetence; IgG, IgM, and IgA levels; antibody plaque assay for a lymphocyte-dependent antigen; and body, thymus, and spleen weights of treated animals.

EPA believes that the cell tissue culture test is probably the most stringent for a viral agent and probably the most sensitive means for testing the capability of a virus to infect and interact with nontarget species. Cell lines of human and nonhuman origin are tested.

Table 4
ECOLOGICAL EFFECTS AND
ENVIRONMENTAL EXPRESSION DATA
REQUIREMENTS

Tier I
- Avian oral
- Avian injection test
- Wild mammal testing
- Freshwater fish testing
- Freshwater aquatic invertebrate testing
- Estuarine and marine animal testing
- Plant studies
- Nontarget insect testing

Tier II
- Terrestrial environmental expression tests
- Freshwater environmental expression tests
- Marine or estuarine environmental expression tests

Tier III
- Terrestrial wildlife and aquatic organism testing
- Avian pathogenicity/reproduction test
- Definitive aquatic animal tests
- Aquatic embryo larvae and life cycle studies
- Aquatic ecosystem test
- Special aquatic tests
- Plant studies

Analysis includes observations for any gross morphological changes, for inhibition of cell division, and for change in infectivity of the bioassay culture fluid. The bioassay must be the most sensitive test available for detecting virus. One of the most valuable aspects of the tissue culture test is the assay for the disappearance of input virus and the synthesis of viral proteins and nucleic acid. Such an assay is most useful in the mouse 3T3 cell line for the assessment of possible slowly developing virus-cell interactions that could lead to virus persistence and/or malignant transformation.[9]

A sensitive, quantitative immunological test (e.g., enzyme-linked immunosorbent assay) and a molecular hybridization test for viral nucleic acid at selected time intervals until termination of the tissue culture test is recommended.

Specific Tier II studies to be conducted are determined on a case-by-case basis depending on the results observed in Tier I and the pesticide use pattern. For example, if persistence of all or part of the viral genome occurs in any of the systems tested, further evaluation would be required to evaluate the implications of this finding relative to potential human hazard. Tier II toxicology studies would usually entail tests of longer exposure and duration, additional species, or perhaps other routes of exposure. Tier III studies consist of long-term (chronic) testing in mammalial test species.

In addition to Tier II toxicology testing, if adverse effects or other problems are observed in Tier I, then studies to assess human exposure via treated food or feed are initiated, if appropriate. These studies are analogous to residue studies for chemical pesticides and could ultimately be used to set a tolerance for the virus in raw agricultural commodities.

C. Ecological Effects

Tests for ecological effects (Table 4) are also organized in a sequential, tier testing scheme, and a similar approach to identifying and evaluating potential hazards is employed. Tier I consists of laboratory studies, using the maximum challenge approach to assess effects in selected species of birds, fish, insects and other invertebrates, and plants. If any of the Tier I effects studies are positive, then the appropriate Tier II studies are indicated. In this scheme,

Tier II studies are designed to assess the environmental expression or fate of the virus in the environment. Once both Tier I and Tier II studies are completed, information on both effects and fate will be available, and a more complete hazard assessment can be made. For example, if environmental fate testing indicates that the virus will not continue to survive and replicate after field application, then certain effects may not be of concern, since nontarget organisms would not be affected under actual conditions of use.

On the other hand, if both the Tier I effects tests and Tier II fate tests are positive, than Tier III studies may be required. Tier III ecological effects studies are analogous to Tier III toxicology tests: both usually consist of longer-term testing in order to fully evaluate effects under conditions that closely approximate actual use situations in terms of extent and duration of exposure.

V. REGISTRATION OF VIRAL PESTICIDES

A. General Organization and Function

EPA's Office of Pesticide Programs (OPP) administers the federal regulation of pesticide products. OPP consists of four divisions (Figure 2). Two of the divisions, Registration (RD) and Hazard Evaluation (HED) are responsible for most of the scientific and administrative aspects of registration with which a pesticide manufacturer is concerned.

Product registration covers new pesticides, existing pesticides for which additional uses are sought (amended registration) or which other manufacturers seek to market (''me-too'' registration), experimental products or uses (experimental use permit), and emergency uses of unregistered products (emergency exemption from registration). OPP is also responsible for two other major programs related to registration; registration standards and special review. The registration standards program evaluates the data base for currently registered products in light of today's testing and data requirements standards. Additional data or label modifications may be required under the registration standards program to ensure product safety and to support continued registration. The special review program administers the process by which EPA conducts an in-depth risk-benefit evaluation of those products whose use (or proposed use) is presumed to cause unreasonable adverse effects to human health or the environment. Products under special review may be denied registration or cancelled, or their use(s) modified or limited, depending on the extent and nature of the hazards they pose compared to the benefits realized from their use.

B. The Registration Process

The following discussion describes the registration process for a ''typical'' agricultural product such as a viral pesticide used to control an insect pest on soybeans. Figure 3 summarizes the registration process.

1. Product Development

Most microbial pesticides have been discovered by insect and plant pathologists during the course of basic research which was not necessarily directed towards development of a pesticide product. Much of the early developmental work on the viral pesticides has been undertaken by government and university laboratories. Once a potential product has been discovered, and initial work on production techniques, formulation and small-scale efficacy testing (e.g., greenhouse tests) has been completed, then field tests are usually needed to refine the formulation, determine the appropriate use parameters, and further evaluate efficacy. It is generally at this point that the product and its developer come under the jurisdiction of FIFRA.

Field testing conducted to accumulate information necessary to register the product normally requires an experimental use permit (EUP). The circumstances under which a permit

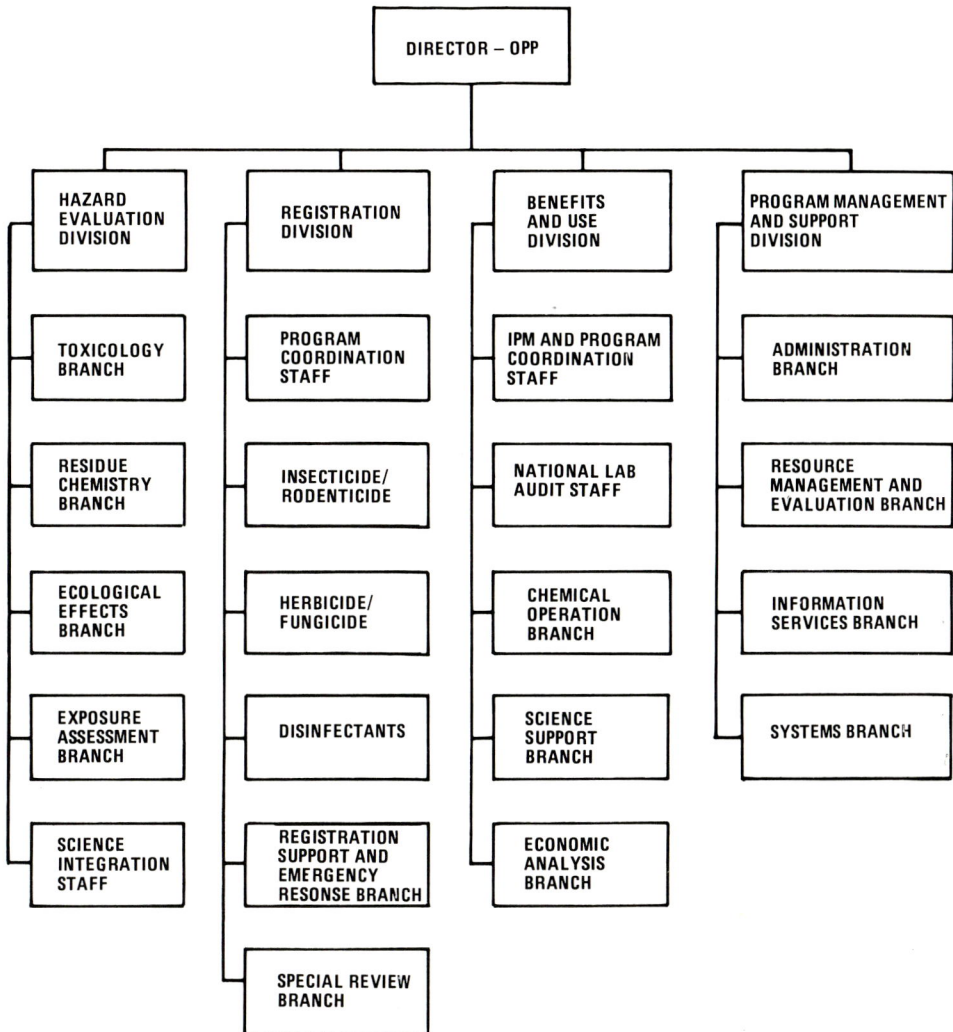

FIGURE 2. Office of Pesticide Programs — organizational chart.

is required and details concerning permit applications, product labeling, and conduct of field trials is described in the Code of Federal Regulations, Part 172.[3] The major elements of an EUP application (EPA Form 8570-17, Figure 4) include identification of the product to be tested, quantity to be tested, test sites and acreage, the proposed testing program, the product label, and the supporting product analysis and safety data. EPA reviews this application to determine whether the EUP is justified and whether issuance of a permit would cause unreasonable adverse effects on human health or the environment. If results of this review and evaluation are favorable, EPA can then issue a permit and give notice of such in the *Federal Register*. Permits are granted for specific time period (usually 1 year), site, use pattern, and quantity of product. Applications are processed as expeditiously as possible and normally require less than 120 days.

2. Registration Application

After a developer has conducted laboratory evaluations and experimental field testing, he has probably developed his prospective product to the point where a specific product formulation and use pattern (e.g., use site, pest, rate and frequency of application) for regis-

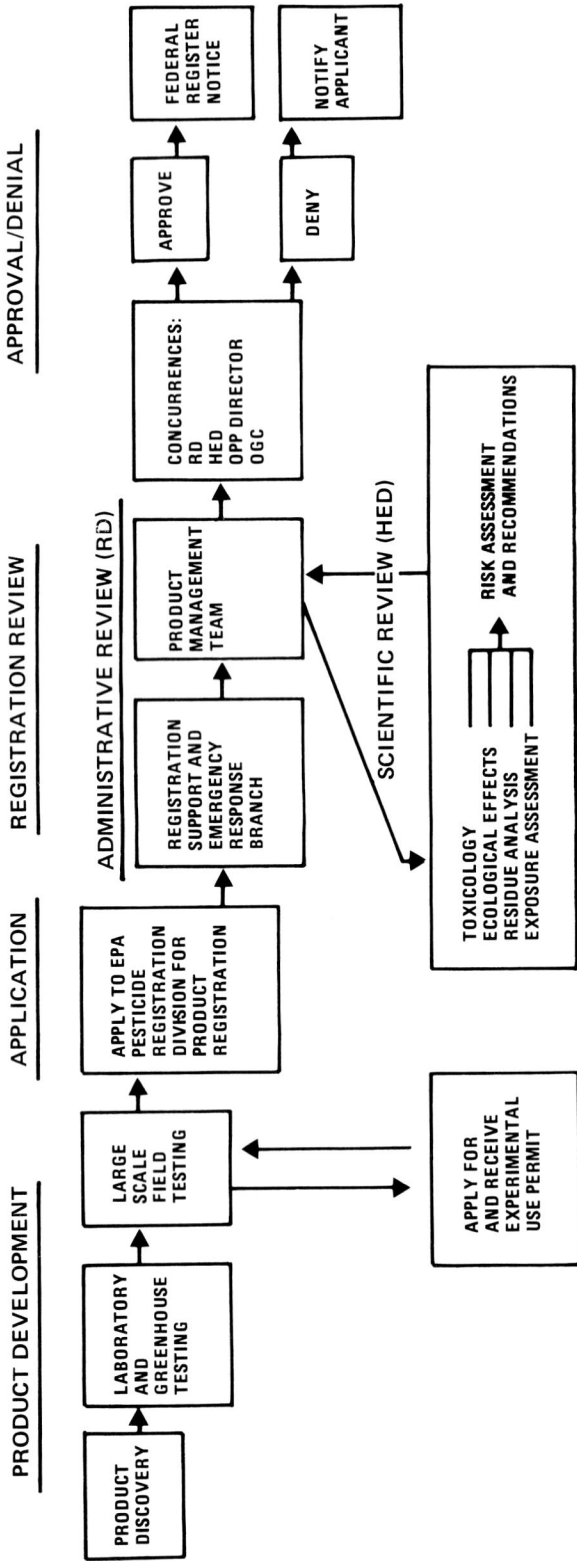

FIGURE 3. The pesticide registration process.

Form Approved OMB No. 158-R0068

U.S. ENVIRONMENTAL PROTECTION AGENCY OFFICE OF PESTICIDE PROGRAMS REGISTRATION DIVISION *(WH-567)* WASHINGTON, D.C. 20460	APPLICATION FOR AN EXPERIMENTAL USE PERMIT TO SHIP AND USE A PESTICIDE FOR EXPERIMENTAL PURPOSES ONLY *(Please read the instructions on reverse before completing)*

1. TYPE OF APPLICATION ☐ NEW ☐ EXTENSION *(Give number below)* PERMIT NO. _____	2. EPA COMPANY NUMBER	3. DATE OF APPLICATION

4. NAME AND ADDRESS OF FIRM/PERSON TO WHOM THE EX-PERIMENTAL USE PERMIT IS TO BE ISSUED *(Include ZIP code)*	5. NAME AND ADDRESS OF SHIPPER *(If shipment is intended and if different from applicant) (Include ZIP code)*

6. NAME OF PRODUCT	7. IS PRODUCT REGISTERED WITH EPA ☐ YES *(If Yes, complete below)* ☐ NO EPA REG. NO. _____

8. TOTAL QUANTITY PROPOSED FOR SHIPMENT/USE PRODUCT: _____ ACTIVE EQUIVALENT _____ POUNDS	9. PROPOSED PERIOD OF SHIPMENT/USE	10. PLACES FROM WHICH SHIPPED

11. WILL TECHNICAL MATERIAL BE IMPORTED ☐ YES *(Give total quantity below)* ☐ NO GALLONS _____ POUNDS _____	12. PLACE WHERE DIRECTIONS FOR USE APPEAR ☐ ON CONTAINER LABEL ☐ IN PRINTED MATTER ACCOMPANYING PRODUCT

CERTIFICATION

This is to certify that food or feed derived from the experimental program will not be used or offered for consumption or sale for consumption except by laboratory or experimental animals if illegal residues are present in or on such food or feed.

13. SPECIFY THE NAME AND TELEPHONE NUMBER OF THE INDIVIDUAL MOST FAMILIAR WITH THIS APPLICATION WHO CAN BE CONTACTED DIRECTLY, IF NECESSARY TO PROCESS THIS APPLICATION	14. SIGNATURE OF APPLICANT OR AUTHORIZED FIRM REPRESENTATIVE	
	15. TITLE	16. DATE SIGNED

BELOW FOR U.S. EPA USE ONLY

In any correspondence on this application, refer to this number: _____ Normal review time indicates that processing of this application is expected to be completed by: Date: _____	RECEIVED BY EPA-OFF REGISTRATION DIVISION, WASHINGTON, D.C. 20460
NAME OF EPA CONTACT	TELEPHONE NO.

EPA Form 8570-17 (2-76)

FIGURE 4.   Experimental use permit application.

tration has been identified. Therefore, the product and its use are sufficiently defined so that production and commercialization are the next major steps. This, then, is the time for the developer to submit an application for registration to EPA.

The information to be submitted in an application for product registration depends on how and where the product will be used. For example, registration of pesticides to be applied on food for human consumption normally requires more extensive toxicology data (than nonfood use products) and data to support a tolerance or an exemption from the requirement of a tolerance under the Federal Food Drug and Cosmetic Act.[17]

Each product is somewhat unique, with its own specific use pattern, formulation and chemical, physical, and biological characteristics. Therefore, the Agency strongly encourages applicants to request preregistration conferences to discuss their specific product, the kinds of data and information that will be necessary to support registration, and appropriate protocols for developing the necessary data. In addition, applicants may wish to discuss the possibility of waiving of modifying certain requirements that may be inappropriate for their specific product. By developing a "dialogue" between EPA and the applicant early in the registration process, unnecessary delays or misunderstandings can usually be avoided.

Regardless of the proposed use and extent of data required, each new product application would include the following materials:

- Application for New Pesticide Product Registration — EPA form 8570-1 (Figure 5)
- Confidential Statement of Formula — EPA form 8570-4 (Figure 6)
- Data Reference Sheet — EPA form 8570-20 (Figure 7)
- Copies of draft labeling
- Copies of any data submitted in support of the application

Registration Division (RD) coordinates the review process for each product registration. Applications submitted to the Registration Division first go to the Registration Support and Emergency Response Branch (RSERB). There, the application is issued a number, a file is created for the application, each volume of data is cataloged, and a determination is made as to the type of registration (e.g., new, amended, "me-too"). Upon completion, RSERB assigns the application to the appropriate Product Management Team in RD.

Product Management Teams are divided among three branches in RD. Each branch is responsible for handling one of the following groups of products: (1) insecticides/rodenticides, (2) fungicides/herbicides, and (3) disinfectants (Figure 2). Each team handles certain active ingredients so that, for example, one product management team would handle all of the viral insecticides. From this point on, the team to which the product is assigned coordinates all registration or other actions (e.g., EUPs) relating to the product. The product manager is, therefore, the applicant's primary contact with the Agency and should be consulted to address questions or problems, to check on the status of the registration review, or to arrange meetings with the scientific staff.

Upon receipt of the application, the Product Manager (PM) performs a detailed administrative review to be sure the application is complete before the review proceeds. The PM publicly announces receipt of new products for registration by publishing a notice of such in the *Federal Register*. A separate *Federal Register* notice is published to announce receipt of applications accompanied by a tolerance petition to support a food or feed use.

3. Concurrent Scientific Reviews

The PM team distributes the applicable data to each science branch on the Hazard Evaluation Division (HED) for review. Scientific review of the data supporting registration is the most technical and time-consuming portion of the review process. The review consists of several concurrent and stepwise reviews by the four HED science branches — Toxicology, Ecological Effects, Exposure Assessment, and Residue Chemistry. Taken together, the scientific reviews comprise a comprehensive risk assessment of the product and its proposed use.

The Toxicology Branch reviews and evaluates the data pertaining to human health assessment (e.g., exposure of mammalian species to assess infectivity, toxicity, and allergenicity). The Residue Chemistry Branch reviews the product analysis data, and for food or

Form Approved OMB No. 2000-0012

EPA

U.S. ENVIRONMENTAL PROTECTION AGENCY	
OFFICE OF PESTICIDE PROGRAM (TS-767)	
WASHINGTON, D.C. 20460	☐ REGISTRATION
APPLICATION FOR PESTICIDE:	☐ AMENDMENT A

Please read instructions on reverse before completing.

SECTION I

1. COMPANY/PRODUCT NO.	2. DATE	3. PRODUCT MANAGER	4. PROPOSED CLASSIFICATION
			☐ GENERAL ☐ RESTRICTED

5. NAME AND ADDRESS OF APPLICANT *(Include ZIP Code)*

☐ CHECK IF THIS IS A NEW ADDRESS

6. PRODUCT NAME

SECTION II

1. SUBJECT OF AMENDMENT

☐ RESUBMISSION IN RESPONSE TO AGENCY LETTER DATED _____

☐ FINAL PRINTED LABEL IN RESPONSE TO AGENCY LETTER DATED _____

☐ OTHER *(explain below)*

SECTION III

1. WILL THIS PRODUCT BE PACKAGED IN:

CHILD-RESISTANT PACKAGING ☐ YES ☐ NO

UNIT PACKAGING ☐ YES ☐ NO

If YES, unit pkg. wt. _____ No. per container _____

WATER-SOLUBLE PACKAGING ☐ YES ☐ NO

If YES, pkg. wt. _____ No. per container _____

2. TYPE OF CONTAINER
☐ METAL
☐ PLASTIC
☐ GLASS
☐ PAPER
☐ OTHER *(Specify)*

3. LOCATION OF NET CONTENTS
☐ LABEL ☐ CONTAINER

4. SIZE(S) OF RETAIL CONTAINER

5. LOCATION OF LABEL DIRECTIONS
☐ ON LABEL
☐ ON MATERIAL ACCOMPANYING PRODUCT

6. MANNER IN WHICH LABEL IS AFFIXED TO PRODUCT
☐ LITHOGRAPH ☐ OTHER *(Specify)*
☐ PAPER GLUED
☐ STENCILED

SECTION IV

1. CONTACT POINT *(Complete items directly below for identification of individual to be contacted, if necessary, to process this application).*

NAME

TITLE

TELEPHONE NO. *(Include Area Code)*

6. DATE APPLICATION RECEIVED *(Stamped)*

2. SIGNATURE

3. TITLE

4. TYPED NAME

5. DATE SIGNED

EPA Form 8570-1 (Rev. 5-81)   PREVIOUS EDITION IS OBSOLETE.

FIGURE 5.   Pesticide product registration — application form.

feed uses, any residue data to support a tolerance or an exemption from requirement of a tolerance. The Ecological Effects Branch evaluates the data on nontarget species such as birds, fish, aquatic and terrestrial insects, and plants. Finally, the Exposure Assessment Branch examines the data pertaining to environmental expression or fate of the viral pesticide in the environment.

CONFIDENTIAL BUSINESS INFORMATION - DOES NOT CONTAIN NATIONAL SECURITY INFORMATION (E.O. 12065)

Form Approved
OMB No. 2000-0483
Approval expires 7-31-84

See instructions on back of last page.

U.S. ENVIRONMENTAL PROTECTION AGENCY
OFFICE OF PESTICIDE PROGRAMS (TS-767)
WASHINGTON, D.C. 20460

CONFIDENTIAL STATEMENT OF FORMULA

☐ BASIC FORMULATION
☐ ALTERNATE FORMULATION   A

1. NAME AND ADDRESS OF APPLICANT/REGISTRANT *(Include ZIP Code)*

2. PAGE ____ OF ____

3. REGISTRATION NUMBER/FILE SYMBOL

4. PRODUCT MANAGER

5. PRODUCT NAME

6. NAME AND ADDRESS OF PRODUCER *(Include ZIP Code)*

7. WEIGHT/GALLON DENSITY

8. pH

9. FLASH POINT/FLAME EXTENSION

10. COUNTRY WHERE FORMULATED

COMMERCIAL COMPONENT *(List each as actually introduced into this formulation. Give Trade Name (if any), Name of Supplier, and EPA Registration Number if applicable.)*

12. SUPPLIER

13. EPA REG. NO.

14. WEIGHT OF EACH COMPONENT*

15. PERCENT BY WEIGHT

16. PURPOSE IN FORMULATION

*If liquid measure, give specific gravity or pounds per gallon.

17. TOTAL WEIGHT OF BATCH

100.00%

18. TYPED NAME OF APPROVING OFFICIAL

19. TITLE

20. TELEPHONE NO. *(Include area code)*

21. DATE

EPA Form 8570-4 (Rev. 10-81)   PREVIOUS EDITION IS OBSOLETE.

ORIGINAL COPY TO EPA
APPLICANT RETAINS SECOND COPY

FIGURE 6.   Confidential statement of formula.

Form Approved OMB No. 2000-0012

EPA

U.S. ENVIRONMENTAL PROTECTION AGENCY
REGISTRATION DIVISION (TS-767)
WASHINGTON, D.C. 20460

DATA REFERENCE SHEET

(See instructions on the back of the last page before completing.)

1. PAGE ___ OF ___

2. EPA REGISTRATION NO./FILE SYMBOL

3. APPLICANT'S NAME AND ADDRESS

4. PRODUCT NAME

5. PRODUCT MANAGER

6. TO ACCOMPANY APPLICATION FOR REGISTRATION DATED:

7. NAME OF STUDY	SOURCE OF STUDY					f. ACCESSION NUMBER *(if known)*
	a. APPLICANT CONDUCTED STUDY *(mark 'X')*	b. OBTAINED FROM EPA *(mark 'X')*	c. OBTAINED FROM ANOTHER FIRM OR SOURCE *(give name and address)*	d. OBTAINED FROM PUBLIC LITERATURE *(give reference)*	e. OTHER *(explain)*	

If you marked "X" in this column, do you wish your name placed on the Data Submitters List? ☐ YES ☐ NO

EPA Form 8570-20 (1-81)

APPLICANT-RETAIN THE LAST COPY

EPA COPY

FIGURE 7. Data reference sheet.

Each science review is conducted concurrently, but the findings of the Residue Chemistry and Exposure Assessment branches are ultimately used by the toxicologists and ecological effects reviewers to complete their risk assessment. As discussed previously, viral pesticides are tested in a step-wise or tier scheme and therefore data on residues and environmental expression are not normally required as part of an initial review. Instead, the initial HED review of viral pesticides is normally based on product analysis, toxicology, and ecological effects data, along with the product label and any other available information on the proposed use pattern.

The completed HED science reviews summarize the adequacy of the submitted data, describe any potential human or environmental risk, specify the need for any additional data or information, and recommend any use limits, restrictions, or label changes as necessary to ensure product safety. The completed reviews are returned to the RD Product Manager.

4. Registration Approval or Denial

Once the PM receives the completed HED science reviews, and notwithstanding other legal and administrative considerations, the PM then makes the regulatory determination as to whether EPA can approve the registration request.

If the PM concludes that the product cannot be registered based on the available data, the applicant is notified of rejection. The reasons for rejection are specified, and any additional information, data, or use restrictions required to complete the application and grant registration are identified.

Applications approved by the PM team are summarized and concurrence is obtained from the involved offices — usually Registration Division, Hazard Evaluation Division, Director of OPP, and the Office of General Counsel (OGC).

Notice of the registration is published in the *Federal Register*. For food and feed uses, a final regulation is published in the *Federal Register* specifying the tolerance or the exemption from the requirement of a tolerance, along with any other provisions or requirements pertaining to registration and quality assurance. For example, the exemption from the requirement of a tolerance for the use of *Heliothis* NPV in or on all raw agricultural products[18] stipulates that the integrity of the seed virus must be assured by periodic checks, and sets forth various specifications to assure the purity and identity of the active ingredient.

VI. REGULATORY STRATEGY FOR GENETICALLY MODIFIED VIRAL PESTICIDES

Looking to the future, EPA believes that the rapidly developing genetic engineering technologies (e.g., recombinant DNA technology) offer the prospect of greatly increasing the numbers and variety of viral pesticides and increasing their market share compared to chemical pesticides. Therefore, EPA is now examining the data and informational requirements for nonengineered viral pesticides and their applicability to the evaluation of engineered microbial pesticides. This was undertaken to determine if these requirements should be modified, and if additional requirements are necessary to properly evaluate genetically modified viral pesticides.[19]

The potential hazards EPA foresees with genetically engineered viruses are, for the most part, the same as those for nonengineered microbes, namely: infectivity, pathogenicity, virulence, toxicity, host range, competitiveness, and survivability. Using genetic engineering techniques, microorganisms could be produced which exhibit a broader host range, a new toxin, enhanced virulence, or greater survivability than their naturally occurring, nonengineered ''parent'' microorganisms.

EPAs major concern is the potential for hazards posed by engineered organisms which would not be identified by the testing scheme for nonengineered viruses discussed in this chapter. The Agency is concerned about the stability of the genetic material in an engineered virus and about the specificity of the inserted gene segment. Current requirements would yield no information about the characteristics the inserted genes are intended to express (e.g., broader host range) and the potential for other characteristics to be unknowingly inserted and expressed. A related concern is the potential for transfer and exchange of genetically engineered characteristics with naturally occurring microorganisms, or with the host or other nontarget organisms, thereby generating new potential hazards.

Using past experience with microbial pesticides as a basis, the Agency has developed a preliminary strategy for addressing the immediate scientific and regulatory issues pertaining to genetically engineered microbial pesticides. The requirements specified in the data requirements regulation (Part 158) will apply to both genetically engineered and nonengineered pesticides. However, additional data or information may be required to address some of the questions discussed above and to support the registration of genetically engineered pesticides. This could include information on the genetic engineering techniques used, the identity of the inserted gene segment (base sequence data of detailed restriction map of the gene), information on the control region of the genes, a description of the "new" traits or characteristics that are intended to be expressed, test to evaluate genetic stability, and some or all of the Tier II environmental expression tests. Additional requirements will be determined on a case-by-case basis depending on the particular virus, its parent virus, the pesticide use pattern, and the manner and extent to which the virus has been "engineered".

VII. SUMMARY

During the past 10 to 20 years, developers of viral pesticides have made a significant commitment to developing viral pathogens of pest organisms and formulating them as effective and practical products for pest control. Similarly, EPA has made considerable progress in the past 10 years in establishing a scientifically sound and practical regulatory program for viral pesticides. As noted previously, the key elements of this program are the data requirements and recommended testing guidelines which specifically address the viral pesticides and their particular characteristics.

The safety record of viral pesticides has, to date, been very favorable. The Agency is not aware of any significant adverse human health or environmental effects resulting from use of these products. Therefore, as advancements are made in formulation and application technology and "new" pesticides are developed using genetic engineering techniques, EPA expects to register many more viral pesticides.

The prospect of more viral pesticides being proposed for registration means the Agency must continue to maintain and upgrade its regulatory program and review staff for registration of viral pesticides. Therefore, the Agency has sponsored, and is continuing to sponsor, workshops, expert panels, and research aimed at providing necessary information to revise and upgrade the guidelines and develop standard, routine test protocols for viral pesticides. The Agency plants to incorporate these research findings into the guidelines within the next 2 to 3 years.

ACKNOWLEDGMENTS

Development of EPAs regulatory program for viral pesticides has been a multidisciplinary task. Thus the efforts of several people should be acknowledged. D. Urban, R. Holst, A. Vaughan, M. Sochard, W. Woodrow, H. Manning, R. Carsel, C. Kawanishi, and W. Nelson. Further, Dr. Kawanishi and Messrs. Urban, Vaughan, and Grable provided helpful comments on this chapter.

REFERENCES

1. The Federal Insecticide, Fungicide, and Rodenticide Act (FIFRA) as amended, (7 U.S.C. 136 *et sec.*), 1978.
2. Code of Federal Regulations, Title 40, Part 162 — Regulations for the Enforcement of the Federal Insecticide, Fungicide and Rodenticide Act, U.S Government Printing Office, Washington, D.C., 1985.
3. Code of Federal Regulations, Title 40, Part 172 — Experimental Use Permits, U.S. Government Printing Office, Washington, D.C., 1985.
4. Code of Federal Regulations, Title 40, Part 158, Data Requirements for Registration, U.S. Government Printing Office, Washington, D.C., 1985.
5. Environmental Protection Agency, *Pesticide Assessment Guidelines, Subdivision M- Biorational Pesticides,* #PB 83-153965, NTIS, Springfield Va., 1983.
6. **Bourquin, A. W., Ahearn, D. G., and Meyers, S. P., Eds.,** Impact of the Use of Microorganisms on the Aquatic Environment, 660-3-75-001, Environmental Protection Agency, Washington, D.C., 1975.
7. **Summers, M. D., Engler, R., Falcon, L. A., and Vail, P., Eds.,** *Baculoviruses for Insect Pest Control: Safety Considerations,* American Society for Microbiology, Washington, D.C., 1975.
8. **Summers, M. D. and Kawanishi, C. Y., Eds.,** Viral Pesticides: Present Knowledge and Potential Effects on Public and Environmental Health, EPA-600/9-78-026, Environmental Protection Agency, Washington, D.C., 1978.
9. American Institute of Biological Sciences, Human Hazard Evaluation Testing for Biorational Pesticides, report to U.S. Environmental Protection Agency under Grant No. R806461, Arlington, Va., 1980.
10. Environmental Protection Agency, Regulation of ''biorational'' pesticides; policy statement and notice of availability of background document, *Fed. Regist.,* 44, 94, 28093, 1979.
11. **Kawanishi, C. Y.,** personal communication, 1981.
12. **Kawanishi, C. Y.,** personal communication, 1984.
13. **Shapiro, M., Bell, R. A., and Owens, C. D.,** *In vivo* propagation of the nuclear polyhedrosis virus of *Lymantria dispar,* in *Characterization, Production and Utilization of Entomopathogenic Viruses,* Ignoffo, C. M., Martignoni, M. E., and Vaughn, J. L., Eds., Proceedings of the Second Conference of Project V, Microbiological Control of Insect Pests, of the US/USSR Joint Working Group on the Production of Substances by Microbial Means, 1980.
14. **Martignoni, M. and Ignoffo, C.,** Biological activity of baculovirus preparations: *in vivo* assay, in *Characterization, Production and Utilization of Entomopathogenic Viruses,* Ignoffo, C. M., Martignoni, M. E., and Vaughn, J. L., Eds., Proceedings of the Second Conference of Project V, Microbiological Control of Insect Pests of the US/USSR Joint Working Group of the Production of Substances by Microbiological Means, 1980.
15. **Vail, P. V.,** Standardization and quantification: insect laboratory studies, in *Baculoviruses for Insect Pest Control: Safety Considerations,* Summers, M. D., Engler, R., Falcon, L. A., and Vail, P., Eds., American Society for Microbiology, Washington, D.C., 1975.
16. **Draize, J. M.,** The Appraisal of Chemicals in Foods, Drugs, and Cosmetics, Association Food and Drug Officials of the U.S., Houston, 1959.
17. Federal Food Drug and Cosmetic Act (FFDCA) as amended, (21 U.S.C. 371,), 1980.
18. Code of Federal Regulations, Title 40, Part 180 — Tolerances and Exemptions from Tolerances for Pesticide Chemicals In or On Raw Agricultural Commodities, U.S. Government Printing Office, Washington, D.C., 1983.
19. **Betz, F. S., Levin, M., and Rogul, M.,** Safety aspects of genetically engineered microbial pesticides, Recombinant DNA Tech. Bull. 6, 4, NIH, Bethesda, Md., 1983.

Chapter 9

PRACTICAL FACTORS INFLUENCING THE UTILIZATION OF BACULOVIRUSES AS PESTICIDES

G. T. Bohmfalk

TABLE OF CONTENTS

I. INTRODUCTION

The first indication that resistance to insecticides was a reality led many leading entomologists to consider alternative ways of controlling inspect pests. Chemical pest control can become methodical and lead to outbreaks of secondary pests, expanding the nature of the pest problem. We know from history that agriculture was productive prior to the development of chemical insecticides, and it should be obvious that there are successful alternatives to them. Biological control through the use of entomopathogens became a reality through the efforts of dedicated individuals in the disciplines of invertebrate pathology and entomology. This reality, however, is reflected in varying degrees of success. Microbial control as it is practiced today is successful in many cases, but to say that it is truly operational and successful to the final degree or in an economic sense stretches the definition. Many factors affect the development and operational use of pathogens in pest control programs, and from a practical standpoint many impediments remain to be overcome. The purpose of this chapter is to review some of the practical factors that influence the use of baculoviruses as pesticides.

II. STATE OF THE ART

The current research on baculoviruses as pesticides places heavy emphasis on basic research. There are many more scientists working on laboratory projects than there are applied scientists working on the practical aspects in the field. And even fewer still are the dedicated practitioners of pest control who use viral pesticides. This is a perplexing situation and cannot continue. Now, more than ever, the industrial sector is scrutinizing its commitments to viral pesticides. This close scrutiny has meant reduced funding for cooperative viral research and even "in-house" cutbacks on projects where candidate baculoviruses are being evaluated. It is sincerely hoped that a demand for products based on baculoviruses will materialize to the point of supporting continued efforts. At this point, however, this is probably a naive expectation. The development of pyrethroids by industry is proceeding rapidly. Each company is trying to get a market share of the crop protection market, which is dominated by several pyrethroids. Pyrethroids are very popular these days and are likely to remain so until resistance becomes more widespread. This popularity bids ill to the development of any alternative strategy to pest control, including microbial control agents such as baculoviruses.

From a practical perspective, there are not many baculoviruses used as pest control agents. Depending on what constitutes practicality, it would be well to consider that there is only one commercialized baculovirus product, Elcar®, the nuclear polyhedrosis virus of *Heliothis* spp. Others which have been studied and developed to some extent, including evaluations in the field, are listed in Table 1. Progress with these is briefly reviewed below.

A. Annual Crops

The *Autographa* nuclear polyhedrosis virus (NPV) has been tested against several pests of annual crops. This virus has a somewhat broader spectrum than is traditional for viruses. It has also been tested under the designation SAN 404 I. The *Trichoplusia ni* NPV has also been tested against several looper pests of annual crops. This virus product has also been studied under the designation of SAN 405 I. *Spodoptera* spp. NPVs have been under scrutiny against several armyworm species for many years on both broadleaved and grass crops. The primary reason that none of these viruses has made it into commercialization is because they were not proven to be better than either Elcar® or *Bacillus thuringiensis* for controlling these pests. Thus, the prospect of returning a reasonable profit after all the costs for development were absorbed did not appear to be a sound expectation.

Table 1
BACULOVIRUSES UTILIZED IN APPLIED RESEARCH PROJECTS

Annual crops	Orchards	Forests
Autographa californica NPV	*Cydia pomonella* GV	*Lymantria dispar* NPV
Heliothis spp. NPV (Elcar®)	*Oryctes rhinoceros* baculovirus	*Choristoneura fumiferana* NPV
Spodoptera spp. NPVs (several)		*Orygia pseudotsugata* NPV
Trichoplusia ni NPV		*Neodiprion* spp. NPVs (several)

B. Orchards

The nonoccluded baculovirus of the palm rhinoceros beetle, *Oryctes rhinoceros,* has been utilized successfully in control projects in tropical regions where palms are important crops. This work has been done by the Food and Agriculture Organization and is supported publicly with no industrial commercialization. The protection afforded the palm industry by this work is very significant and speaks well of governmental contributions to agriculture in developing countries.

The granulosis virus (GV) of the codling moth, *Cydia pomonella,* is nearing commercialization in California and abroad. The results on large-scale testing of SAN 406 I, another designation for this virus, continue to be impressive. It is highly virulent against a primary pest. This fact makes it unique and when available commercially, promises to change substantially the pest management practices in apples, pear, and walnuts. The main reason it has not been given product status is that the microbial control industry is uncertain of the market size and, therefore, of the expected revenue that might be generated to cover development costs.

C. Forests

The *Lymantria dispar* NPV has been studied extensively in Central Europe and the U.S.[1] Although the studies done by the U.S. Department of Agriculture (USDA) Accelerated Gypsy Moth Research, Development, and Application Program, have led to registration by the Environmental Protection Agency (EPA), no company has entered into production of the product. It is generally thought that this virus has not been commercialized because the use of *B. thuringiensis* currently offers satisfactory large-scale population management of the gypsy moth.

The NPVs of the sawflies represent an interesting situation where there is very little expectation of commercialization. Not only does there not appear to be a demand for large amounts of standardized product, but the cost to produce these viruses is expected to be high because artificial media (synthetic diet) for rearing sawfly larvae have not been developed. This limits the possibilities for mechanization and the maintenance of sawfly colonies becomes a costly endeavor. There are reports that a farmer cooperative in Indiana produced and marketed *Neodiprion sertifer* NPV, but stopped after regulatory procedures were promulgated. There is also some commercial production and marketing of partially purified suspensions of *N. sertifer* viral polyhedra. This is being done by state owned Kemira Ox, a chemical company in Norway.[2]

The NPVs of the spruce budworm, *Christoneura fumiferana,* and the Douglas-fir tussock moth, *Orygia pseudotsugata,* have been developed to some extent, but commercialization is not expected in the near future.

III. GENERAL CONSIDERATIONS

Any discussion of the current state of the art for microbial control must consider current pest control strategies in general. Also helpful would be a consideration of the likely direction of pest control strategies in the near future. In this context, the pyrethroid insecticides are such an overbearing influence that they affect both. The overwhelming majority of all insecticide research and development is in the area of the pyrethroids. At present, industry is focusing on pyrethroids almost exclusively. By 1988, it is expected that ten pyrethroids will be fully commercialized. These ten are all closely related, and, given the current lack of regulatory scrutiny, all are expected to be registered for use in the great majority of areas treated with any insecticides. They are all cheap to produce, and mechanization would lead to a volume response in profit. The greater the amount of material sold, the higher the demand, and in production this means a lower unit cost, which results in a higher unit profit.

The paradox here is that resistance to pyrethroids is certain to occur, and it too will follow a volume response. Few members of the chemical insecticide industry contemplate a "post-pyrethroid" era. This could well result in disastrous consequences because the pyrethroids are so similar. Resistance to one will no doubt mean resistance to them all as a result of cross resistance. And then where will microbial control be? Hopefully, along with other efforts in the development of alternative strategies, at a position to be rapidly implemented and deployed to prevent very substantial economic losses to farmers. Therefore, the increased popularity of chemical synthetics should not lead to the abandonment of research on other areas. The needs are obvious. Products and strategies must be developed that will make alternatives such as microbial pesticides more practical.

A. The Practical Aspects of Pest Control

The use of any biological insecticide, baculoviruses included, is currently seen as "a good idea but impractical." This is admittedly a generalization, for in some areas such as gypsy moth management in the forests, technology that employs *B. thuringiensis* as a control agent is advancing rapidly. However, in agriculture, where over one billion dollars is spent annually in insecticides, chemical insecticides are used almost exclusively. It is a complicated state of affairs because so many diverse interests are represented in the course of pest control.

It is noted above that most work on baculoviruses emphasizes basic research. It is a long way from basic research to operational technology. This results from a paucity of effort given to the implementation of any "new" control strategy. It is almost as if we expect the knowledge of baculovirus-based insecticides to move to the site of need through some divine guidance, and all of a sudden be practical. This is not true, of course, so some acknowledgment by the basic researchers that useful technology is dependent on a multiplicity of effort is definitely in order. Concomitantly it should be acknowledged by all researchers involved in applied studies that basic research will provide an improvable technology. Once this gap is bridged, a unified effort can proceed. A quick review of the scientific literature will identify the basic scientists. But who are the others who will bring to practice any alternative pest control strategies?

B. The Consumer

The ultimate consumer of a pest control strategy is the grower. Although this point is debatable, it is based on the premise that a grower can do whatever he pleases on his own land. Most growers are businessmen. They profit from production. And most are keen on analyses of profits/losses, income/expense, and so on. Because of the demand on his system, any decision in regard to whether to use a chemical insecticide or an alternative depends only slightly on sociobiological ramifications such as environmental quality. Most growers are not rewarded for environmental consideration and cannot, therefore, consider it as pivotal in a decision-making process. The farmers are influenced heavily by industry representatives, consultants, agricultural extension service specialists, and representatives of financial institutions.

Industry representatives are most visible in their field sales force. Many times they will have advanced degrees in agriculture and speak with an authoritative voice. Because of their allegiance to the company they represent, one should only expect that they would represent that company's products in a most favorable light. And the more frequent the farmer's thought processes are influenced by a certain line of reasoning (represented by the industry spokesman), the more likely he is to respond to his own reasoning with due influence considered. So large companies with many representatives in the field seem to be better able to influence the farmers than smaller companies with correspondingly fewer representatives in the field. This again is a broad generalization because there are always individual farmers who are not influenced so easily. But suffice it to say that a farmer is influenced considerably by industry representation.

Consultants are utilized as advisors in areas of heavy insecticidal use. Although many are experienced and well-educated, no particular qualifications are needed to engage in the business of consulting. Many times consultants are specialists in some particular area outside of their realm of influence. If a specialist becomes a consultant, he will likely engage in "total agricultural consulting." This sometimes results in shortcomings when a recommendation is needed outside their expertise. So the recommendation will likely become either whatever is perceived or the easiest to utilize and be accepted, or whatever the other consultants are using to handle similar situations. At this time, very few specialists trained in microbial control are functioning as consultants to agricultural producers.

Agricultural Extension Service specialists are subject matter specialists trained in many fields and charged with the education of the agriculture producers. Many function as integrated pest management (IPM) specialists. IPM, by its very nature, involves many diverse areas of expertise and the interrelationships of IPM can be complex. IPM specialists are generally educated traditionally and most have little experience in the specifics of pest control. These specifics of host plant/pest/ecosystem in their physical reality may be encountered by the IPM specialist for the first time after he arrives at his position of extension specialist. Once on the job, the social pressures of his position, coupled with the fact that there is a tremendous amount of money invested in the crop developing under his stewardship, can be burdensome.[3] Because pest outbreaks, by their very nature, occur in a relatively narrow time window and the response must be intense, time becomes important, and alternative strategies and creativity are inhibited. Due to negative perceptions about biological control in general and microbial control specifically, it is easier to respond by recommending a chemical control agent.

The processes one goes through, whether he is a consultant or an extension specialist, seem to be the root of the problem in getting recommendations for alternative strategies made. It is the responsibility of all of us involved in the science of pest control through innovation to see that the influencers (consultants and extension specialists) understand and believe in alternative strategies. The thought processes are vitally important and must be exercised in the course of choosing a control method. It is indeed unfortunate that the few stimulating and creative individuals involved in pest management do not have an open forum for their ideas, free from the influence of traditional pest control.

Financial institution representatives are mentioned here only because they have taken on a highly influential role in agriculture production. Because most production is done on borrowed money, the lenders often influence significantly the kinds of decisions made in the course of production; even to the point of pest control decisions. It is important to realize that a banker, not usually trained in any area of pest management, must be included when considering who to educate in pest control principles.

C. The Universities

Land grant universities, and others involved in agriculture, are tremendous resources for implementing change and innovation. Their charge is the generation, transmittal, and application of knowledge. This is done by teaching, research, and extension. Ideally, a trilogy of catalyses is desirable. But unfortunately, the three areas do not always communicate as well as necessary for any wholesale implementation of their own collective strategies. Traditionally, the universities have been responsible for the generation of knowledge which in turn moves out to implementation through the cooperative extension services. While this is a distinguished and time-honored tradition, one need only consider the state of affairs in applied pest control today (i.e., ''squirt and count'' pyrethroid technology) and determine how relevant this is today.

Teaching — Interdisciplinary systems approaches are relevant now. The unilateral thrust for excellence of the past may not be appropriate today. The complex environmental problems are sound examples of the complexity of the situation today. There are few individuals on university faculties who have experience in the integration of several disciplines. This is not unique to entomology. Other disciplines tend to approach problems unilaterally also. Notable, the direction taken by endeavors such as ''The Consortium for Integrated Pest Management,'' and other large-scale interdisciplinary efforts, mark a response in the right direction.

Research — There is a tremendous gap between the abstract theory of the basic scientist and the intuitive responses of the practitioner. In some ways, this is good and allows for a continuum of activity. But the links between them are weak, and upsets caused by fiscal austerity, reassignment, and reorganization can have a devastating effect. In demanding times, realizing goals of research efforts are more easily accomplished when under the control of an individual. This also holds true for teaching and extension. In general, little is done to ensure a logical and relevant bridge between those involved in different disciplines. Perhaps an intradisciplinary *team* cooperation should be considered before undertaking interdisciplinary efforts, especially during times of limited resources. A quick review of a trade publication reflects the scarcity of resources spent in researching and promoting alternative strategies. There simply is not enough spent on alternatives to mount an intensive effort that will allow reasonable progress.

Extension — Ideally, the extension arm of the land grant system allows for information purveyal from the universities to a local setting. There also is a response to *local* need inherent in the extension systems. This system depends primarily on a network of generalists who conduct programs at a county (parrish or burrough) level. Their success is generally a reflection of the capacity of the county agent to speak the language of the researchers in terms that can be understood by the farmers. In alternative strategies, such as the use of baculovirus insecticides, the complexities seem to be more than the generalist can grasp and employ successfully. Viruses can, however, complement chemical control strategies because they are easy to use and unilateral in approach. Also, growers respond favorably to rapid results which are easily evaluated and attributable to a recommendation. In contrast, the abstractness of insect ecology is lost between the basic area and useful technology for lack of an effective information delivery system.

In some states where agriculture is a dominant economic force, there are local specialists. As previously mentioned, their time has been dedicated to solving the complexities of transferring scientific information on a local basis. They have an influential role in determining which tactics are used in the pest control practices in their sphere of influence. This works well in theory, but often the bureaucracies are stifling. An oversimplified example of the bureaucratic pecularities can be seen in a large state well known for innovation in pest management, where a cadre of local specialists in entomology were given new titles. They were changed from ''County Extension Entomologist'' to ''County Extension Agent-Pest Management.'' This was done to accommodate a thriving bureaucracy which meant only to ''clarify'' the position of local specialists. Many of the entomologists held advanced degrees, but were not the true generalists they were meant to be. It is difficult to see how such a change could make a positive contribution to the development of alternative strategies in these highly critical times.

D. The Legislators

There will always be the search for the panacea. And this will result in continued pressure by the public for more effective pest control agents. Hopefully, we will move to an economic condition that will result in a demand for environmental compatibility. Currently, environmental compatibility is regarded as expensive. Public support will continue to play the most important role in legislative activity in this area. Since so much of this activity will be done with governmental support, the political responsibilities must be met. As unpopular as it may be today, regulatory enhancement must be brought about. Tighter restrictions on the sale, use, and development of control agents should be implemented. Most public figures involved in legislation have backgrounds in law or economics, and can usually grasp the importance of developing environmentally sound agents like baculoviruses. It follows then that someone must teach or influence them. This brings about either changes in responsibilities, or new roles for members of the scientific community. It will not be enough to promote alternatives without displaying unarguably the futility of wholesale dependence on chemical control. There must be a reconciliation of the economic and environmental factors that influence pest control strategies. This must be made with a clear voice towards policies affecting both the science and practice of baculovirus technology.

IV. ECOSYSTEM CONSIDERATIONS

A baculovirus used as an insecticide can be introduced into the pest's environment in several ways. For instance, annual crops are usually void, or nearly so, of a pathogen which might offer some control of a pest. Thus, the virus must be introduced. This is done in several ways, but usually parallels those types of approaches used for chemical control. For example, microbial control agents in use today are available in formulations similar to those of chemical insecticides. They are meant to be generally applied over the agricultural crop to be protected. Although this approach can offer good results, there are many extrinsic factors which can have a negative impact on this approach. Most pathogens are stomach poisons; that is they must be ingested to be effective. Thus, to kill the insect, a lethal dose must be eaten. This requires a lethal dose distribution over areas where the pest will be eating. In cotton it is relatively easy to distribute a pathogen over the foliage and control leaf-feeding pests. It is quite another matter when fruit feeders, like *Heliothis* spp. which feed in protected areas at isolated sites such as the floral buds or developing bolls, are to be controlled.

The growth character of the plant to be protected also influences the degree of efficacy expected from a pathogen. Again, cotton is a good example. Its habit of continual growth consistently exposes new growth in the course of normal phenological growth. Orchard crops tend to set fruit and then mature them. Very little growth continues after the fruit is set. As a result, even application of a stomach poison can offer high levels of protection, especially with multiple applications which may have some cumulative effect. Coverage is obviously of critical importance, yet there are trends in agricultural application technology which tend toward low volume applications and even ultra low volume applications. Chemicals with their different mode of action may very well be useful at reduced volumes, but to expect a stomach poison to be evenly distributed over the critical area through applications of lower volumes might be unrealistic.

The sensitivity of baculoviruses to ultraviolet light is well known, and affects significantly the time window during which they are effective. Chemical controls are no less sensitive. In fact, the newer pyrethroid insecticides are sensitive to light and are substantially more sensitive to high ambient temperatures (not characteristic of baculoviruses). However, because a baculovirus must persist until eaten, rather than just persist until physically contacted, it may appear that they are less robust and cannot be relied upon for longer-term control.

Baculovirus insecticides are frequently characterized by pest control advisors as being less effective at controlling large insect populations. While this may be true from the standpoint of rapid population reductions, quite the opposite if long-term population suppression is the goal. This impression is persistent because many evaluations simply use acute subtraction as the method of appraisal. Because baculoviruses do not kill the target insect until one or more days after ingestion, one could easily get the impression that the viruses were not killing insects, if evaluations were made too soon after an application. Often a 24-hr post application count on the insect population is all that is used to determine efficacy. Obviously, a mistaken impression would be given through such an evaluation.

Seldom are ecological factors, such as larval cohort or development cycle, overlayed onto the phenological development of the crop being protected. This is seen as too costly in terms of time and complication. Most pest control advisors (consultants or extension specialists) will point out that these are the chief causes for not utilizing a baculovirus insecticide in more pest management programs.

An educational program to overcome these inadequacies would be a tremendous undertaking. Basic biological relationships are often not well understood. Moreover, the key personnel involved in making pest control decisions are so diverse in their training and interests that it is impossible to find common areas upon which agreements might arise. Obviously, no one group could undertake the task which would result in a more acceptable state of affairs in biorational pest management. However, certainly a stronger commitment by all involved would be beneficial.

V. THE INDUSTRIAL POSITION

Industrial commercialization is a powerful and necessary force in pest management. It has the ability to influence the impact of science on the consumer. All the scientific progress available today will count for little if it is not transferred from the developer to the user. And this reality impacts on the science itself. The capitalistic society prevailing today is driven by consumer desires and needs, and the integrated effort to satisfy these at a profit. It matters not whether the consumer is a farmer, consultant, or any other person likely to be influenced by the practice of pest management.

Commercialization is a complicated state of affairs, but can be simplified to reflect the goal of marketing a product. Successful marketing depends on:

- Having the appropriate *product* to meet a need
- Being positioned in the right *place*
- Giving the definitive *promotion*
- Being competitively *priced*

The four Ps, as they are called, are necessary for making a reasonable profit in any marketplace, whether one is dealing with baculovirus pesticides or widgets.[4] These components must be present collectively to ensure that a company will succeed in marketing. If one or more of the components is missing, any group charged with the responsibility of marketing a product will undoubtedly fail. The study of marketing is a science which is, in its own way, no less complicated than baculovirology. The history of marketing abounds with failures that resulted because the essential components of marketing were missing. To think that a baculovirus product used in pest management would not be subject to the same kinds of commercial demands would be unrealistic. So with the four Ps in mind, several questions arise:

A. Can a Baculovirus become an Appropriate Product?

The answer here is an unquestionable yes. To answer this question, which is not very demanding, all that is necessary is to have a candidate viral product that offers some degree of efficacy against a definable population of pests. The literature is replete with citations where some formulation of a product has offered some degree of mortality when used against a pest population (i.e., Elcar®, codling moth virus).

In a more realistic sense, a product consists of more than a set of physical ingredients. Although technical details are very important to those involved in the science of invertebrate pathology, they have very little bearing in a user's conception of the product. The user is content with something that has the capacity to give satisfaction (real or not!). This capacity can and does include the actual physical constituents of the product, plus use directions, efficacy warranties, application techniques, and perceived attractiveness. If a product simply meets the use for which it is intended, then technical details have little meaning. A farmer wants something that will "kill bugs". He does not really care about the intricate nature by which baculoviruses can accomplish this. So the criteria for a successful product remains straightforward, but can get quite complicated.

B. Can a Viral Product be Positioned in the Right Place?

Here again the answer is yes. Simply recognizing where there is a market niche into which the kind of control offered by a baculovirus product will fit is all that is required.

There is some uncertainty here in regard to whether this has been done properly in the past. For example, Elcar® was determined as a product which should be used for the management of *Heliothis* spp. in cotton. It was not until 1982 that a broader label for Elcar® was obtained, thereby expanding the place where the product would fit. This came after several years of unsuccessful experience in the highly competitive and confusing agrochemical marketing system that prevails in the U.S. cotton agroecosystem. The expectation of a sophisticated product like Elcar® succeeding in such a limited market was probably ill conceived at best. Naturally, the tremendous revenues generated by agrochemical companies in cotton pest control would be a source of interest to any group, and so it was with Elcar®. But to neglect the identity and character of the market and its needs can prove a serious and painful mistake.

C. Can a Product Based on a Baculovirus be *Promoted* Definitively?

Because product promotion does not depend on the actual composition of the product itself, it then becomes something that *must* be done to establish a position of a product in the marketplace. So a more important question is not whether a product *can* be promoted properly but whether it *will* be. This can depend entirely on a company's commitment to that product. The commitment is often a reflection of its sales capacity which immediately sets up a "Catch-22" situation. In a developing market into which *any* new product fits, the financial requirements, as well as other contributions to promotion, can be larger than potential revenue, at least in the short-term. In the longer-term, the contribution the product made back to the company must balance, and then surpass promotional costs, or it will be an unsuccessful endeavor. This must be understood because to underemphasize promotion will result in incomplete positioning, and ultimately result in a negative impact on the product.

Promotional questions can be truly frustrating, and most company managers "play it by ear". Few firms involved in agrichemical businesses use valid theories or research data for determining promotional commitments. Here, again, it is a difficult task to bridge the gap between the abstract analysis of the marketing theorist and the intuitive reactions of a practitioner, both of which hold keys to successful promotion management.

D. Can a Pesticide Based on a Baculovirus be Competitively *Priced*?

This question is perhaps the most difficult one to address. Not only because it is so involved, but because of its paramount importance. It can be answered only one way. It *must* be priced competitively or failure is certain. One way to determine pricing is to look at the perceived competition and its price. The problem with this is that the competition must be identified correctly, and there must also be some level of incentive contained in the price. In the past, a consensus on what the competition for baculovirus is has not been met. Bacterial pesticides, organochlorines, organophosphates, carbamates, nematodes, pyrethroids, or any number of other agents (or combinations of them) have been considered as competition. The true identity of any competition for baculovirus products is still poorly characterized. Whatever it is, it must be identified more clearly before we can carry on with the business of the commercialization of baculoviruses. For in any pricing of a product, there must be enough compensation provided to cover the investment in the product and a fair return in the form of profit. The lack of profit in industry yields the discontinuation of the product. And any product discontinued will be difficult, if not impossible, to replace in with a similar kind.

So the marketing function in pest management must be performed. It probably does not matter who, what, where, when, or how the intricacies are performed, just as long as they are not neglected and are realized as essential. Industry has the ability and responsibility to do this. Input from the scientific community is sought, appreciated, and used often. Cooperation can result in a viable industry which will be in place to continue the practical development of the basic discoveries of the bench. In this way, mutual benefit can and will be derived from baculoviruses as pesticides.

VI. REGULATORY CONSIDERATIONS

The registration of products like baculoviruses has been done in three distinct historical phases. The three phases mark times when procedures involved in the regulatory process differed considerably.

Pre-Elcar® phase — Elcar® (*Heliothis* NPV) was the first viral product registered for commercial use. The registration effort for Elcar® was a pioneering effort. It was very expensive and rigorously done. The reason for the high cost of registration was that the viral candidate was put through all the testing required for a chemical insecticide and additional specific infectivity tests. And, since both the EPA and the regulatory personnel representing industry were engaged in an effort new to all, there were numerous changes made during the testing and registration process. Many requirements were altered as the process proceeded. Industry was asked to present many kinds of data. In retrospect, many of the data were not particularly relevant. Nevertheless, the testing necessary to meet the data requirements was done. This was costly, and resulted in much time lost in registration. The EPA simply did not have the expertise to properly and expeditiously register the ''biorationals'', as they were called. After Elcar® was registered, the precedent was set, and the procedures and intricacies of the registration process were evaluated. They were made more realistic and, as a result, it became easier to register baculoviruses (see Chapter 8).

Post-Elcar® phase — In 1978, after Elcar® was fully registered, the U.S. Congress amended the federal laws which dealt with the registration of pesticides, the Federal Insecticide, Fungicide, and Rodenticide Act (FIFRA). Generally, the results brought about a relaxation of the requirements for superfluous data. And, since the procedures were more realistic with regard to basic biology, the process presented to baculoviruses (and other biologicals) an advantage over chemical insecticides.

But in 1978, the ''Cite-All'' regulations were promulgated by the EPA. This meant that the registrant was required, for the first time, to cite all the data in EPAs files which contained testing results for the active ingredient being considered. And the registrant had to offer to pay compensation to the owner of the data for the use of that data (since it would be used in support of a new registration). Also at that time, conditional registrations began to be issued. Conditional registrations were granted on the condition that all or some of the additional data requirements on the risks involved were waived. The result here was a drastic reduction in the time required for new registrations. The EPA also determined that efficacy data for a product should not affect whether a product received a label or not. The most obvious result of the relaxation of efficacy data was the reduction in the amount of field testing for an active ingredient. Theoretically, a product could be registered for a use for which it had never been tested. And several were!

After 1979 with "Cite-All", conditional registrations and efficacy data waivers, a rather odd state of affairs resulted. Those rulings meant that a registrant was required to cite data of others even if they themselves had a complete data package. And another registrant could utilize the data package of a previously registered active ingredient without the permission of the owner of the data. This meant that companies were using their competitor's data packages to register their own products. So similar registrations were easier to obtain. For example, a company choosing to register a baculovirus could easily register a product if it was the same virus as another which had already been registered. And, since the data had already been acepted by the agency, no reviews were considered necessary. Pertinent here is the fact that "living organisms" are not patented, and, therefore, not protected through normal commercial patent law. These "Me-Too" (as they are referred to by the EPA) registrations became commonplace. Several registrations for *Bacillus thuringiensis* (a successful bacterial insecticide) were obtained as "Me-Too" registrations.

The period from 1978 to 1982 was considered a time of regulatory bliss for biological insecticides. This period also saw an administration which, through EPA policy, believed that it was very important to bring new biologicals (baculoviruses included) to the marketplace as expeditiously as possible because of concern about the effects of chemical insecticides on the environment. Registration applications involving biologicals were even given priority reviews, and tolerance filing fees were waived since biologicals are generally exempt from tolerance anyway and their registrations was seen to be in the public interest.

Current Policy — The change in the administration that occurred in 1982 has brought sweeping changes to the regulatory process. This was expected since the EPA was put under the direction of its critics. The pesticide industry, in general, was not completely convinced that "Cite-All" was a valid policy, especially the large and powerful companies who had the ability to generate more complete data packages. The original owners of the data were given 10 years of exclusive use after which that data could be used without permission. Most data compensations, as determined by the EPA, were not seen as sufficient because of (1) the risk taken to develop the compound, (2) the time required to develop the data which would meet registration requirements, and (3) the time from submission to registration was not considered.

As a result of these procedures, a second registrant could gain quick and easy (not to mention less costly) access to the marketplace. Anyone with a complete data package did not see the value in "Cite-All" nor the need to comply with it. It was the dissatisfaction with the imbalance that led to litigation. There were two important rulings in court cases involving "Cite-All". These resulted in the declaration of a moratorium by EPA, and the issuance of pesticide registrations came to an abrupt halt. The court cases in question were the *National Agricultural Chemical Association* v. *U.S. Environmental Protection Agency*, No. 79-2063 (D.D.C., Jan. 20, 1983) ("NACA") and *Monsanto Co.* v. *Acting Administrator*, No. 79-366C(1) (E.D. Mo., May 9, 1983) ("Monsanto"). The "NACA" ruling required that the EPA discontinue *requiring* applicants to "Cite-All" and to develop an optional registration procedure. The "Monsanto" decision disallowed EPA from using data in support of registrations without the original submitter's permission. For a 90-day period, only a few *new* proprietary chemicals were registered. Only in cases where one applicant could prove conclusively that they were the only data submitter would registration be considered.

During this period, the EPA developed interim procedures for satisfying registration requirements under the "NACA" and "Monsanto" rulings. The procedures, which became effective on June 30, 1983, allowed two methods by which an applicant could achieve registration. Under the first option, the applicant was required to provide the agency with a listing of the data requirements for a given registration and a list of data references (or actual data) to fulfill the requirements. The major stipulation here was that the applicant could not cite another's data without the data owner's permission. Where data gaps appear to exist, the applicant was to inform all other data owners of the perceived gap. If *one* data owner had the data necessary to fill a gap in the data base in the EPAs files, all others were required to either obtain permission to use it, or produce the data themselves. Registration was not to be granted until the EPA agreed with the data requirements, the data references (or data) were supplied, the data gaps were resolved, and the data review, if any, was accomplished.

The second option was referred to as "Cite-All Data with Permission". Under this option, the applicant was required to cite-all, but must have the permission of *all* data submitters. One advantage of this option is that the applicant was not required to go through the elaborate indexing systems.

Under either option, biologicals were at a disadvantage. For the older biologicals, the registrant was required to comply with the elaborate indexing procedures which would no doubt be time comsuming and cause registration delays. But, more importantly, new biologicals, including baculoviruses currently under development, are adversely affected. Under the old regulations, the first registrant was given 10 years exclusive use for supporting data. Since others were required to "cite-all" and not allowed to use anyone else's data without permission, competitors were effectively kept out of the marketplace for 10 years. Under the present regulations, if an applicant is willing to develop the relatively inexpensive data package, a second registration is fairly easily attainable. With patent protection being very limited in the case of biologicals, the only remaining incentive to industry to develop new biologicals, under an "exclusive use" system, was eliminated.

The "Monsanto" decision has been appealed to the Supreme Court and the ramifications of the decision are yet to be realized. For the present, little incentive is available for an industry to make a commitment to the development of biologicals on strictly economic reasoning.[5] Despite this, there are indications that a new area of concentration is opening with specific emphasis on genetically engineered biological organisms.[6] This area is quite similar to conventional microbial control and much is common here.

VII. CONCLUSION

The use of baculoviruses and other microbial control agents has not progressed significantly over the last decade due largely to the development of effective pyrethroid insecticides. Pyrethroids are simply too attractive to the practitioners of pest control. Nonetheless, many scientists and pest management personnel understand the futility of total reliance on *any* chemical pesticide. Therefore, the need for the development of alternative strategies continues. Perhaps something as simple as pyrethroid resistance will result in a united effort to develop alternatives. Similarly "new" microbial control agents based on genetically engineered organisms could offer sufficient incentives to industry and practitioners to push the products through to field use.

REFERENCES

1. **Lewis, F. B., Rollison, W. D., and Yendol, W. B.,** Gypsy moth: research toward integrated pest management, *U.S. Dep. Agric. For. Serv. Tech. Bull.,* No. 1584, 1979.
2. **Cunningham, J. C. and Entwhistle, P. F.,** Control of sawflies by baculovirus, in *Microbial Control of Pests and Plant Diseases, 1970-1980,* Burges, H. D., Ed., Academic Press, New York, 1981, 380.
3. **Frisbie, R. E.,** personal communication, 1983.
4. **Ryan, W. T.,** Product as a variable, in A Guide To Marketing, Learning Systems Company, Homewood, Ill., 1981, 66.
5. **Brotherton, S.,** personal communication, 1983.
6. **Kirschbaum, J. B.,** Potential implication of genetic engineering and other biotechnologies to insect control, *Annu. Rev. Entomol.,* 30, 51, 1985.

Chapter 10

INSECT RESISTANCE TO BACULOVIRUSES

D. T. Briese

TABLE OF CONTENTS

I. INTRODUCTION

The efficient use of baculoviruses as agents for biological control necessitates a knowledge of interactions between virus and host-insect in the field. An essential requirement for this is an understanding of the dosage-mortality responses of an insect to virus infection.[1] Within a given insect population, those responses do not remain a constant property of the species but may be expected to change with time and circumstance. For practical purposes, therefore, the problem of insect resistance to baculoviruses concerns essentially the consequences of such changes in susceptibility to virus infection on the efficiency of control measures.

Changes in the level of susceptibility shown by populations are due to changes in the frequencies of individuals showing particular responses to virus infection. Laboratory bioassays have demonstrated that there is considerable variability in the response of individual insects to given dosages of virus.[2] In a recent review, Briese and Podgwaite[3] described several factors which act to determine the particular response of an insect, and showed that these factors could be grouped into three categories: developmental, environmental, and genetic. Developmental factors concern changes in the level of resistance shown by many insect species as they grow and progress through successive stages of maturation. Environmental factors involve those external influences that might modify the expression of resistance through the actual defense mechanisms of the insect (as distinct from influences that act through selection of controlling gene systems). Finally, genetic factors concern changes in the frequency of genes controlling the defense mechanisms of insects to virus infection.

This chapter examines how these factors influence the expression of resistance to baculoviruses. A feature common to all the factors is that they operate through the physiological and biochemical defense mechanisms of the insect; consequently, those mechanisms will be examined in some detail. With regard to practical control procedures, developmental and environmental factors affect mainly the short-term expression of resistance, whereas genetic factors may lead to longer-term changes in levels of susceptibility. Thus, an understanding of the effects of and interplay between these factors under field conditions is a prerequisite for the systematic usage of baculoviruses as pest control agents.

II. DEVELOPMENTAL FACTORS AFFECTING RESISTANCE

The relationships between the age of an insect and its response to virus infection have now been documented in many instances, based mainly on comparisons between larvae of different instars. The majority of cases, including all instances in which the amount of virus ingested by each larva was known, have demonstrated an increase in resistance with larval age (Table 1). The few exceptions[4,5] involved the exposure of larvae to different concentrations of virus, the actual amount ingested remaining unknown. Therefore, although the lethal concentration may have remained similar, the older larvae would have consumed more food and hence ingested more virus particles.[6,7] Appropriately analyzed, these cases would probably also show an age-related increase in resistance.

Although it is difficult to compare the data in Table 1, due to the differences in the age spans of the insects examined and in bioassay techniques, it is still possible to see that there is a wide variability in age-related resistance between species — the rate at which resistance to particular baculoviruses develops being several orders of magnitude greater in some species than in others.

Further, this increase in resistance has been found to be directly proportional to the weight of the larvae in the case of *Pieris rapae, P. brassicae,*[1] and *Mamestra brassicae.*[18] Data from several other studies listed in Table 1 were analyzed and a linear regression was found between log LD_{50} values and log weight. In all cases, involving both nuclear polyhedrosis virus (NPV) and granulosis virus (GV), body weight accounted for most of the variation in

Table 1

INSECTS WHICH SHOW INCREASING RESISTANCE TO BACULOVIRUS INFECTION WITH AGE (SPECIES LISTED APPROXIMATELY IN ORDER OF LEAST TO GREATEST DIFFERENCE IN RESPONSE[a])

Species	Virus	Instars tested	Method of bioassay[b]	LD50 or LC50		Relative susceptibility[c]	Ref.
				Most susceptible stage	Most resistant stage		
Pseudoplusia includens	NPV	1—6	LC50	2.1	74.5	35	8
Ceramica picta	NPV	1 & 3	LC50	63	3,300	52	9
Adoxophyes orana	NPV	1—5	LC50	ca. 1,200	ca. 100,000	ca. 80	10
Gilpinia hercyniae	NPV	1—4	LD50	20	800	40	11
Anticarsia gemmetalis	NPV	2—5	LD50	103	4,100	40	12
Heliothis zea	NPV	3 day—8 day	LD50	9.5	2,294	240	13
Mamestra configurata	NPV	1—6	LD50	18	14,130	780	14
Operophthera brumata	NPV	1—5	LD50	2.5	1,780	710	15
Hyphantria cunea	NPV	2, 4, 5	LD50	479	188,000	390	16
Malacosoma neustria	NPV	3 & 4	LD50	1,405	12,320	9	17
Mamestra brassicae	NPV	1—5	LD50	7	238,370	34,000	18
Lymantria dispar	NPV	1—4	LD50	265	2,540,000	9,600	7
Malacosoma disstria	NPV	1, 3, 4	LD50	50	3,400,000	68,000	6
Laspeyresia pomonella	GV	1 & 3	LC50	17	175	10	19
Adoxophyes orana	GV	1, 3, 4	LC50	ca. 3,000	ca. 500,000	ca. 160	10
L. pomonella	GV	1 & 5	LD50	5	49	10	20
Pieris rapae	GV	1—4	LD50	5	662	130	1
Lacanobia oleraceae	GV	2—5	LD50	19,900	3,980,000	200	21
H. cunea	GV	2, 4, 5	LD50	70,600	187,000,000	2,600	16
P. brassicae	GV	1—4	LD50	66	23,000,000	350,000	1

[a] Order is approximate because different sets of instars have been compared for each species.

[b] LD50 indicates actual dosage given to each larva is known or was calculated, and LC50 indicates a concentration of virus was given and actual dosage is not known. The results of these two treatments are not comparable.

[c] Ratio of highest to least resistance stage, rounded to two significant figures.

the regression (Table 2). The regression slopes differed markedly between species, reflecting the interspecific variability in age-related resistance suggested by Table 1.

This relationship does not appear to hold over the full span of larval development, however, for Evans[18] reported a deviation from the linear regression, due to a sharp increase in the resistance to GV of midfifth-instar and older larvae of *M. brassicae*. Similar results have also been recorded for late-instar larvae of *Operophthera brumata*,[15] *M. configurata*,[14] and *Heliothis armigera*[22] exposed to their respective NPVs. Evans[18] analyzed the relationship between larval development time and LT_{50} (time required for 50% virus-induced mortality), and found that the point at which full "maturation resistance" developed should occur when the time needed to produce a full cycle of virus infection exceeded the time remaining for a larva to reach pupation (Figure 1). As weight is also related to development time, full maturation resistance should occur at approximately 30% of maximum larval weight,[18] a useful indicator for estimating its occurrence in the field. In a subsequent investigation of maturation resistance in *M. brassicae*, Evans[23] found that these highly resistant late-instar larvae could still be infected when given very high dosages. Two points were identified at which resistance deviated sharply from the previous linear relationship with weight. These corresponded to the times when larvae were just entering their final instar and just prior to prepupation. The reasons for these sharp increases to very high levels of resistance in late-instar larvae could stem from shifts in the balance of juvenile and molting hormones that occur at these stages.[23] Moreover, in Lepidoptera, NPV infections are known to be interrupted during periods of rapid cell proliferation at metamorphosis, and the embryonic regenerative cells only become susceptible to infection once they have begun to differentiate, essentially restarting the infection cycle.[24]

Changes in susceptibility to infection also occur during larval molts. Watanabe[25] found that synthesis of polyhedral protein in cytoplasmic polyhedrosis virus (CPV)-infected cells of *Bombyx mori* was suppressed during the molt, although synthesis of viral ribonucleic acid (RNA) was unaffected. Moreover, Kobayashi et al.[26] found changes in the susceptibility to NPV within instars of *B. mori*. Susceptibility to peroral infection was greatest immediately before and immediately after a molt, but decreased in between. Within-instar differences in susceptibility to a GV have also been observed for *P. brassicae*.[27] Thus, the true pattern of resistance in developing larvae probably follows that shown in Figure 2, and the linear regression represents an averaging of the changes taking place.

This direct relationship with weight over the early stages of development appears to be a general property of age-related resistance, although Payne[28] could see no *a priori* reason for it, since an infectious agent should be able to produce an infection regardless of the size of the host. The problem then is whether the effect is merely a "dilution" of the dosage required to initiate a lethal infection by increasing weight, or whether weight correlates with changes in some underlying defense mechanism. An analysis of the regressions of LD_{50}/mg on log weight for several species (Table 3) suggests that both effects might be involved.

In four of the seven species infected by an NPV and in one of the three infected by a GV, regressions slopes were no longer significantly different from zero. This indicates that the dosage per unit body weight remained relatively constant (Figure 3). In the remaining five cases, the regressions still produced significant positive slopes (Table 2, Figure 3) indicating that viral resistance was increasing more rapidly than larval weight. Thus, these nine species responded as though, for the one group *(Anticarsia gemmetalis, M. configurata, O. brumata, H. zea, and P. rapae)*, age-related increases in resistance were due primarily to a nonspecific "dilution" effect correlated with increasing weight, whereas in the other group *(M. brassicae, Lymantria dispar, Hyphantria cunea, and P. brassicae)* this effect was supplemented by the development of specific defense mechanisms against viral infection.

The weight gain over the period for which this relationship holds (i.e., before full "maturation resistance" occurs) ranges up to several hundred-fold depending on the species.

Table 2
PARAMETERS FOR REGRESSION OF LD$_{50}$ ON BODY WEIGHT OF LARVAE EXPOSED TO VIRUS AT DIFFERENT AGES[a]

Species — virus interaction	Slope	Intercept	Variance accounted for by regression (%)	Degrees of freedom	Significance of regression	Ref.
Anticarsia gemmetalis — NPV	0.70	1.98	92.3	3	$p < 0.05$	12
Mamestra configurata — NPV	0.76	2.05	86.1	5	$p < 0.05$	14
Operophthera brumata — NPV	1.09	1.78	99.8	4	$p < 0.01$	15
Heliothis zea — NPV	1.15	1.08	97.5	5	$p < 0.01$	13
M. brassicae — NPV	1.56	2.40	99.8	4	$p < 0.01$	18
Hyphantria cunea — NPV	2.02	3.03	98.0	2	$p < 0.05$	16
Lymantria dispar — NPV	2.16	1.35	99.7	3	$p < 0.01$	7
Pieris rapae — GV	1.15	1.71	98.8	3	$p < 0.01$	1
P. brassicae — GV	2.63	3.77	99.0	3	$p < 0.01$	1
H. cunea — GV	2.68	5.38	95.7	2	$p < 0.05$	16

[a] Based on data sets in which the amount of virus ingested by each larva was known.

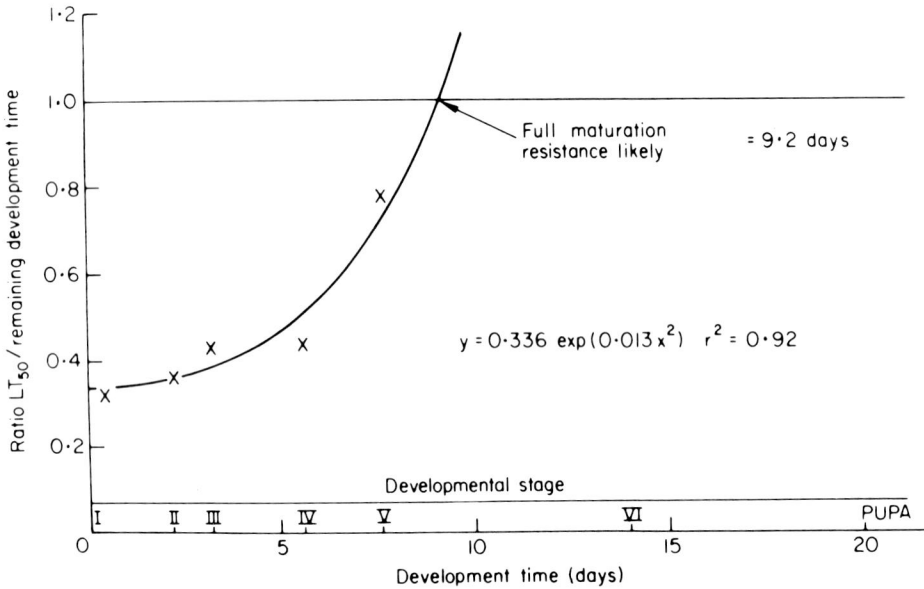

FIGURE 1.   Relationship between *Mamestra brassicae* larval development time and a parameter (y) based on the ratio of instar LT_{50} to remaining time to pupation. Maturation resistance is expected when the ratio (y) exceeds 1. (From Evans, H. F., *J. Invertebr. Pathol.*, 37, 101, 1981. With permission.)

Therefore, other insects, in which weights are unknown, may also be categorized, depending on whether the increase in resistance shown by them is of this order of magnitude or greater. Thus, from Table 1, *Laspeyresia pomonella* and *Gilpinia hercyniae* would probably belong to the group showing only a "dilution" response, whereas *Malacosoma disstria* could be placed in the group considered to possess additional defenses against viral attack. Interestingly, this latter group contains two species *(Lymantria dispar* and *P. brassicae)* in which between-population differences in response to a virus have been found.

This nonspecific "dilution" effect could involve passive factors, such as a reduction in surface-volume ratio of the midgut with larval growth. Watanabe[29] found that larvae of *B. mori* infected with CPV produced highly pathogenic feces for a short period immediately after peroral inoculation. No polyhedra were present, suggesting that uncoated virus particles had passed directly through the insect. A reduced surface-volume ratio in older larvae would increase the probability of a virus particle passing through the midgut without attaching to susceptible epithelial cells. Other weight-related physiological changes might also be involved, such as an increase in gut pH with age.[30,31] The underlying specific defense mechanisms against viral attack are poorly understood, although several potential mechanisms have been identified.[32] These will be discussed in greater detail in Section V.

III. ENVIRONMENTAL FACTORS AFFECTING RESISTANCE

Many environmental factors can modify the relationship between a virus and its host insect. Some may act directly on the virus and affect its prevalence in the field (e.g., ground cover, soil type, cultural practices).[33] Others may act either directly on the insect to alter its response to viral attack or affect the subsequent insect-virus interaction. This section will consider those factors which appear to change the levels of susceptibility to a virus shown by certain insects (Table 4), rather than those which may affect the actual dosage of virus received by the insect.

Early evidence of the role played by such factors arose from commercial silkworm-rearing

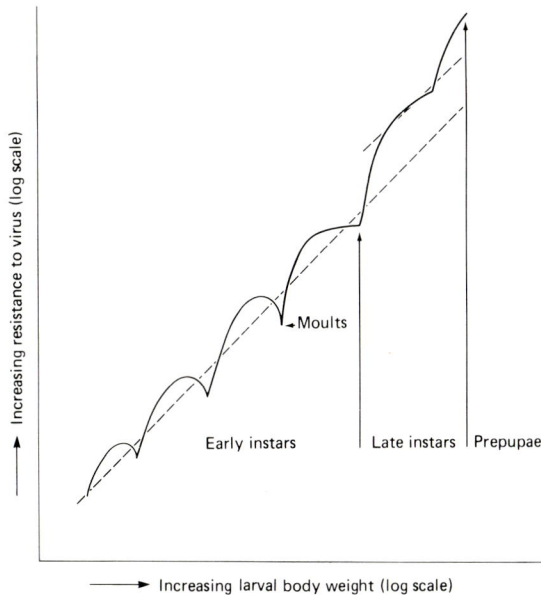

FIGURE 2.   Pattern of increasing resistance to virus infection during the development of insect larvae, showing increased susceptibility at early larval moults (after Kobayashi[26]) and the occurrence of maturation resistance during late instars (after Evans[23]). Dashed lines indicate linear relationships of dosage to weight.

operations, where it was observed that the incidence of viral infections, including NPV, was greater in *B. mori* larvae in autumn than in spring.[25,34] These changes were associated with the quality of the mulberry leaves upon which the larvae fed.[25] More recent experiments have shown that specific nutritional components may have an effect, because changes in the sucrose, protein, and cellulose levels of artificial diets all affected the susceptibility of *B. mori* to viral infection.[39] Protein levels appeared to directly affect antiviral activity and protease activity of larval digestive fluids, whereas cellulose merely acted as a feeding stimulant.[40] David and Taylor[37] also found that a sucrose-deficient diet caused increased susceptibility in *P. brassicae* to peroral infection with GV. In this case it appeared that the absence of sucrose led directly to greater uptake of the virus by midgut epithelial cells, since the effect was only observed when sucrose was withheld at the time of inoculation, not before or after. A several-fold increase in susceptibility has also been noted when larvae of *B. mori* were shifted from natural to synthetic foods,[44] this probably due to the absence of substances essential to the synthesis of antiviral agents in the digestive fluid.[46] Moreover, stress caused by switching to unfamiliar natural food sources may also lower resistance in other Lepidoptera.[36]

Physical factors may also be important (Table 2), since short exposures to extremes of temperature can increase the susceptibility of larvae to viral infection.[41] By contrast, though, very high temperature treatments have been used to raise the resistance levels of late-instar larvae of *B. mori*[42] to infection by infectious flacherie virus (IFV), and of *Pseudaletia unipuncta*[43] to infection by NPV. In the case of *B. mori*, high temperatures stimulated a discharge of infected midgut cells, followed by regeneration, whereas in the case of *P. unipuncta* the production of polyhedral protein was inhibited. It should be noted though, that the temperatures used in such treatments were generally beyond the normal tolerance range of the insects concerned.[43] Such effects would, therefore, not be significant in the field.

Table 3

PARAMETERS FOR REGRESSION OF LD_{50}/mg ON BODY WEIGHT OF LARVAE EXPOSED TO VIRUS AT DIFFERENT AGES

Species — virus interaction	Slope	Intercept	Variance accounted for by regression (%)	Degrees of freedom	Significance of regression	Source of raw data
Anticarsia gemmetalis — NPV	−0.29	1.98	69.1	3	N.S.	12
Mamestra configurata — NPV	−0.24	2.05	36.5	5	N.S.	14
Operophthera brumata — NPV	0.07	1.82	64.8	4	N.S.	15
Heliothis zea — NPV	0.15	1.08	38.6	5	N.S.	13
M. brassicae — NPV	0.56	2.40	100.0	4	$p < 0.01$	18
Hyphantria cunea — NPV	1.02	3.03	93.3	2	$p < 0.05$	16
Lymantria dispar — NPV	1.17	1.34	99.0	3	$p < 0.01$	7
Pieris rapae — GV	0.15	1.72	38.8	3	N.S.	1
P. brassicae — GV	1.63	3.79	99.7	3	$p < 0.01$	1
H. cunea — GV	1.68	5.38	90.0	2	$p < 0.05$	16

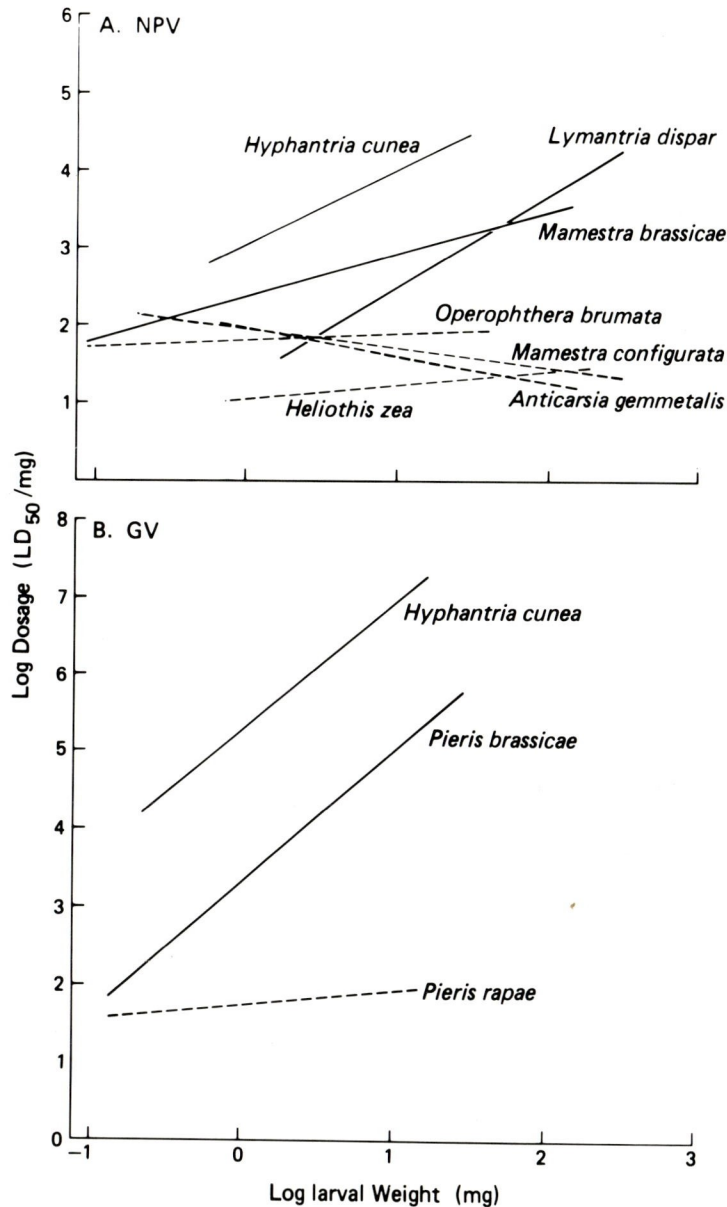

FIGURE 3.   Relationship between viral dosage/unit weight and larval weight for nine species of Lepidoptera infected perorally with (A) NPV and (B) GV (dashed regression lines have slopes which are not significant, based on Table 3).

Light, too, can apparently directly affect the susceptibility of insects to viral infection. *B. mori* larvae reared in constant darkness were found to be approximately ten times more susceptible to NPV than those reared in light.[44] Two factors appear to be involved — a reduction in the ratio of susceptible columnar to nonsusceptible goblet cells in the midgut epithelium[44] and an inability to synthesize the antiviral red fluorescent protein in the absence of light.[45]

Finally, Watanabe[46] found that sublethal doses of DDT and organophosphate insecticides can increase the susceptibility of *B. mori* to peroral infection with NPV. Reduced dosages

Table 4
ENVIRONMENTAL FACTORS WHICH MAY INFLUENCE THE LEVEL OF SUSCEPTIBILITY SHOWN BY INSECTS TO VIRAL INFECTION

Insect — virus interaction	Environmental factor	Specific influence	Effect	Possible underlying mechanism	Ref.
Bombyx mori — NPV	Undefined	Seasonal changes	More susceptible in autumn than in spring	Not known	34
B. mori — NPV	Nutritional	Synthetic vs. natural food	More susceptible on synthetic food	Possibly due to absence of substances required to synthesize antiviral agents in digestive fluid	35
Pieris brassicae — GV	Nutritional	Unfamiliar natural food	More susceptible than when reared on "usual" food	Not known	36
P. brassicae — GV	Nutritional	Absence of sucrose at time of infection	Increased susceptibility	Absence of sucrose apparently results in greater penetration of virus into midgut epithelium and subsequent persistence	37
P. brassicae — GV	Nutritional	Casein deficient synthetic diet	Increased susceptibility	Not known	38
B. mori — NPV	Nutritional	Low sucrose, low protein or high cellulose synthetic diets	Increased susceptibility	Protein levels directly affect antiviral activity and protease activity of digestive fluid	39, 40
B. mori — CPV[a]	Temperature stress	Exposure to extremes	Increased susceptibility	Not known	41
B. mori — IFV[a]	Temperature stress	Exposure to very high temperature after infection	Reduced susceptibility	Insects discharged infected cells; regenerated cells were not susceptible to infection	42

					Ref.
Pseudaletia unipuncta — NPV	Temperature stress	Exposure to high temperature after infection	Reduced susceptibility	Possibly affects adsorption/penetration of virus into midgut cells and inhibits subsequent replication of polyhedra	43
B. mori — NPV	Light	Reared in constant dark or constant light	More susceptible when reared in constant cark	Reduction in proportion of susceptible columnar cells observed in gut of larvae reared in dark	44
B. mori — NPV	Light	Reared in constant darkness	Increased susceptibility	Light required for synthesis of antiviral red fluorescent protein in larval digestive fluids	45
B. mori — NPV	Chemical stress	Sublethal dosages of insecticides	Increased susceptibility	Not known	46

a Not baculoviruses; CPV = polyhedrosis virus; IFV = infectious flacherie virus.

of chemical insecticides and baculoviruses used in combination have produced additive effects in several field trials.[47]

It appears that there are many nutritional, physical, or chemical stressors, which may influence the level of susceptibility shown by insects at a particular time. Generally, it appears that insects exposed to abnormal conditions are more susceptible to infection by baculoviruses. Thus, although the changes in susceptibility are relatively small when compared to changes in age-related resistance (Section II) or to changes in resistance levels between populations of a species (Section IV), the environmental factors causing them must be considered, or controlled, in any analysis of insect resistance to baculoviruses.

IV. GENETIC FACTORS AFFECTING RESISTANCE

A. Population Differences in Resistance and their Genetic Bases

Most baculoviruses being studied as potential biological control agents are endemic and, therefore, would have coevolved with their host. Therefore, from time to time genes conferring greater resistance to the virus may have arisen and become fixed in discrete populations of the host insect, such as those genes known to confer resistance to chemical insecticides.[48] Numerous studies have reported differences in response to peroral infection with a baculovirus between geographically distinct populations of a species.[2,3,32] Furthermore, in the case of species such as *L. dispar* and *Zeiraphera diniana*, which are known to undergo periodic fluctuations in density, such changes in response may occur within a population at different phases of the population cycle.[3,49,50] A gradual, but significant increase in LD_{50} was observed over 11 generations in a laboratory population of *L. dispar* exposed to NPV.[51] Unfortunately, though, much of the earlier work was based either on single-dosage comparisons or incomplete dosage-mortality analyses (see Briese[32] for examples), leaving interpretation of the data open to speculation. In several more recent studies, the dosage-mortality data has been adequately quantified, and these indicate differences in LD_{50} values of up to 800-fold between populations (Table 5). Moreover, comparisons of the regression slopes (Table 5) indicate that the differences are not merely due to changes in variability of the response, but to a real shift in the overall level of resistance shown by each population. In fact, in three of the cases, *Spodoptera frugiperda*,[54] *Epiphyas postvittana*,[57] and *Phthorimaea operculella*,[61] the resistance shown by particular populations has been shown to have a genetic basis.

These studies, plus other investigations involving different groups of viruses indicate that shifts in resistance may be controlled either by complex genetic mechanisms or, in some instances, by single autosomal genes (Table 6). In the latter case, both dominant and recessive alleles determining resistance have been reported. In two cases, involving small nonoccluded viruses infecting *B. mori*, the genes involved conferred total resistance to infection (Table 6). However, these viruses follow different invasion pathways than the baculoviruses, and therefore different defense mechanisms are probably involved.[3] In baculoviruses and cytoplasmic polyhedrosis viruses, total resistance has not been observed. Rather, where single genes have been identified, they appear to control relatively large shifts in susceptibility to the virus (Table 6). However, when large samples have been examined, LD_{50} values of individual populations have been found to be widely dispersed rather than grouped at a few extremes:[34,60] this cannot be explained by single gene mechanisms. It seems likely that other background genes contribute to this additional variation either by modifying the expression of the major gene or by acting through some other defense mechanisms which affect susceptibility to a lesser degree (see Section V). A marked F_1 heterosis in resistance to viral infection has been observed in hybrid populations of *B. mori* exposed to both CPV[62] and NPV,[34] suggesting that the genetic background to major resistance genes also affects the phenotypic expression of resistance to a virus, as has been found in the case of chemical insecticides.[67]

Table 5
INSECT SPECIES IN WHICH POPULATIONS HAVE BEEN EXAMINED FOR DIFFERENCES IN SUSCEPTIBILITY TO BACULOVIRUSES

Species	Virus	No. of populations	Origin of populations	Instar tested	Range of resistance factors	Range of slopes	Ref.
Bombyx mori	NPV	14	Commercial culture	4th	2—630	Not given	52
Lymantria dispar	NPV	4	Laboratory	2nd	1—129	Not given	53
Spodoptera frugiperda	NPV	2	Laboratory	1st	3—5	Not given[a]	54
Phrygandia californica	NPV	5	Field	5th	<2	0.77—1.21	55
S. littoralis	NPV	2	1 Laboratory 1 Field	5th	12	0.84—1.09	56
Epiphyas postvittana	NPV	3	2 Laboratory 1 Field	3rd	50—160	1.77—2.56	57
Pieris rapae	GV	2	Field	4th	No difference		55
P. brassicae	GV	4	Laboratory 1 Field	4th	2—12	1.25—1.59	30
P. brassicae	GV	2	1 Laboratory	3rd	No difference		1
Plodia interpunctella	GV	2	Laboratory	1st	7	1—1.2	58
P. interpunctella	GV	2	Laboratory	3rd	ca. 800	Not given[b]	59
Phthorimaea opercullella	GV	6	Field	3rd	1—10	0.60—0.92	60
P. opercullella	GV	6	Field	3rd	1—11	0.69—1.03	60
P. opercullella	GV	4	1 Laboratory 3 Field	3rd	4—30	0.70—0.97	61

[a] Actual values not given, but genetic basis for difference described.
[b] Actual values not given, but slope similar for both susceptible and resistant strains.

Table 6
GENETIC MECHANISMS CONTROLLING INTRASPECIFIC DIFFERENCES IN RESISTANCE TO A VIRUS

Species	Virus	Difference in susceptibility	Genetic mechanism	Ref.
Bombyx mori	CPV[a]	>400×	Dominant autosomal gene	63
B. mori	IFV[a]	Nonsusceptible	Recessive autosomal gene	64
B. mori	DNV[a]	Nonsusceptible	Recessive autosomal gene	65, 66
Spodoptera frugiperda	NPV	5×	One or more genes lacking dominance	54
Epiphyas postvittana	NPV	160×	Polygenic	57
Phthorimaea operculella	GVV	30×	Dominant autosomal gene	61

[a] Not baculoviruses.

B. Selection for Resistance

Studies on the selection of resistance genes within a population have produced somewhat ambivalent results (Table 7). There is considerable indirect evidence, however, that such selection has occurred unintentionally in the laboratory, for several of the examples of increased resistance involve inbred laboratory populations that had been exposed to outbreaks of their associated viruses: *Pieris brassicae*[68] and *Phthorimaea operculella*[60] to GV, and *S. frugiperda*[56] and *E. postvittana*[58] to NPV. In the case of *P. operculella,* selection experiments by Briese and Mende[69] support this suggested course of events. They obtained a 140-fold increase in LD_{50}, without a significant change in regression slope, after six generations of selection in a previously susceptible population. On the other hand, selection in the population of *P. operculella,* known to possess a gene conferring resistance, produced only a slight additional increase in LD_{50}, but a significant increase in slope values. Briese and Mende interpreted this to mean that the former population included a small proportion of individuals with the resistance gene, the frequency of which increased rapidly under selection pressure from the virus, whereas the latter population was already homogeneous for this gene, and therefore the changes observed were due mainly to an elimination of more susceptible individuals and a subsequent reduction in variability of response to the virus. As there was no heterosis between the two populations, it would appear that the same mechanism is involved in both cases.[69]

This is not the only way in which the expression of resistance may be altered, for selection may also act on the arrangement of background genes. Watanabe[71] was able to produce a 10- to 20-fold increase in the resistance to CPV shown by two selected populations of *B. mori*. Interestingly, hybrid crosses of these populations did not show heterosis for resistance, whereas hybrid crosses of the control lines from which they were derived did. Watanabe[71] interpreted this to mean that selection had acted on different polygenic backgrounds in the respective populations to produce in them genetic structures which were more homogeneous for resistance and hence did not exhibit strong heterosis.

Several attempts to select for resistance have produced no changes, even after many generations (Table 7), but Briese[32] has suggested there may have been no gene initially present upon which selection could act. In all successful attempts a shift in response occurred within relatively few generations. The rapidity with which changes in frequency of phenotypes can take place under laboratory selection by a virus is illustrated by Ignoffo and Garcia's[75] experimental manipulation of a mixed population of *Heliothis zea* and *H. virescens*. *H. virescens* is five times more tolerant of *Autographa* NPV than *H. zea,* and after three consecutive generations of exposure to the virus the ratio of a mixed population of the two species changed from 1:1 to 366:1.

Table 7
CASES IN WHICH ATTEMPTS HAVE BEEN MADE TO SELECT FOR INCREASED RESISTANCE TO A VIRUS IN LABORATORY POPULATIONS OF LEPIDOPTERA

Species	Virus	Instar selected	No. of generations selected	Selection pressure (% mortality)	Resistance factor	Ref.
Bombyx mori	IFV[a]	3rd	5	Not stated	11—28	70
B. mori	CPV[a]	5th	8	530—92%	14—20	71
Heliothis zea	NPV	3 or 4 days	25	50—70%	No difference	72
H. armigera	NPV	3rd	22	40—80%	No difference	73
H. armigera	GV	3rd	22	15—90%	No difference	73
Cydia pomonella	GV	4th	7	59—96%	No difference	74
C. pomonella	GV	1st	7	51—90%	7—8	74
Phthorimaea operculella	GV	3rd	6	34—71%	140	69

[a] Not baculoviruses; IFV = infectious flacherie virus, CPV = cytoplasmic polyhedrosis virus.

The real problem that selection for resistance poses, though, is not whether it can be produced in inbred laboratory populations, but whether it will occur in the field to such an extent as to compromise the use of baculoviruses in control programs. To date, none of the programs involving the use of such viruses in large-scale trials (e.g., sawflies,[76,77] *Orgyia pseudotsugata*,[76] *Heliothis* spp.,[78] *L. dispar*,[79] *S. littoralis*,[80] *Laspeyresia pomonella*,[81] *Oryctes rhinoceros*[82]) have encountered problems due to selection for increased resistance. However, very few studies have actually addressed the problem of resistance in the field.

In some extreme cases, such as the densonucleosis virus (DNV) infecting *B. mori* in which total resistance exists and where insect populations are confined to discrete localities, it is relatively easy to establish a pattern between the epizootiology of the virus and the resistance of particular populations.[83] In the case of baculoviruses, however, observed differences in the levels of resistance observed are not so well-defined. In the mid1950s, Martignoni[84] reported an increase in LD_{50} and in the regression slope for *Z. diniana* infected with GV, following an outbreak of the virus which had caused increasingly high mortality in the field during the previous year. While the data obtained was not sufficient to permit a fuller interpretation, it is possible that resistant individuals were present in the population, for Benz[50] subsequently found that a population of *Z. diniana* collected from the same area showed greater heterogeneity in virus susceptibility than one collected from farther afield. Benz considered that this could have occurred because the former population contained both susceptible and resistant forms selected for during previous GV outbreaks.

Furthermore, it has been hypothesized that patterns in the response of populations of *Phryganidia californica* to NPV,[55] and of *P. operculella* to GV,[60] reflected the recent history of exposure of the insect to its virus. During a virus outbreak, selection for resistant phenotypes would occur. Meanwhile, in the period between epizootics, there would again be an increase in the frequency of susceptible phenotypes which, it is suggested, are then at a selective advantage. Although data indicating that susceptible individuals might be intrinsically more fit have not yet been obtained, there is strong evidence for this where resistance to chemical insecticides is concerned.[85,86]

Mitchell and Fuxa[87] are currently investigating the effects of exposure to NPV on field populations of *S. frugiperda*. Experimental data are still being collected and definite conclusions cannot yet be drawn, but their findings indicate a trend toward reduced susceptibility and increased heterogeneity in response after exposure to the virus.

Finally, mention should be made of a closely related aspect: the induction of latent virus infections. Aizawa et al.[88] were able to increase resistance to the induction of NPV through physical and chemical stressors in *B. mori* by selection over 11 generations. Interestingly, they found no correlation between resistance to virus induction and resistance to peroral infection. This is in contrast to Huber's[74] later findings that a strain of *L. pomonella*, selected for resistance to peroral infection with GV, was also less prone to "spontaneous" outbreaks of latent GV. This is not necessarily contradictory because a latent virus infection does not have to overcome an insect's defenses against invasion by the pathogen. If the major defense mechanisms were directed against invasion, they would not be correlated with resistance to induction. However, if they were directed against subsequent development of the viral infection, they possibly could be correlated. More will be said of such insect defense mechanisms in the following section.

V. DEFENSE MECHANISMS CONTRIBUTING TO RESISTANCE

A. The Physiological and Biochemical Nature of Defense Mechanisms

Before considering the defense mechanisms of an insect against viral attack, it is useful to provide a synopsis of the pathway of invasion and infection followed by baculoviruses (for greater detail see Granados[89]). Although viruses may also be acquired transovarially, transovum, through spiracles or through parasitism, the most common route of entry is per os. In larvae of the order Lepidoptera, following ingestion the occlusion bodies are dissolved, releasing virions into the midgut lumen. Free virions then attach to, and enter, the columnar cells of the midgut epithelium. The other major midgut cell type, goblet cells, appears to be resistant to baculovirus infection,[89] although they provide the site of entry for some of the nonoccluded groups of virus.[90] Having entered the cell, virions uncoat either at the nuclear pore (GV) or within the nucleus (NPV) before replication commences in the nucleus of the columnar cells. Replicated nucleocapsids may acquire an envelope and become occluded or, more commonly, bud through the cell membrane as nonoccluded virus into the hemocoel where they proceed to cause the systemic infection of the host. Infection of secondary tissues results in the production of further nonoccluded virus and the formation of occluded virus which is released into the environment through cell lysis and larval death, eventually to continue the infection cycle in another host.

This process can essentially be divided into three phases: preinvasion, including the period during which the baculovirus is present in the gut of the host insect; invasion, involving the attachment, entry, and establishment of infection in the midgut columnar cells; and postinvasion, when virus particles enter the hemocoel and set up secondary infections. There are a number of potential defense mechanisms which can act at each phase of the infection pathway (Table 8). The examples given in Table 8 concern defenses against other virus groups as well as baculoviruses. This is because most mechanisms do not appear to be highly virus-specific, but rather are general defenses against certain pathways of invasion. This is indicated by the significant positive correlation between the resistance shown by different strains of *B. mori* to NPV and CPV,[52] both of which invade the midgut columnar cells.

From the available evidence, it seems that the major defense mechanisms operate during the invasion phase. While considerable antiviral activity has been detected in the digestive juices of *B. mori*, including the isolation of specific agents (see Table 8), this could not be correlated to the resistance shown by various strains of the insect to peroral infection, and therefore seems to be only a minor factor.[62] Of the other potential preinvasion defense mechanisms, dissolution of the protein occlusion bodies is a relatively nonspecific process and hence is probably not very important,[89] and the role of the peritrophic membrane is still somewhat unclear,[89] although it is considered capable of preventing virus particles from reaching the midgut epithelium.[25,93]

The invasion phase can be experimentally bypassed by injecting virus particles directly into the hemocoel. This has resulted in a reduction in LD_{50} values of up to 10,000-fold in larvae of *B. mori* injected with CPV, compared to those given the virus perorally.[25] Similarly, the LD_{50} of *Galleria melonella* given an intrahemocoelic injection of NPV was some 40 to 50 times less than that of larvae fed the virus.[105] Likewise, final-instar larvae of *Manduca sexta*, which were immune to very high peroral dosages of *G. melonella* NPV, could be infected by the injection of nonoccluded virus.[106] This suggests that resistance to peroral infection depends more on the prevention of successful invasion of the susceptible midgut cells than on prevention of postinvasion multiplication and spread of the virus.

One important defense mechanism against invasion involves the discharge of infected midgut cells into the gut lumen at each larval molt and their replacement by newly formed cells.[95-97] In the case of *B. mori* infected with CPV, the regenerated columnar cells became reinfected,[96] whereas, in *Hyphantria cunea* larvae infected with the same *B. mori* CPV, the regenerated cells were immune to subsequent infection.[97] A similar discharge and regeneration occurs in the goblet cells of resistant strains of *B. mori* infected with the nonoccluded IFV, but not in susceptible strains.[95] In the case of the nonoccluded DNV, strains of *B. mori* showing genetically determined nonsusceptibility to the virus[66] cannot be infected by either peroral or intrahemocoelic inoculation. This suggests that host-cell receptors or host-cell enzymes involved in viral replication may be blocked in this instance.[91]

Table 8
MECHANISMS BY WHICH INSECTS MAY DEFEND AGAINST VIRAL ATTACK

Potential defense mechanism	Evidence for such a mechanism	Type of variability reported for mechanism
Preinvasion Phase of Viral Infection Pathway		
Prevention of dissolution of occlusion bodies	Probably not important, as dissolution appears to be a nonspecific process[89]	None reported
Degradation, inactivation, or agglutination of released virions in the gut lumen	Increasing pH in midgut could lead to greater inactivation of GV in *Pieris brassicae*[30]	Between larval instars[30]
	Digestive fluid of *Bombyx mori* showed antiviral activity toward NPV[91]	Between populations[62]
	Isolation of red fluorescent protein from *B. mori*, which agglutinates NPV[92]	None reported
Virions cannot penetrate the peritrophic membrane	Electron microscopy indicates potential of membrane as barrier to microbial infection in *Orgyia pseudotuga*,[93] but probably not very important for baculoviruses[89]	None reported
Invasion Phase of Viral Infection Pathology		
Virions unable to attach to midgut epithelial cells of host	Nonsusceptibility of *B. mori* strains to DNV may involve interference to host-cell receptor synthesis[90]	Between populations[90]
Midgut cells do not support viral replication	Nonsusceptibility of *B. mori* strains to DNV may involve interference to virus replication[90]	Between populations[90]
	GV replication is partially blocked or inhibited in midgut epithelial cells of *Archips*[94]	Between individuals[94]
Discharge of infected midgut cells — regenerated cells susceptible to subsequent infection	Midgut cells of *B. mori* infected with IFV were discharged at a molt and replaced with new cells which subsequently became infected — cycle repeated each molt[95]	Between populations — only resistant strains showed cell regeneration[95]
	Midgut cells of *B. mori* infected with CPV were discharged at a molt and replaced with new cells which were subsequently reinfected — cycle repeated at each molt[96]	
Discharge of infected midgut cells — regenerated cells immune to subsequent infection	Midgut cells of *Hyphantria cunea*, infected with *Bombyx* CPV were discharged and replaced with new cells which were not susceptible to reinfection[97]	None reported

Table 8 (continued)
MECHANISMS BY WHICH INSECTS MAY DEFEND AGAINST VIRAL ATTACK

Potential defense mechanism	Evidence for such a mechanism	Type of variability reported for mechanism
Postinvasion Phase of Viral Infection Pathway		
Phagocytosis and lysis of virions in the hemolymph	Phagocytosis of NPV by granular cells in *B. mori* hemolymph, followed by digestion by lysosymal enzymes[98,b]	None reported
	Increase in susceptibility of *Galleria mellonella* to NPV following blockage of phagocytes[99]	None reported
	Lysis of NPV in *Lymantria dispar* hemolymph[100]	Between populations[100]
Chemical inhibition of virions in the hemolymph	Viral inhibition factor found in NPV-infected hemolymph of *B. mori*[101]	None reported
Resistance tissue cells following invasion by virions	Selection of *Spodoptera frugiperda* ovarian cell clones highly resistant to infection by *S. frugiperda* NPV and *Autographa californica* NPV[102]	Between cell clones[102]
Reduction in the rate of virus replication within cells	Increased metabolism of cells can lead to more rapid infection by NPV of *B. mori* embryos[103] and pupae[104]	Between individuals[103,104]

ᵃ No direct evidence.
ᵇ This did not prevent subsequent systemic infection.
ᶜ Activity of the viral inhibitory factor blocked by healthy hemolymph.

That such resistance can occur at the cellular level is demonstrated in the case of selected cell clones of *S. frugiperda* which showed a high level of resistance to both *Spodoptera* NPV and *Autographa* NPV, in spite of the fact that they adsorbed virus equally as well as the more susceptible cell clones.[102] The only observed difference between the two types of cells was that the resistant cells grew more slowly[102] suggesting that the rate of cell metabolism may be important in influencing resistance. This is supported by experiments which showed that the stimulation of cell metabolism by ecdysone analogues in diapausing embryos and pupae of *B. mori* resulted in more rapid infection by NPV.[103,104] Such changes in hormone levels may explain in part the increase in susceptibility to viral infection immediately before and after molts (Figure 2), and have been linked to the onset of maturation resistance in late-instar larvae.[23]

Cellular resistance could occur during the invasion phase if midgut cells were involved and/or in the postinvasion phase during the systemic infection of the insect. Other postinvasion defense mechanisms (Table 8) seem to play only a minor role. For example, Stairs[99] found a mere 13-fold increase in the susceptibility to NPV of *G. melonella* larvae whose phagocytes had previously been blocked by injections of India ink.

B. Relationship Between Developmental, Environmental, and Genetic Determination of Resistance Levels

Table 8 indicates that several of the lines of defense against viral attack vary between populations suggesting dependence on genetically determined differences in response to the virus. The midgut barrier seems to be the primary mechanism involved, and is augmented by other processes occurring in the preinvasion (antiviral activity of digestive fluids) and postinvasion (phagocytosis and lysis of virions in the hemolymph) phases.[32] This is consistent with the polygenic nature of resistance described in Section IV.

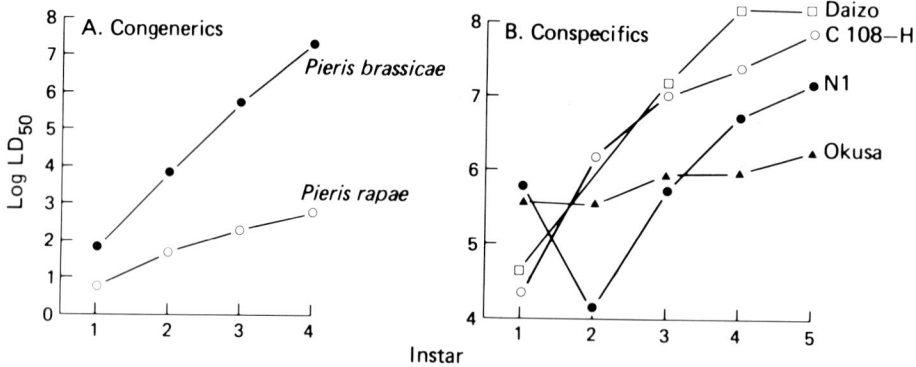

FIGURE 4.   Differences in the rate of age-related increases in resistance to infection between (A) two species of *Pieris* infected with the same GV (data from Payne et al.[1]) and (B) four strains of *Bombyx mori* infected with the same CPV (data from Aruga and Watanabe[62]).

Environmental factors also seem to act on the same set of mechanisms (see Table 4). It has been shown that light can influence the cell composition of the midgut epithelium,[44] that temperature extremes can induce cellular discharge,[43] and that both light and nutrient levels can affect the synthesis of antiviral agents.[35,45] Moreover, temperature and photoperiod can influence cell metabolic rates, either directly or mediated through hormone production, and this has been shown to modify rates of virus infection (Table 8). In *Pieris brassicae* larvae subjected to nutritional stress, one population showed increased susceptibility to GV while the response of a more resistant population remained unchanged.[36] This suggests that the genetic mechanisms conferring resistance on the latter population may control not only the physiological defenses against the viral attack, but also may determine the degree to which such defenses are affected by environmental stimuli.

Developmental factors seem to affect the same group of defense mechanisms. Thus, the reduction in susceptibility to NPV with age in *B. mori* larvae has been related to the inhibitory properties of the intestine[26] and to the enhanced antiviral activity produced by the digestive fluids of older larvae.[93] This suggests that the same defense mechanisms are involved in the development of differential resistance between instars and between populations. In fact, in *B. mori,* the increase in resistance to NPV with age was more rapid in resistant than in susceptible larvae.[26]

Such an accentuated development of age-related tolerance of virus in resistant populations has been found in several other cases. A sharp increase in the resistance to GV of one particular strain of *Plodia interpunctella,* relative to another more susceptible one, has been observed between the second and third instars.[61] The same trend was noted between populations of different species as well as between conspecific populations (Figure 4). These patterns probably represent part of a range of insect-virus interactions which culminates in virtual immunity at all stages of development. An example of an interaction at the other limit of the range involves the infection of the sphingid *Manduca sexta* by NPV from the pyralid *G. melonella.*[106] While older larvae could not be infected, neonates were susceptible when fed extremely large dosages of the virus. It should perhaps be pointed out that this range of responses does not necessarily include the phenomenon of host-specificity, which may well involve properties of the virus, but rather the response of an insect to virus which is capable of infecting it.

In summary, the three groups of factors affecting the expression of resistance are closely interrelated. As an insect grows, there is an apparent dilution effect, in that more virus is required to initiate a lethal infection. However, in addition to this, there exists a set of defense mechanisms whose activities are genetically determined and which may be subject to modification by environmental factors. These seem to become more effective as the insect matures, and thus increase the rate at which age-related resistance develops in the populations concerned.

VI. RESISTANCE AND THE USE OF BACULOVIRUSES FOR PEST CONTROL

This and previous reviews[3,32] have demonstrated that individual insect species show variability in their response to a particular virus, and that part of this variability is under genetic control, making it heritable and subject to natural selection. The present chapter has also considered how developmental and environmental factors can contribute to the expression of resistance. It is apparent that, as insects grow and pass through successive developmental stages, their resistance to viral infection increases, and the rate at which this occurs is probably determined by the genetic mechanisms mentioned above. Differences in susceptibility between instars and between populations appear to involve similar physiological and biochemical defense mechanisms. Environmental factors also seem to act through those mechanisms, although they tend to produce smaller shifts in response to a virus than do the developmental or genetic factors.

How then do such changes in susceptibility affect the decisions required to use a baculovirus successfully for pest control in the field? The importance of each set of factors affecting resistance depends largely on the time scale being considered. In the short-term, age-related factors are probably the most important because of the very large differences in susceptibility that can occur between age classes (up to 350,000, see Table 1). Their importance is dependent on the behavior of the insect, since all stages may not be vulnerable to field applications of the virus. For example, if the larvae burrow, only neonates are exposed to a virus-contaminated surface.[20] However, many insect pests are vulnerable at all stages of their larval life, and in such cases it will be important to consider the age-structure of a target population when adopting control procedures.

The relevance of age-related resistance to decision-making will also depend on the rate at which resistance develops in the insect, a property which apparently has a strong genetic component. In those species which show a relatively constant LD_{50} per weight relationship (Figure 3) it may have a minor effect. In one such species, *Pieris rapae*, the increased feeding rates of older larvae largely compensated for increases in their resistance to the virus, which might equalize the dosage level at all stages.[28,107] In the field, however, other factors such as virus inactivation might reduce the effective dosage eventually consumed by older larvae, for *Heliothis* spp. larvae, which also show a flat LD_{50} per weight relationship (Figure 3), showed differential mortality between larval instars when given a single dosage level of NPV in the field.[108] Such an effect would obviously be more evident in species that show a strong positive regression of LD_{50} on weight; in practice such species might prove more difficult to control with baculoviruses. In fact, LD_{50} per weight relationships might prove a useful criterion for identifying species that might be amenable to control by a particular virus.

Finally, the reason for using the virus will also influence the way in which age-related resistance is manipulated. When the virus is used as an insecticide, it might be desirable in many cases to apply it when only early instars are present in order to obtain a high degree of protection with minimum dosage.[10,109] As the more susceptible early instars show a shorter LT_{50} when infected (see Figure 1), the damage inflicted by such larvae before they succumb to the virus will also be minimized. When life stages show wide overlap, dosages may have to be modified to achieve an acceptable compromise between the amount, and hence the cost, of virus applied, and the level of mortality achieved. On the other hand, where the intention is to initiate a long-lasting epizootic, it might be desirable to choose a time for virus application when the age-structure of the population favors survival of some individuals (greater yield of virus particles obtained from older larvae[28]) to continue transmission of the virus. The possible effect of virus-induced changes in the age-structure of target populations might influence the outcome of such control attempts.

In the field, most environmental factors which directly affect an insect's response to viral attack will probably have little bearing on control procedures. They appear to cause only small changes in response, which would probably lie within any safety margins incorporated into dosage levels. Moreover, since they mainly involve aspects of nutrition or climate which are largely uncontrollable, there is little scope for deliberate manipulation of environmental factors to alter susceptibility. The one exception might be the increased susceptibility caused in certain cases by sublethal dosages of insecticide. This could lend itself to integrated control procedures which would exploit this apparent synergism to achieve a more cost-efficient control.[47]

It has already been pointed out that genetic factors are involved in age-related resistance. The same factors may also influence short-term control procedures since most insect-virus interactions have evolved over long periods and differences already exist between field populations of a species.[60] Thus, a dosage suitable for one population may be less effective against another. Where differences are small, they may lie within the dosage safety margins as has been suggested for relevant environmental factors. However, as shown in Section IV, quite large differences can occur between populations, which are often of the same order as economically unacceptable increases in resistance to chemical insecticides.[60] There will be a need to systematically monitor the response of target populations in order to ensure that proposed dosage rates remain adequate. It is in the development of suitable bioassay techniques for this that the control of environmental factors will be important, to ensure the accuracy and comparability of data.

Probably the most critical aspect of these genetic factors is that they are subject to selection, which can occur quite rapidly (Section IV). This can affect long-term procedures for the application of baculoviruses since even selection for relatively small changes in response could seriously alter the cost-effectiveness of viral control. The greatest risk of increased resistance will occur when baculoviruses are used as insecticides with frequent applications at high dosage. This situation could destabilize any coevolved balance in the pattern of resistance buildup and reversion and lead to the promotion of widespread resistance.[69] Fortunately this has not yet been recognized as a problem in field trials, and numerous modifying factors, such as immigration and the differential fitness of resistant and susceptible phenotypes under different conditions, should influence the outcome of selection.

From models concerned with insect resistance to chemical insecticides[48,110,111] it seems that many of these factors can be manipulated to minimize the risk of resistance. In this regard, the suggestion that age-related resistance develops more rapidly in a "resistant" individual than in a susceptible one has important implications. Thus, while increased resistance may develop in early-instar larvae,[74] selection pressure will be stronger if older larvae are exposed to a virus. Consequently, the age-structure of a population when treated may not only influence the short-term effectiveness of control procedures, but may also influence the development of long-term shifts in the response of the population to that virus. Hence, the direction of virus applications against the early-instar larvae of pests, which in many cases is the most cost-efficient method and affords the greatest protection to the crop, appears to have the added advantage of reducing the risks of selection for increased resistance.

Other factors, too, distinguish viral control agents from chemical insecticides. Viruses can reproduce themselves, they can occur in latent forms which are activated by stresses, including other viruses,[112] and they are genetically variable[113,114] and hence are themselves amenable to selection for changes in virulence.[115,116] Even though there are sufficient analogies between viral and chemical insecticides to enable models for resistance to chemical insecticides to serve as useful guidelines,[3] the role played by these viral properties in modifying the development of insect resistance to viruses will eventually have to be understood. For example, two GVs isolated from *Pieris* spp. showed only very small structural differences (97.7% homology in deoxyribonucleic acid (DNA) sequence). Though highly homologous, large differences in the virulence of these viruses were noted for the relatively resistant *P. brassicae,* but no significant change was found in the response of *P. rapae,* which was highly susceptible to both viruses.[117]

There has recently been a trend toward the modeling of insect-pathogen interactions, to provide an insight into the population dynamics of both insect and pathogen, and into the effect of control procedures.[118-120] These models will also need to take into account the variability shown by the insect in its response to the pathogen. At the moment there is insufficient data to undertake this. More information is required on changes in susceptibility that occur under field conditions. Mitchell and Fuxa[87] are studying this using *S. frugiperda* and its NPV.

The relationship between developmental and genetic factors controlling resistance needs to be explored more fully from a quantitative viewpoint; we also need to determine the more fundamental aspects of the formal genetics and underlying physiology of defense mechanisms.

Obviously, much needs to be learned before the effects on control procedures of differential susceptibility to baculoviruses within a species can be predicted. This chapter has provided evidence that such studies are as important as the other problem areas facing the development of viral control agents (e.g., virus biology, mass production, delivery systems, safety considerations).[121,122] An understanding of the nature of resistance should permit improvements in control techniques and enable measures to be taken to reduce the risks of unwanted changes occurring in susceptibility to virus infection.

REFERENCES

1. **Payne, C. C., Tatchell, G. M., and Williams, C. F.,** The comparative susceptibilities of *Pieris brassicae* and *Pieris rapae* to a granulosis virus from *P. brassicae, J. Invertebr. Pathol.,* 38, 273, 1981.
2. **Burges, H. D.,** Possibilities of pest resistance to microbial control agents, in *Microbial Control of Insects and Mites,* Burges, H. D. and Hussey, N. W., Eds., Academic Press, New York, 1971, 445.
3. **Briese, D. T. and Podgwaite, J. D.,** Development of viral resistance in insect populations, in *Viral Insecticides for Biological Control,* Maramorosch, K. and Sherman, K. E., Eds., Academic Press, New York, 1985, 361.
4. **Smith, O. J., Hudges, K. M., Dunn, P. H., and Hall, I. M.,** A granulosis virus disease of the western grape leaf skeletonizer and its transmission, *Can. Entomol.,* 88, 507, 1956.
5. **Cunningham, J. C.,** Pathogenicity tests of nuclear polyhedrosis viruses infecting the eastern hemlock looper, *Lambdina fiscellaria fiscellaria* (Lepidoptera: Geometridae), *Can. Entomol.,* 102, 1534, 1970.
6. **Stairs, G. R.,** Quantitative differences in susceptibility to nuclear-polyhedrosis virus among larval instars of the forest tent caterpillar, *Malacosoma disstria* (Hubner), *J. Invertebr. Pathol.,* 7, 427, 1965.
7. **Burgerjon, A., Biache, G., Chaufaux, J., and Petre, Z.,** Sensibilité comparée en fonction de leur age, des chenilles de *Lymantria dispar, Mamestra brassicae* et *Spodoptera littoralis* aux virus de la polyedrose nucleaire, *Entomophaga,* 26, 47, 1981.
8. **Livingston, J. M., McLeod, P. J., Yearian, W. C., and Young, S. Y., III,** Laboratory and field evaluation of a nuclear polyhedrosis virus of the soybean looper, *Pseudoplusia includens, J. Ga. Entomol. Soc.,* 15, 194, 1980.
9. **Capinera, J. L. and Kanost, M. R.,** Susceptibility of the zebra caterpillar to *Autographa californica* nuclear polyhedrosis virus, *J. Econ. Entomol.,* 72, 570, 1979.
10. **Flückiger, C. R.,** Untersuchungen über drei Baculovirus-Isolate des Schalenwicklers, *Adoxophyes orana* F. V. R. (Lep., Tortricidae), dessen Phänologie und erste Feldversuche, als Grundlagen zur Microbiologischen Bekämpfung diese Obstschadlings, *Mitt. Schweiz. Entomol. Ges.,* 55, 241, 1982.
11. **Cunningham, J. C. and Entwistle, P. F.,** Control of sawflies by baculovirus, in *Microbial Control of Pests and Plant Diseases, 1970-1980,* Burges, H. D., Ed., Academic Press, New York, 1981, 379.

12. **Boucias, D. G., Johnson, D. W., and Allen, G. E.,** Effects of host age, virus dosage, and temperature on the infectivity of a nucleopolyhedrosis virus against velvetbean caterpillar, *Anticarsia gemmatalis,* larvae, *Environ. Entomol.,* 9, 59, 1980.

13. **Allen, G. E. and Ignoffo, C. M.,** The nucleopolyhedrosis virus of *Heliothis:* quantitative in vivo estimates of virulence, *J. Invertebr. Pathol.,* 13, 378, 1969.

14. **Bucher, G. E. and Turnock, W. J.,** Dosage responses of the larval instars of the Bertha armyworm, *Mamestra configurata* (Lepidoptera: Noctuidae), to a native nuclear polyhedrosis, *Can. Entomol.,* 115, 341, 1983.

15. **Wigley, P. J.,** The Epizootiology of a Nuclear Polyhedrosis Virus Disease of the Winter Moth, *Operophtera brumata* L. at Wistman's Wood, Dartmoor, Ph.D. thesis, University of Oxford, Oxford, 1976.

16. **Boucias, D. G. and Nordin, G. L.,** Interinstar susceptibility of the fall webworm, *Hyphantria cunea,* to its nucleopolyhedrosis and granulosis viruses, *J. Invertebr. Pathol.,* 30, 68, 1977.

17. **Magnoler, A.,** Bioassay of nucleopolyhedrosis virus against larval instars of *Malacosoma neustria, J. Invertebr. Pathol.,* 25, 343, 1975.

18. **Evans, H. F.,** Quantitative assessment of the relationships between dosage and response of the nuclear polyhedrosis virus of *Mamestra brassicae, J. Invertebr. Pathol.,* 37, 101, 1981.

19. **Sheppard, R. F. and Stairs, G. R.,** Dosage-mortality and time-mortality studies of a granulosis virus in a laboratory strain of the codling moth, *Laspeyresia pomonella, J. Invertebr. Pathol.,* 29, 216, 1977.

20. **Laing, D. R. and Jacques, R. P.,** Codling moth: techniques for rearing larvae and bioassaying granulosis virus, *J. Econ. Entomol.,* 73, 851, 1980.

21. **Crook, N. E., Brown, J. D., and Foster, G. N.,** Isolation and characterization of a granulosis virus from the tomato moth, *Laconobia oleracea,* and its potential as a control agent, *J. Invertebr. Pathol.,* 40, 221, 1982.

22. **Whitlock, V. H.,** Effect of larval maturation on mortality induced by a nuclear polyhedrosis and granulosis virus infections of *Heliothis armigera, J. Invertebr. Pathol.,* 30, 80, 1977.

23. **Evans, H. F.,** The influence of larval maturation on responses of *Mamestra brassicae* L. (Lepidoptera: Noctuidae) to nuclear polyhedrosis virus infection, *Arch. Virol.,* 75, 163, 1983.

24. **Stairs, G. R.,** The effect of metamorphosis on nuclear polyhedrosis virus infection in certain Lepidoptera, *Can. J. Microbiol.,* 11, 509, 1965.

25. **Watanabe, H.,** Resistance of the silkworm to cytoplasmic polyhedrosis virus, in *The Cytoplasmic-Polyhedrosis Virus of the Silkworm,* Aruga, H. and Tanada, Y., Eds., University of Tokyo Press, Tokyo, 1971, 169.

26. **Kobayashi, M., Yamaguchi, S., and Yokoyama, Y.,** Influence of the larval development on the susceptibility of the silkworm, *Bombyx mori* L., to the nuclear polyhedrosis virus, *J. Sericult. Sci. Jpn.,* 38, 431, 1969 (in Japanese).

27. **David, W. A. L., Clothier, S. G., Woolner, M., and Taylor, G.,** Bioassay on insect virus on leaves. II. The influence of certain factors associated with the larvae and the leaves, *J. Invertebr. Pathol.,* 17, 178, 1971.

28. **Payne, C. C.,** Insect viruses as control agents, *Parasitology,* 84, 35, 1982.

29. **Watanabe, H.,** Pathogenic changes in the faeces from the silkworm, *Bombyx mori* L., after peroral inoculation with a cytoplasmic-polyhedrosis virus, *J. Sericult. Sci. Jpn.,* 37, 385, 1968 (in Japanese).

30. **Ripa, R.,** Studies of the Susceptibility of *Peiris brassicae* (L.) to a Granulosis Virus, Ph.D. thesis, University of London, London, 1978.

31. **Stiles, B. and Paschke, J. D.,** Midgut pH in different instars of three *Aedes* mosquito species and the relation between pH and susceptibility of larvae to a nuclear polyhedrosis virus, *J. Invertebr. Pathol.,* 35, 58, 1980.

32. **Briese, D. T.,** Resistance of insect species to microbial pathogens, in *Pathogenesis of Invertebrate Microbial Diseases,* Davidson, E. W., Ed., Allanheld Osmun, Totowa, N.J., 1981, 511.

33. **Fuxa, J. R. and Geaghan, J. P.,** Multiple-regression analysis of factors affecting prevalence of nuclear polyhedrosis virus in *Spodoptera frugiperda* (Lepidoptera: Noctuidae) populations, *Environ. Entomol.,* 12, 311, 1983.

34. **Aratake, Y.,** Strain differences of the silkworm, *Bombyx mori* L., in the resistance to a nuclear polyhedrosis virus, *J. Sericult. Sci. Jpn.,* 42, 230, 1973 (in Japanese).

35. **Hayashiya, K., Nishida, J., and Matsubara, F.,** Inactivation of nuclear polyhedrosis virus in the digestive juice of silkworm larvae, *Bombyx mori* L., I. Comparison of antiviral activities in the digestive juices of larvae reared on natural and artificial diets, *Jpn. J. Appl. Entomol. Zool.,* 12, 189, 1968.

36. **David, W. A. L. and Gardiner, B. O. C.,** The incidence of granulosis deaths in susceptible and resistant *Pieris brassicae* (Linnaeus) larvae following changes of population density, food, and temperature, *J. Invertebr. Pathol.,* 7, 347, 1965.

37. **David, W. A. L. and Taylor, C. E.,** The effect of sucrose content of diets on susceptibility to granulosis virus disease in *Pieris brassicae, J. Invertebr. Pathol.,* 30, 117, 1977.

38. **David, W. A. L.**, The granulosis virus of *Pieris brassicae* (L.) and its relationship with its host, *Adv. Virus Res.*, 22, 112, 1978.

39. **Watanabe, H. and Imanishi, S.**, The effect of the content of certain ingredients in an artificial diet on the susceptibility to virus infection in the silkworm, *Bombyx mori*, *J. Sericult. Sci. Jpn.*, 49, 404, 1980 (in Japanese).

40. **Watanabe, H.**, personal communication, 1983.

41. **Watanabe, H.**, Temperature effects on the manifestation of susceptibility to peroral infection with cytoplasmic polyhedrosis in the silkworm, *Bombyx mori* L., *J. Sericult. Sci. Jpn.*, 33, 286, 1964 (in Japanese).

42. **Inoue, H. and Tanada, Y.**, Thermal therapy of the flacherie virus disease in the silkworm, *Bombyx mori*, *J. Invertebr. Pathol.*, 29, 63, 1977.

43. **Watanabe, H. and Tanada, Y.**, Infection of a nuclear-polyhedrosis virus in armyworm *Pseudaletia unipuncta* Haworth (Lepidoptera: Noctuidae), reared at high temperature, *Appl. Entomol. Zool.*, 7, 43, 1972.

44. **Watanabe, H. and Takimiya, K.**, Susceptibility of the silkworm larvae, *Bombyx mori*, reared under different light conditions to polyhedrosis viruses, *J. Sericult. Sci. Jpn.*, 43, 403, 1976 (in Japanese).

45. **Hayashiya, K., Nishida, J., and Uchida, Y.**, The mechanism of formation of red fluorescent protein in the digestive juice of silkworm larvae: the formation of chlorophyllide-A, *Jpn. J. Appl. Entomol. Zool.*, 20, 37, 1976 (in Japanese).

46. **Watanabe, H.**, Susceptibility to virus infection in the silkworm, *Bombyx mori*, applied topically with sublethal dosages of insecticide, *J. Sericult. Sci. Jpn.*, 40, 350, 1971 (in Japanese).

47. **Jacques, R. P. and Morris, O. N.**, Compatibility of pathogens with other methods of pest control and with different crops, in *Microbial Control of Pests and Plant Diseases 1970-1980*, Burges, H. D., Ed., Academic Press, New York, 1981, 695.

48. **Whitten, M. J. and McKenzie, J. A.**, The genetic basis for pesticide resistance, in *Proc. 3rd Australasian Conf. Grass. Invertebr. Ecol.*, Lee, K. E., Ed., South Australia Government Printer, Adelaide, 1982, 1.

49. **Vasiljevic, L.**, Influence of various phases of gradation upon the susceptibility of gypsy moth caterpillars to polyhedrosis, *J. Sci. Agric. Res.*, 14, 1, 1961.

50. **Benz, G.**, Untersuchungen über die Pathogenität eines Granulosis-Virus des grauen Larchenwicklers *Zeiraphera diniana* (Guenee), *Agron. Glasn.*, broj 5-6-7, 566, 1962.

51. **Rollinson, W. D.**, unpublished data cited in **Briese, D. T.** and **Podgwaite, J. D.**, Development of viral resistance in insect populations, in *Viral Insecticides for Biological Control*, Maramorosch, K. and Sherman, K. E., Eds., Academic Press, New York, 1985, 361.

52. **Watanabe, H.**, Relative virulence of polyhedrosis viruses and host-resistance in the silkworm, *Bombyx mori* L. (Lepidoptera: Bambycidae), *Appl. Entomol. Zool.*, 1, 139, 1966.

53. **Rollinson, W. D. and Lewis, F. B.**, Susceptibility of gypsy moth larvae to *Lymantria* spp. nuclear and cytoplasmic polyhedrosis virus, *Zastita Bilja*, 24, 163, 1973.

54. **Reichelderfer, D. F. and Benton, C. V.**, Some genetic aspects of the resistance of *Spodoptera frugiperda* to a nuclear polyhedrosis virus, *J. Invertebr. Pathol.*, 23, 378, 1974.

55. **Martignoni, M. E. and Schmid, P.**, Studies on the resistance to virus infection in natural populations of Lepidoptera, *J. Invertebr. Pathol.*, 3, 62, 1961.

56. **Klein, M. and Podoler, H.**, Studies on the application of a nuclear polyhedrosis virus to control populations of the Egyptian cotton worm *Spodoptera littoralis*, *J. Invertebr. Pathol.*, 32, 244, 1978.

57. **Briese, D. T., Mende, H. A., Grace, T. D. C., and Geier, P. W.**, Resistance to a nuclear polyhedrosis virus in the light-brown apple moth *Epiphyas postvittana* (Lepidoptera: Tortricidae), *J. Invertebr. Pathol.*, 36, 211, 1980.

58. **Hunter, D. K. and Hoffman, D. F.**, Susceptibility of two strains of Indian meal moth to a granulosis virus, *J. Invertebr. Pathol.*, 21, 114, 1973.

59. **Hunter, F. R. and Vigneswaren, K.**, personal communication, 1983.

60. **Briese, D. T. and Mende, H. A.**, Differences in susceptibility to a granulosis virus between field populations of the potato moth, *Phthorimaea operculella* (Zeller) (Lepidoptera: Gelechiidae), *Bull. Entomol. Res.*, 71, 11, 1981.

61. **Briese, D. T.**, Genetic basis for resistance to a granulosis virus in the potato moth *Phthorimaea operculella*, *J. Invertebr. Pathol.*, 39, 215, 1982.

62. **Aruga, H. and Watanabe, H.**, Resistance to per os infection with cytoplasmic polyhedrosis virus in the silkworm, *Bombyx mori* (Linnaeus), *J. Insect Pathol.*, 6, 387, 1964.

63. **Watanabe, H.**, Resistance to peroral infection by the cytoplasmic-polyhedrosis virus in the silkworm, *Bombyx mori* (Linnaeus), *J. Invertebr. Pathol.*, 7, 257, 1965.

64. **Funada, T.**, Genetic resistance of the silkworm, *Bombyx mori* L., to an infection of a flacherie virus, *J. Sericult. Sci. Jpn.*, 37, 281, 1968 (in Japanese).

65. **Furuta, Y.**, On the heredity of the resistance to small flacherie virus in the silkworm *Bombyx mori*, *J. Sericult. Sci. Jpn.*, 47, 241, 1978 (in Japanese).

66. **Watanabe, H. and Maeda, S.,** Genetically determined non-susceptibility of the silkworm, *Bombyx mori,* to infection with a densonucleosis virus (Densovirus), *J. Invertebr. Pathol.,* 38, 370, 1981.

67. **McKenzie, J. A., Dearn, J. M., and Whitten, M. J.,** Genetic basis of resistance to diazinon in Victorian populations of the Australian sheep blowfly, *Lucilia cuprina, Aust. J. Biol. Sci.,* 33, 85, 1980.

68. **David, W. A. L. and Gardiner, B. O. C.,** A *Pieris brassicae* (Linnaeus) culture resistant to a granulosis virus, *J. Insect Pathol.,* 2, 106, 1960.

69. **Briese, D. T. and Mende, H. A.,** Selection for increased resistance to a granulosis virus in the potato moth, *Phthorimaea operculella* (Zeller) (Lepidoptera: Gelechiidae), *Bull. Entomol. Res.,* 73, 1, 1983.

70. **Uzigawa, K. and Aruga, H.,** On the selection of resistant strains to the infectious flacherie virus in the silkworm, *Bombyx mori* L., *J. Sericult. Sci. Jpn.,* 35, 23, 1966 (in Japanese).

71. **Watanabe, H.,** Development of resistance in the silkworm, *Bombyx mori,* to peroral infection of a cytoplasmic polyhedrosis virus, *J. Invertebr. Pathol.,* 9, 474, 1967.

72. **Ignoffo, C. M. and Allen, G. E.,** Selection for resistance to a nuclear polyhedrosis virus in laboratory populations of the cotton bollworm, *Heliothis zea, J. Invertebr. Pathol.,* 20, 187, 1972.

73. **Whitlock, V. H.,** Failure of a strain of *Heliothis armigera* (Hubn.) (Noctuidae: Lepidoptera) to develop resistance to a nuclear polyhedrosis virus and a granulosis virus, *J. Entomol. Soc. S. Afr.,* 40, 251, 1977.

74. **Huber, J.,** Selektion einer Resistenz gegen perorale Infektion mit einem Granulosisovirus bei einem Laborstamm des Apfelwicklers, *Laspeyresia pomonella* L., Diss. No. 5044, Eidgenössische Technische Hochschule Zürich, Switzerland, 1974.

75. **Ignoffo, C. M. and Garcia, C.,** Shifts in ratio of laboratory populations of *Heliothis zea* and *H. virescens* surviving exposure to a non-specific and specific nucleopolyhedrosis virus, *Environ. Entomol.,* 8, 1102, 1979.

76. **Cunningham, J. C.,** personal communication, 1983.

77. **Entwistle, P. F.,** personal communication, 1983.

78. **Luttrell, R. G.,** personal communication, 1983.

79. **Lewis, F. B.,** personal communication, 1983.

80. **McKinley, D. J.,** personal communication, 1983.

81. **Huber, J.,** personal communication, 1983.

82. **Bedford, G. O.,** Control of the rhinoceros beetle by baculovirus, in *Microbial Control of Pests and Plant Diseases 1970—1980,* Burges, H. D., Ed., Academic Press, New York, 1981, 409.

83. **Watanabe, H. and Shimizu, T.,** Epizootiological studies on the occurrence of densonucleosis in the silkworm, *Bombyx mori,* reared at sericultural farms, *J. Sericult. Sci. Jpn.,* 49, 485, 1980.

84. **Martignoni, M. E.,** Contributo alla conscenza di una granulosi di *Eucosma griseana* (Hubner) (Tortricidae, Lepidoptera) quale fattore limitante il pullulamento dell'insetto nella Engadina alta, *Mitt. Schweiz. Anst. Forstl. Versuchstierkd.,* 32, 371, 1957.

85. **Curtis, C. F., Cook, L. M., and Wood, R. J.,** Selection for and against insecticide resistance and possible methods of inhibiting the evolution of resistance in mosquitoes, *Ecol. Entomol.,* 3, 273, 1978.

86. **Roush, R. T. and Plapp, F. W., Jr.,** Effects of insecticide resistance on biotic potential of the house fly (Diptera: Muscidae), *J. Econ. Entomol.,* 75, 708, 1982.

87. **Mitchell, F. and Fuxa, J. R.,** personal communication, 1983.

88. **Aizawa, K., Furuta, Y., and Nakamura, K.,** Selection of a resistant strain to virus induction in the silkworm, *Bombyx mori, J. Sericult. Sci. Jpn.,* 30, 405, 1962.

89. **Granados, R. R.,** Infectivity and mode of action of baculoviruses, *Biotech. Bioeng.,* 22, 1377, 1980.

90. **Watanabe, H.,** Characteristics of densonucleosis in the silkworm, *Bombyx mori, Spn. Agric. Res. Q.,* 15, 183, 1981.

91. **Aratake, Y. and Ueno, H.,** Inactivation of a nuclear polyhedrosis virus by the gut juice of the silkworm, *Bombyx mori* L., *J. Sericult. Sci. Jpn.,* 42, 279, 1973.

92. **Hayashiya, K., Uchida, Y., and Himeno, M.,** Mechanism of anti-viral action of red fluorescent protein on nuclear polyhedrosis virus in silkworm larvae, *Jpn. J. Appl. Entomol. Zool.,* 22, 238, 1978 (in Japanese).

93. **Brandt, C. R., Adang, M. J., and Spence, K. D.,** The peritrophic membrane: ultrastructural analysis and function as mechanical barrier to microbial infection in *Orgyia pseudotsuga, J. Invertebr. Pathol.,* 32, 12, 1978.

94. **Pinnock, D. E. and Hess, R. T.,** Electron microscope observations on granulosis virus replication in the fruit tree leaf roller, *Archips argyrospila:* infection of the midgut, *J. Invertebr. Pathol.,* 30, 354, 1977.

95. **Inoue, H.,** Multiplication of an infectious flacherie virus in the resistant and susceptible strain of the silkworm, *Bombyx mori, J. Sericult. Sci. Jpn.,* 43, 318, 1974.

96. **Inoue, H. and Miyagawa, M.,** Regeneration of the midgut epithelial cells in the silkworm, *Bombyx mori,* infected with ciruses, *J. Invertebr. Pathol.,* 32, 373, 1978.

97. **Yamaguchi, K.,** Natural recovery of the fall webworm, *Hyphantria cursa,* to infection by a cytoplasmic polyhedrosis virus of the silkworm, *Bombyx mori, J. Invertebr. Pathol.,* 33, 126, 1979.

98. **Inoue, K.,** Phagocytosis of nuclear polyhedra and the localization of the acid phosphatase in the granular cell of the silkworm, *Bombyx mori* L., *J. Sericult. Sci. Jpn.,* 43, 394, 1974 (in Japanese).

99. **Stairs, G. R.,** Changes in the susceptibility of *Galleria mellonella* (Linneus) larvae to nuclear polyhedrosis virus following blockage of the phagocytes with India ink, *J. Insect Pathol.,* 6, 363, 1964.
100. **Larionov, G. V. and Baranovskii, V. I.,** Effect of virus infection on the lytic activity of hemolymph of *Ocneria dispar* and *Dendrolimus sibiricus, Izv. Sib. Otd. Akad. Navk SSR Ser. Biol. Nauk,* 0(2), 131, 1979 (in Russian).
101. **Aizawa, K.,** Defense reactions of the silkworm, *Bombyx mori,* against the nuclear polyhedrosis, *Proc. 4th Int. Colloq. Insect Pathol.,* 352, 1970.
102. **Crawford, A. M. and Sheehan, C.,** Persistent baculovirus infections: *Spodoptera frugiperda* NPV and *Autographa californica* NPV in *Spodoptera frugiperda* cells, *Arch. Virol.,* 78, 65, 1983.
103. **Kobayashi, M.,** Factors affecting the susceptibility of cultured silkworm embryo, *Bombyx mori* L., to nuclear polyhedrosis virus infection, *Jpn. J. Appl. Entomol. Zool.,* 26, 82, 1982 (in Japanese).
104. **Kobayashi, M. and Kawase, S.,** Effect of alteration of endocrine mechanism on the development of nuclear polyhedrosis in the isolated pupal abdomens of the silkworm, *Bombyx mori, J. Invertebr. Pathol.,* 36, 6, 1980.
105. **Stairs, G. R.,** Dosage-mortality response of *Galleria mellonella* (Linnaeus) to a nuclear polyhedrosis virus, *J. Invertebr. Pathol.,* 7, 5, 1965.
106. **Fraser, M. and Stairs, G. R.,** Susceptibility of *Trichoplusia ni, Heliothis zea* (Noctuidae) and *Manduca sexta* (Sphingidae) to a nuclear polyhedrosis virus from *Galleria mellonella* (Pyralidae), *J. Invertebr. Pathol.,* 40, 255, 1982.
107. **Tatchell, G. M.,** The effects of a granulosis virus infection and temperature on the food consumption of *Pieris rapae, Entomophaga,* 26, 291, 1981.
108. **Luttrell, R. G., Yearian, W. C., and Young, S. Y.,** Effects of Elcar® (*Heliothis zea* nuclear polyhedrosis virus) treatments on *Heliothis* spp., *J. Ga. Entomol. Soc.,* 17, 211, 1982.
109. **Stacey, A. L., Young, S. Y., and Yearian, W. C.,** Effect of larval age and mortality level on damage to cotton by *Heliothis zea* infected with *Baculovirus heliothis, J. Econ. Entomol.,* 70, 383, 1977.
110. **Comins, H. N.,** The management of pesticide resistance: models, in *Genetics in Relation to Insect Management,* Hoy, M. A. and McKelvey, J. J., Jr., Eds., Rockefeller Foundation, New York, 1979, 55.
111. **Tabashnik, B. E. and Croft, B. A.,** Managing pesticide resistance in crop-arthropod complexes: interactions between biological and operational factors, *Environ. Entomol.,* 11, 1137, 1982.
112. **McKinley, D. J., Brown, D. A., Payne, C. C., and Harrap, K. A.,** Cross-infectivity and activation studies with four baculoviruses, *Entomophaga,* 26, 79, 1981.
113. **Shapiro, M. and Ignoffo, C. M.,** Nucleopolyhedrosis of *Heliothis:* activities of isolates from *Heliothis zea, J. Invertebr. Pathol.,* 16, 107, 1970.
114. **Hughes, P. R., Gettig, R. R., and McCarthy, W. J.,** Comparison of time-mortality response of *Heliothis zea* to 14 isolates of *Heliothis* nuclear polyhedrosis virus, *J. Invertebr. Pathol.,* 41, 256, 1983.
115. **Stairs, G. R., Fraser, T., and Fraser, M.,** Changes in growth and virulence of nuclear polyhedrosis virus from *Choristoneura fumiferana* after passage in *Trichoplusia ni* and *Galleria mellonella, J. Invertebr. Pathol.,* 38, 230, 1981.
116. **Wood, H. A., Hughes, P. R., Johnston, L. B., and Langridge, W. H. R.,** Increased virulence of *Autographa californica* nuclear polyhedrosis virus by mutagenesis, *J. Invertebr. Pathol.,* 38, 236, 1981.
117. **Crook, N. E.,** A comparison of the granulosis viruses from *Pieris brassicae* and *Pieris rapae, Virology,* 115, 173, 1981.
118. **Anderson, R. M.,** Theoretical basis for the use of pathogens as biological control agents of pest species, *Parasitology,* 84, 3, 1982.
119. **Pinnock, D. E. and Brand, R. J.,** A quantitative approach to the ecology of the use of pathogens for insect control, in *Microbial Control of Pests and Plant Diseases 1970—1980,* Burges, H. D., Ed., Academic Press, New York, 1981, 655.
120. **Brand, R. J. and Pinnock, D. E.,** Application of biostatistical modelling to forecasting the results of microbial control trials, in *Microbial Control of Pests and Plant Diseases 1970—1980,* Burges, H. D., Ed., Academic Press, New York, 1981, 667.
121. **I.P.R.C.,** Recommendations, in *Proc. of Workshop on Insect Pest Management with Microbial Agents: Recent Achievements, Deficiencies, and Innovations,* Insect Pathology Resource Center, Ithaca, N.Y., 1980, 53.
122. **Burges, H. D.,** Strategies for microbial control of pests in 1980 and beyond, in *Microbial Control of Pests and Plant Diseases 1970—1980,* Burges, H. D., Ed., Academic Press, New York, 1981, 797.

INDEX

A

Abiotic persistence, 113—114
Abnormalities, 138
Abundance of virus, 101—102, 120
Acephate, 146
Additivity, 139, 144
Adjuvant mixture, 144
Adoxophyes orana, 38, 141, 239
Adsorption to clay colloids, 114
Adventitious viral agents, 65
Agarose gel strength, 23—24
Agarose overlay, 23
Age of host insect, 34—35, 99, 141
  multiple infection in relation to, 138
Age-related resistance, 238, 240, 256—258
Age-structure of target population, 257—258
Agricultural application technology, 229
Agricultural Extension Service specialists, 227
Agriculture, 167—168, 170—171
Agrotis c-nigrum, 135
Agrotis segetum, 135, 137, 142
Air flow, production environment, 41
Alfalfa butterfly, see *Colias eurytheme*
Alfalfa looper, see *Autographa californica*
Alfalfa pests, 110
Alternate hosts, in vivo production, 36—38
Alternative pest control strategies, 226
Aluminum chloride, 72
Animal sera, 65
Annual crops, 224—225
Antagonism, 139, 141, 144
Anticarsia germmetalis, 239—241, 244
Apanteles melanoscelus, 112
Apodemus, 111
Apple leafroller, see *Adoxophyes orana*
Apple moth, see *Hyponomeuta malinellus*
Application of baculoviruses, 157—158, 164—179
  agriculture, 167—168, 170—171
  compatibility with other pesticides, 172
  equipment, 165—168
  forests, 165—166, 169—170
  tank mixtures, 168—171
  timing, 172—174
Arctia caja, 37
Arctiid, see *Diacrisia virginica*
Argyresthia conjugella, 37
Artificial diets, 42
Artificially introduced pathogens, 123
Autodissemination of virus, 124
Autographa californica, 36, 134—135
Autographa californica nuclear polyhedrosis virus
    (AcNPV), 22, 36
  agarose gel strength, 24
  alternate hosts, 37
  applied research projects, 224—225
  commercial and semicommerical preparation, 183
  formulation method, 160

host range, 102
infection of *Trichoplusia ni* with, 92

B

Bacillus cereus, 142
Bacillus sphaericus, 142
Bacillus thuringiensis
  compatibility with other pesticides, 172
  controlled laboratory studies of interactions, 141
  dosage-mortality response curves, 140
  dual infection mortality rate, 143
  insect control, 224
  interactions with baculoviruses, 143
  interactions with chemicals, 145
  interference with baculovirus infection, 149
  mechanical disruption, 137—138
  standardization, 159
Bacteria, 141
Bacterium paracoli, 142
Baculovirus-based insecticides, 226
Baculovirus-chemical interactions, see Interactions
Baculovirus heliothis, 34, 36, 38—39, 146
Baculovirus-parasitoid interactions, see Interactions
Baculovirus-pathogen interactions, see Interactions
Baits, 143
Balsam fir, 109, 163
Beet armyworm, see *Spodoptera exigua*
Bidrin, 146
Bioassay, 2—5, 16
  in vitro, see In vitro assay
  in vivo, see In vivo assay
Biological control, 224
Biological pesticides, 158, 206
Biology of host insect, 34
Biotic persistence, 110—113
Bird-cherry moth, see *Hyponomeuta evonymellus*
Birth rate, 125
Bivert, 169
Blarina, 111
BMD 03S, 17
Body weight, 93
Bollworm, see *Baculovirus heliothis*
Bolograms, 140
Bombyx iridescent virus, 142
Bombyx mori, 33
  chemical stress and resistance to viral infection,
      247
  cytoplasmic polyhedrosis virus, 240, 242
  genetic mechanisms in resistance to, 250
  light and resistance to viral infection, 247
  nutritional factor vs. resistance to viral infection,
      246
  population differences in susceptibility to, 249
  production temperature, 41
  selection of resistance genes in laboratory, 250—
      252